PROBABILITY

For the Enthusiastic Beginner

David Morin

Harvard University

ISBN-10: 1523318678
ISBN-13: 978-1523318674

Printed by CreateSpace

Additional resources located at:
www.people.fas.harvard.edu/~djmorin/book.html

Cover image:
Faience polyhedron inscribed with letters of the Greek alphabet
The Metropolitan Museum of Art
Fletcher Fund, 1937
www.metmuseum.org

Contents

Preface

This book is written for high school and college students learning about probability for the first time. Most of the book is very practical, with a large number of concrete examples and worked-out problems. However, there are also parts that are a bit theoretical (at least for an introductory book), with many mathematical derivations. All in all, if you are looking for a book that serves as a quick reference, this may not be the one for you. But if you are looking for a book that starts at the beginning and derives everything from scratch in a comprehensive manner, then you've come to the right place. In short, this book will appeal to the reader who has a healthy level of enthusiasm for understanding how and why the standard results of probability come about.

Probability is a very accessible (and extremely fun!) subject, packed with challenging problems that don't require substantial background or serious math. The examples in Chapter 2 are a testament to this. Of course, there are plenty of challenging topics in probability that *do* require a more formal background and some heavy-duty math. This will become evident in Chapters 4 and 5 (and the latter part of Chapter 3). However, technically the only math prerequisite for this book is a comfort with algebra. Calculus isn't relied on, although there are a few problems that do involve calculus. These are marked clearly.

All of the problems posed at the ends of the chapters have solutions included. The difficulty is indicated by stars; most problems have two stars. One star means plug and chug, while three stars mean some serious thinking. Be sure to give a solid effort when solving a problem, and don't look at the solution too soon. If you can't solve a problem right away, that's perfectly fine. Just set it aside and come back to it later. It's better to solve a problem later than to read the solution now. If you do eventually need to look at a solution, cover it up with a piece of paper and read one line at a time, to get a hint to get started. Then set the book aside and work things out for real. That way, you can still (mostly) solve it on your own. You will learn a great deal this way. If you instead head right to the solution and read it straight through, you will learn very little.

For instructors using this book as the assigned textbook for a course, a set of homework exercises is posted at *www.people.fas.harvard.edu/~djmorin/book.html*. A solutions manual is available to instructors upon request. When sending a request, please point to a syllabus and/or webpage for the course.

The outline of this book is as follows. Chapter 1 covers combinatorics, which is the study of how to count things. Counting is critical in probability, because probabilities often come down to counting the number of ways that something can happen. In Chapter 2 we dive into actual probability. This chapter includes a large number of examples, ranging from coins to cards to four classic problems presented in Section 2.4. Chapter 3 covers expectation values, including the variance and standard deviation. A section on the "sample variance" is included; this is rather mathematical and can be skipped on a first reading. In Chapter 4 we introduce the concept of a continuous distribution and then discuss a number of the more common probability distributions. In Chapter 5 we see how the binomial and Poisson distributions reduce to a Gaussian (or normal) distribution in certain limits. We also discuss the law of large numbers and the central limit theorem. Chapter 6 is somewhat of a stand-alone chapter, covering correlation and regression. Although these topics are usually found in books on statistics, it makes sense to include them here, because all of the framework has been set. Chapter 7 contains six appendices. Appendix C deals with approximations to $(1 + a)^n$ which are critical in the calculations in Chapter 5, Appendix E lists all of the main results we derive in the book, and Appendix F contains a glossary of notation; you may want to refer to this when starting each chapter.

A few informational odds and ends: This book contains many supplementary remarks that are separated off from the main text; these end with a shamrock, "♣." The letters N, n, and k generally denote integers, while x and t generally denote continuous quantities. Upper-case letters like X denote a random variable, while lower-case letters like x denote the value that the random variable takes. We refer to the normal distribution by its other name, the "Gaussian" distribution. The numerical plots were generated with Mathematica. I will sometimes use "they" as a gender-neutral singular pronoun, in protest of the present failing of the English language. And I will often use an " 's" to indicate the plural of one-letter items (like 6's on dice rolls). Lastly, we of course take the frequentist approach to probability in this introductory book.

I would particularly like to thank Carey Witkov for meticulously reading through the entire book and offering many valuable suggestions. Joe Swingle provided many helpful comments and sanity checks throughout the writing process. Other friends and colleagues whose input I am grateful for are Jacob Barandes, Sharon Benedict, Joe Blitzstein, Brian Hall, Theresa Morin Hall, Paul Horowitz, Dave Patterson, Alexia Schulz, and Corri Taylor.

Despite careful editing, there is essentially zero probability that this book is error free (as you can show in Problem 4.16!). If anything looks amiss, please check the webpage *www.people.fas.harvard.edu/~djmorin/book.html* for a list of typos, updates, additional material, etc. And please let me know if you discover something that isn't already posted. Suggestions are always welcome.

David Morin
Cambridge, MA

Chapter 1

Combinatorics

TO THE READER: This book is available as both a paperback and an eBook. I have made a few chapters available on the web, but it is possible (based on past experience) that a pirated version of the complete book will eventually appear on file-sharing sites. In the event that you are reading such a version, I have a request:

If you don't find this book useful (in which case you probably would have returned it, if you had bought it), or if you do find it useful but aren't able to afford it, then no worries; carry on. However, if you do find it useful and are able to afford the Kindle eBook (priced below $10), then please consider purchasing it (available on Amazon). If you don't already have the Kindle reading app for your computer, you can download it free from Amazon. I chose to self-publish this book so that I could keep the cost low. The resulting eBook price of around $10, which is very inexpensive for a 350-page math book, is less than a movie and a bag of popcorn, with the added bonus that the book lasts for more than two hours and has zero calories (if used properly!).

– David Morin

Combinatorics is the study of how to count things. By "things" we mean the various combinations, permutations (different orderings), subgroups, and so on, that can be formed from a given set of objects/people/etc. For example, how many different outcomes are possible if you flip a coin four times? How many different full-house hands are there in poker? How many different committees of three people can be chosen from five people? What if we additionally designate one person as the committee's president? Knowing how to count these types of things is critical for an understanding of probability, because when calculating the probability of a given event, we often need to count the number of ways that the event can happen.

The outline of this chapter is as follows. In Section 1.1 we introduce the concept of *factorials*, which are ubiquitous in the study of probability. In Section 1.2 we learn how to count the number of possible permutations (orderings) of a set of objects. Section 1.3 covers the number of possible combined outcomes of a repeated experiment, where each repetition has an identical set of possible results. Examples include rolling dice and flipping coins. In Section 1.4 we learn how to count the number of subgroups that can be formed from a given set of objects, where the order within the subgroup matters. An example is choosing a committee of people in which all of the positions are distinct. Section 1.5 covers the related question of the number of subgroups that can be formed from a given set of objects, where the order within the subgroup *doesn't* matter. An example is a poker hand; the order of the cards in the hand is irrelevant. We find that the answer takes the form of a *binomial coefficient*. In Section 1.6 we summarize the various results we have found so far. We discover that one result is missing from our counting repertoire, and we remedy this in Section 1.7. In Section 1.8 we look at the binomial coefficients in more detail.

After learning in this chapter how to count all sorts of things, we'll see in Chapter 2 how the counting can be used to calculate probabilities. It's usually a trivial step to obtain a probability once you've counted the relevant things, so the work we do here will prove well worth it.

1.1 Factorials

Before getting into the discussion of actual combinatorics, we first need to look at a certain quantity that comes up again and again. This quantity is called the *factorial*. We'll see throughout this chapter that when dealing with a situation that involves an integer N, we often need to consider the product of the first N integers. This product is called "N factorial," and it is denoted by "$N!$".[1] For the first few integers, we have:

$$1! = 1,$$
$$2! = 1 \cdot 2 = 2,$$
$$3! = 1 \cdot 2 \cdot 3 = 6,$$
$$4! = 1 \cdot 2 \cdot 3 \cdot 4 = 24,$$
$$5! = 1 \cdot 2 \cdot 3 \cdot 4 \cdot 5 = 120,$$
$$6! = 1 \cdot 2 \cdot 3 \cdot 4 \cdot 5 \cdot 6 = 720. \tag{1.1}$$

As N increases, $N!$ gets very large very fast. For example, $10! = 3,628,800$, and $20! \approx 2.43 \cdot 10^{18}$. In Chapter 2 we will introduce an approximation to $N!$ called *Stirling's formula*. This formula makes it clear what we mean by the statement, "$N!$ gets very large very fast."

We should add that $0!$ is defined to be 1. Of course, $0!$ doesn't make much sense, because when we talk about the product of the first N integers, it is understood that we start with 1. Since 0 is below this starting point, it is unclear what $0!$ actually means. However, there is no need to try too hard to make sense of it, because as we'll see below, if we simply define $0!$ to be 1, then a number of formulas turn out to be very nice.

---

[1] I don't know why someone long ago picked the exclamation mark for this notation. But just remember that it has nothing to do with the more common grammatical use of the exclamation mark for emphasis. So try not to get too excited when you see "$N!$"!

Having defined $N!$, we can now start counting things. With the exception of the result in Section 1.3, all of the main results in this chapter involve factorials.

1.2 Permutations

A *permutation* of a set of objects is a way of ordering them. For example, if we have three people – Alice, Bob, and Carol – then one permutation of them is Alice, Bob, Carol. Another permutation is Carol, Alice, Bob. Another is Bob, Alice, Carol. It turns out that there are six permutations in all, as we will see below. The goal of this section is to learn how to count the number of possible permutations. We'll do this by starting off with the very simple case where we have only one object. Then we'll consider two objects, then three, and so on, until we see a pattern. The route we take here will be a common one throughout this book: Although many of the results can be derived in a few lines of reasoning, we'll take the longer route where we start with a few simple examples and then generalize until we arrive at the desired results. Concrete examples always make it easier to understand a general result.

One object

If we have only one object, then there is clearly only one way to "order" it; there is no ordering to be done. A list of one object simply consists of that one object, and that's that. (If we use the notation where P_N stands for the number of permutations of N objects, then we have $P_1 = 1$.)

Two objects

With two objects, things aren't completely trivial like they are in the one-object case, but they're still very simple. If we label our two objects as 1 and 2, then we can order them in two ways:

$$1\ 2 \qquad \text{or} \qquad 2\ 1$$

So we have $P_2 = 2$. At this point, you might be thinking that this result, along with the above $P_1 = 1$ result, suggests that $P_N = N$ for any positive integer N. This would mean that there should be three different ways to order three objects. Well, not so fast...

Three objects

Things get more interesting with three objects. If we call them 1, 2, and 3, then we can list out the possible orderings. The permutations are shown in Table 6.1.

$$
\begin{array}{ccc}
1\ 2\ 3 & 2\ 1\ 3 & 3\ 1\ 2 \\
1\ 3\ 2 & 2\ 3\ 1 & 3\ 2\ 1
\end{array}
$$

Table 1.1: Permutations of three objects.

So we have $P_3 = 6$. Note that we've grouped these six permutations into three subgroups (the three columns), according to which number comes first. It isn't necessary to group them this way, but we'll see below that this method of organization has definite advantages. It will simplify how we think about the case where the number of objects is a general number N.

REMARK: There is no need to use the numbers 1, 2, 3 to represent the three objects. You can use whatever symbols you want. For example, the letters A, B, C work fine, as do the letters H, Q, Z. You can even use symbols like ⊗, ♠, ♡. Or you can mix things up with ⊙, W, 7. The point is that the numbers/letters/symbols/whatever simply stand for three different things, and they need not have any meaningful properties except for their different appearances when you write them down. However, having said this, there is certainly something simple about the numbers 1, 2, 3, . . ., or the letters A, B, C, . . ., so we'll generally work with these. In any case, it is usually a good idea to be as economical as possible and not write down the full names, such as Alice, Bob, Carol, etc. ♣

Four objects

The pattern so far is $P_1 = 1$, $P_2 = 2$, and $P_3 = 6$. Although you might be able to guess the general rule from these three results, it will be easier to see the pattern if we look at the next case with four objects. Taking a cue from the above list of six permutations of three objects, let's organize the permutations of four objects (labeled 1, 2, 3, 4) according to which number comes first. We end up with the 24 permutations shown in Table 1.2.

1 2 3 4	2 1 3 4	3 1 2 4	4 1 2 3
1 2 4 3	2 1 4 3	3 1 4 2	4 1 3 2
1 3 2 4	2 3 1 4	3 2 1 4	4 2 1 3
1 3 4 2	2 3 4 1	3 2 4 1	4 2 3 1
1 4 2 3	2 4 1 3	3 4 1 2	4 3 1 2
1 4 3 2	2 4 3 1	3 4 2 1	4 3 2 1

Table 1.2: Permutations of four objects.

If we look at the last column, where all the permutations start with 4, we see that if we strip off the 4, we're simply left with the six permutations of the three numbers 1, 2, 3 that we listed in Table 6.1. A similar thing happens with the column of permutations that start with 3. If we strip off the 3, we're left with the six permutations of the numbers 1, 2, 4. Likewise for the columns of permutations that start with 2 or 1. The 24 permutations listed in Table 1.2 can therefore be thought of as four groups (the four columns), each consisting of six permutations.

Five objects

For five objects, you probably don't want to write down all the permutations, because it turns out that there are 120 of them. But you can *imagine* writing them all down. And for the present purposes, that's just as good as (or even better than) actually writing them down for real.

Consider the permutations of 1, 2, 3, 4, 5 that start with 1. From the above result for the $N = 4$ case, the other four numbers 2, 3, 4, 5 can be permuted in 24 ways. So there are 24 permutations that start with 1. Likewise, there are 24 permutations that start with 2. And similarly for 3, 4, and 5. So we have five groups (columns, if you want to imagine writing them that way), each consisting of 24 permutations. The total number of permutations of five objects is therefore $5 \cdot 24 = 120$.

General case of N objects

Collecting the above results, we have

$$P_1 = 1, \qquad P_2 = 2, \qquad P_3 = 6, \qquad P_4 = 24, \qquad P_5 = 120. \qquad (1.2)$$

Do these numbers look familiar? Yes indeed, they are simply the $N!$ results in Eq. (1.1). Does this equivalence make sense? Yes, due to the following reasoning.

- $P_1 = 1$, of course.

- $P_2 = 2$, which can be written in the suggestive form, $P_2 = 2 \cdot 1$.

- For P_3, Table 6.1 shows that $P_3 = 6$ can be thought of as three groups (characterized by which number appears first) of the $P_2 = 2$ permutations of the second and third numbers. So we have $P_3 = 3P_2 = 3 \cdot 2 \cdot 1$.

- Similarly, for P_4, Table 1.2 shows that $P_4 = 24$ can be thought of as four groups (characterized by which number appears first) of the $P_3 = 6$ permutations of the second, third, and fourth numbers. So we have $P_4 = 4P_3 = 4 \cdot 3 \cdot 2 \cdot 1$.

- Likewise, the above reasoning for $N = 5$ shows that $P_5 = 5P_4 = 5 \cdot 4 \cdot 3 \cdot 2 \cdot 1$. And so on and so forth. Therefore:

- At each stage, we have $P_N = N \cdot P_{N-1}$. Since the sequence of numbers starts with $P_1 = 1$, this relation is easily seen to be satisfied by the general formula,

$$\boxed{P_N = N!} \qquad (1.3)$$

Basically, you just need to tack on a factor of N at each stage, due to the fact that the permutations can start with any of the N numbers (or whatever objects you're dealing with). The number of permutations of N objects is therefore $N!$.

The strategy of assigning seats

An equivalent way of thinking about the $P_N = N!$ result is the following. For concreteness, let's say that we have four people, Alice, Bob, Carol, and Dave. And let's assume that they need to be assigned to four seats arranged in a line. The $N!$ result tells us that there are $4! = 24$ different permutations they can take. We'll now give an alternative derivation that shows how these 24 orderings can be understood easily by imagining the seats being filled one at a time. We'll get a lot of mileage out of this type of "seat filling" argument throughout this chapter and the next.

- There are four possibilities for who is assigned to the first seat.

- For each of these four possibilities, there are three possibilities for who is assigned to the second seat (because we've already assigned one person, so there are only three people left). There are therefore $4 \cdot 3 = 12$ possibilities for how the inhabitants of the first two seats are chosen.

- For each of these 12 possibilities, there are two possibilities for who is assigned to the third seat (because there are only two people left). There are therefore $4 \cdot 3 \cdot 2 = 24$ possibilities for how the inhabitants of the first three seats are chosen.

- Finally, for each of these 24 possibilities, there is only one possibility for who is assigned to the fourth seat (because there is only one person left, so we're stuck with him/her). There are therefore $4 \cdot 3 \cdot 2 \cdot 1 = 24$ possibilities for how the inhabitants of all four seats are chosen. The 1 here doesn't matter, of course; it just makes the formula look nicer.

You can see how this counting works for the $N = 4$ case in Table 1.2. There are four possibilities for the first entry, which stands for the person assigned to the first seat if we label the people by 1, 2, 3, 4. Once we pick the first entry, there are three possibilities for the second entry. And once we pick the second entry, there are two possibilities for the third entry. And finally, once we pick the third entry, there is only one possibility for the fourth entry. You can verify all these statements by looking at the table.

If you want to think in terms of a picture, the above process is depicted in the branching tree in Fig. 1.1. We've changed the numbers 1, 2, 3, 4 to the letters A, B, C, D, with the different possibilities at each branch being listed in alphabetical order, left to right. We've listed the four possibilities in the first stage and the twelve possibilities in the second stage. However, we haven't listed the 24 possibilities in each of the last two stages, because there isn't room in the figure. But one possibility in each stage is shown.

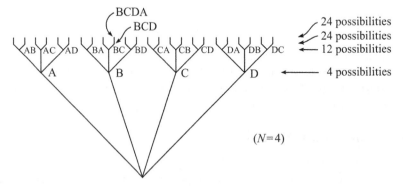

Figure 1.1: The branching tree for permutations of four objects. The number of branches in each fork decreases by one at each successive stage.

It should be emphasized that when dealing with situations that involve statements such as, "There are a possibilities for Outcome 1, and for each of these there are b possibilities for Outcome 2, and for each of these there are c possibilities for Outcome 3, and so on...," the total number of different possibilities when all of the outcomes are listed together is the *product* (not the sum!) of the numbers of possibilities for the different outcomes, that is, $a \cdot b \cdot c \cdots$. You should stare at Table 1.2 and Fig. 1.1 until you're comfortable with this. The reason for the product boils down to the words, "...for *each* of these...," in the above statement. As a simple analogy, if 7 people are each carrying 3 books, then there are $7 \cdot 3 = 21$ (not $7 + 3 = 10$) books in all.

Example (Five plus four): Nine people are to be assigned to nine seats in a row, with the stipulation that five specific people go in the left five seats, and the remaining four people go in the right four seats. How many different assignments can be made?

Solution: There are five ways to put someone (from the five specific people) in the leftmost seat, and then for each of these five ways there are four ways to put someone (from the remaining four of the five specific people) in the next seat, and so on. So there are $5! = 120$ ways to assign the five specific people to the left five seats. For *each* of these 5! ways, there are $4! = 24$ ways to assign the remaining four people to the right four seats (by the same reasoning as above). The total number of ways of assigning the nine people (with the given restriction) is therefore $5! \cdot 4! = 120 \cdot 24 = 2,880$.

Note that this result is much smaller than the $9! = 362,880$ result in the case where there is no restriction, that is, where any person can sit in any seat. The ratio of these two results is $9!/(5! \cdot 4!) = 126$. This sort of number (a quotient involving three factorials) will play a huge role in Section 1.5.

1.3 Ordered sets, repetitions allowed

In this section we'll learn how to count the number of possible outcomes of repeated identical processes/trials/experiments, where the order of the individual results matters. This scenario is the first of four related scenarios we'll discuss in this chapter. These are summarized later in Tables 1.11 and 1.12, with the present scenario being the upper-left one in the table. Two common examples are repeated rolls of a die and repeated flips of a coin. We'll discuss these below, but let's start off with an example that involves drawing balls from a box.

Let's say that we have a box containing five balls labeled A, B, C, D, E. We reach in and pick a ball and write down the letter. We then *put the ball back in the box*, shake the box around, and pick a second ball (which might be the same as the first ball) and write down this letter next to the first one, to the right of it (so the order matters). Equivalently, we can imagine having *two* boxes (a left one and a right one) with identical sets of balls labeled A, B, C, D, E, and we pick one ball from each box. We can think about it either way. The point is that the process of picking a ball is identical each time. We'll refer to this kind of setup in various equivalent ways, but you should remember that all of the following phrases mean the same thing:

- identical trials

- with replacement

- repetitions allowed

Basically, *identical* trials can be constructed by *placing* the ball you just drew back in the box, which means that it's possible for a future ball to be a *repeat* of a ball you've already drawn. Of course, with things like dice and coins, the trials are inherently identical, which means that repetitions are automatically allowed. So we don't need to talk about replacement. You don't remove the dots on a die after you roll it!

How many possible pairs of letters (where repetition is allowed and where the order matters) can we pick in the above five-ball example? More generally, how many different ordered sets of letters can we pick if we do n trials instead of only two? Or if we have N balls instead of five?

In the case of $N = 5$ balls and $n = 2$ trials, the various possibilities are shown in Table 1.3. There are five possibilities for the first pick (represented by the five columns in the table), and then for each of these there are five possibilities for the second pick (represented by the five different entries in each column, or equivalently by the five rows). The total number of possible pairs of letters is therefore $5 \cdot 5 = 25$. Remember that the order matters. So AC is different from CA, for example.

A A	B A	C A	D A	E A
A B	B B	C B	D B	E B
A C	B C	C C	D C	E C
A D	B D	C D	D D	E D
A E	B E	C E	D E	E E

Table 1.3: Drawing two balls from a box containing five balls, with replacement.

If we do only $n = 1$ trial instead of two, then there are of course just $5^1 = 5$ possibilities. Instead of the square in Table 1.3, we simply have one column (just looking at the second letter in each pair), or one row (just looking at the first letter in each pair).

If we increase the number of trials to $n = 3$, then the square in Table 1.3 becomes a cube, with the third axis (pointing into the page) representing the third pick. For each of the 5^2 possibilities in Table 1.3 for the first two letters, there are five possibilities for the third, yielding $5^2 \cdot 5 = 5^3 = 125$ possible triplets in all. Again remember that the order matters. So AAD is different from ADA, for example.

Similarly, $n = 4$ trials yield $5^3 \cdot 5 = 5^4 = 625$ possibilities. In this case the corresponding geometrical shape is a 4-dimensional hypercube – not exactly an easy thing to visualize! Now, the point of listing out the possibilities in a convenient geometrical shape is that it can help you do the counting. However, if the geometrical shape is a pain to visualize, then you shouldn't bother with it. Fortunately there is no need to visualize higher-dimensional cubes. The above pattern of reasoning tells us that there are 5^n different possible results when doing n trials of picking a letter from a 5-letter box, with replacement and with the order mattering.

More generally, if we do n trials involving a box that contains N letters instead of the specific number 5, then the total number of possible results is N^n. This is true because there are N possible results for the first pick. And then for each of these N results, there are N possible results for the second pick, yielding N^2 results for the first two picks. And then for each of these N^2 results for the first two picks, there are N possible results for the third pick, yielding N^3 results for the first three picks. And so on. Remember (as we noted near the end of Section 1.2) that the total number of results of n trials here is the *product* (not the sum) of the N possible results for each trial. So we obtain N^n (and not nN).

Our main result in this section is therefore: The number of possible outcomes when picking n objects from a box containing N distinct objects (with replacement after each stage, and with the order mattering) is:

$$\boxed{\text{Number of possible outcomes} = N^n} \qquad (1.4)$$

This N^n "power-law" result is demonstrated pictorially for $N = 3$ in the branching tree in Fig. 1.2. At each vertex, we have a choice of three paths. A diagonally leftward path corresponds to picking the letter A, an upward path corresponds to the letter B, and a diagonally rightward path corresponds to the letter C.

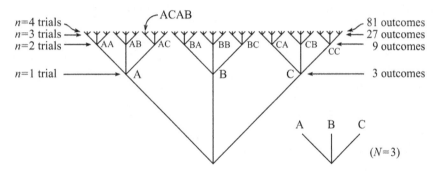

Figure 1.2: The branching tree for ordered lists chosen from three objects, with replacement.

After $n = 1$ trial, there are 3 possibilities: A, B, C. After $n = 2$ trials, there are $3^2 = 9$ possibilities: AA, AB, AC, BA, BB, BC, CA, CB, CC. After $n = 3$ trials, there are $3^3 = 27$ possibilities. We haven't labeled them in the figure, because they wouldn't fit, but they are listed in Table 1.4 (grouped in a reasonable manner).

After $n = 4$ trials, there are $3^4 = 81$ possibilities: AAAA, AAAB, etc. We have indicated one of these in Fig. 1.2, namely ACAB. (The arrow points at the middle branch of the relevant top-level triplet.) If you want to list out all 81 possibilities, you can put an A in front of all 27 entries in Table 1.4, and then a B in front of all of them, and then finally a C in front of all of them.

After another trial or two, the branches in Fig. 1.2 become too small to distinguish the different outcomes. But as with the hypercubes mentioned above, there is fortunately no need to write down the entire branching tree. The tree simply helps in understanding the N^n result.

(There are two differences between the present N^n result and the $N!$ permutation result in Eq. (1.3). First, the factors in N^n are all N's, because there are N possible

AAA	BAA	CAA
AAB	BAB	CAB
AAC	BAC	CAC
ABA	BBA	CBA
ABB	BBB	CBB
ABC	BBC	CBC
ACA	BCA	CCA
ACB	BCB	CCB
ACC	BCC	CCC

Table 1.4: The 27 ordered lists of three objects chosen from a set of three objects, with replacement.

outcomes for each of the identical trials (because we put the ball back in the box after each trial), whereas the factors in $N!$ start with N and decrease to 1 (because there is one fewer possibility at each stage; once a letter/number is used up, we can't use it again). This difference is evident in Figs. 1.1 and 1.2. In the latter, the number of branches is always the same at each stage, whereas in the former, the number of branches decreases by one at each stage. The second difference is that the N^n result involves your choice of the number n of trials (which may very well be larger than n; there is no restriction on the size of n), whereas $N!$ involves exactly the N factors from N down to 1, because we're looking at orderings of the entire set of N objects.)

Let's now look at two classic examples, involving dice and cards, where the N^n type of counting comes up.

Example 1 (Rolling dice): If you roll a standard six-sided die twice (or equivalently, roll two dice), how many different possible ordered outcomes are there?

Solution: There are six possibilities for what the first die shows, and six for the second. So there are $6^2 = 36$ possibilities in all. If you want to list them out, they are shown in Table 1.5.

1, 1	2, 1	3, 1	4, 1	5, 1	6, 1
1, 2	2, 2	3, 2	4, 2	5, 2	6, 2
1, 3	2, 3	3, 3	4, 3	5, 3	6, 3
1, 4	2, 4	3, 4	4, 4	5, 4	6, 4
1, 5	2, 5	3, 5	4, 5	5, 5	6, 5
1, 6	2, 6	3, 6	4, 6	5, 6	6, 6

Table 1.5: The 36 possible ordered outcomes for two dice rolls.

Since we are assuming that the order matters, a $2, 5$ is different from a $5, 2$. That is, rolling a 2 on the first die (or, say, the left die if you're rolling both at once) and then a 5 on the second die (or the right die) is different from rolling a 5 and then a 2. All 36 outcomes in Table 1.5 are distinct.

REMARK: This remark gets a little ahead of things, but as a precursor to our discussion of probability in the next chapter, we can ask the question: what is the probability of obtaining a sum of 7 when rolling two dice? If we look at Table 1.5, we see that six different outcomes yield a sum of 7. They are 1,6; 2,5; 3,4; 4,3; 5,2; 6,1. Since all 36 possibilities are equally likely (because the probability of any number showing up on any roll is the same, namely 1/6), and since six of the possibilities yield the desired sum of 7, the probability of rolling a sum of 7 is $6/36 = 1/6 \approx 16.7\%$. From the table, you can quickly verify that 7 is the sum that has the most outcomes corresponding to it. So 7 is the most probable sum. We'll discuss all the various nuances and subtleties about probability in the next chapter. For now, the lesson to take away from this is that the ability to count things is extremely important in calculating probabilities! ♣

Example 2 (Flipping coins): If you flip a coin four times (or equivalently, flip four coins), how many different possible ordered outcomes are there?

Solution: There are two possibilities (Heads or Tails) for what the first coin shows, and two for the second, and two for the third, and two for the fourth. So there are $2 \cdot 2 \cdot 2 \cdot 2 = 2^4 = 16$ possibilities in all. If you want to list them out, they are shown in Table 1.6.

HHHH	THHH
HHHT	THHT
HHTH	THTH
HHTT	THTT
HTHH	TTHH
HTHT	TTHT
HTTH	TTTH
HTTT	TTTT

Table 1.6: The 16 possible ordered outcomes for four coin flips.

We have grouped the various possibilities into two columns according to whether the first coin shows a Heads or a Tails. Each column has eight entries, because $2^3 = 8$ is the number of possible outcomes for three coins. (Just erase the first entry in each four-coin outcome, and then each column gives the eight possible three-coin outcomes.) Similarly, it's easy to see why five coins yield $2^5 = 32$ possible outcomes. We just need to take all 16 of the four-coin outcomes and tack on an H at the beginning, and then take all 16 again and tack on a T at the beginning. This gives $2 \cdot 16 = 32$ possible five-coin outcomes.

REMARK: As another probability teaser, we can ask: What is the probability of obtaining exactly two Heads in four coin flips? Looking at Table 1.6, we see that six outcomes have two Heads. They are HHTT, HTHT, HTTH, THHT, THTH, and TTHH. Since all 16 possibilities are equally likely (because the probability of either letter showing up on any flip is the same, namely 1/2), and since six of the possibilities yield the desired outcome of two Heads, the probability of obtaining two Heads is $6/16 = 3/8 = 37.5\%$. As with the sum of 7 in the previous example, you can quickly verify by looking at Table 1.6 that two Heads is the most likely number. ♣

1.4   Ordered sets, repetitions not allowed

In this section we will answer the question: How many different sets of n objects can be chosen from a given set of N objects, where the order matters and where repetitions are *not* allowed. This is the second of the four related scenarios summarized in Tables 1.11 and 1.12. The present scenario is the upper-right one in the table. In Section 1.5 we will answer the question for the case where the order *doesn't* matter (and where repetitions are again not allowed). Note that in both of these cases (unlike in Section 1.3 where repetitions were allowed), we must of course have $n \leq N$, because we can't use a given object/person more than once.

When dealing with situations where repetitions are *not* allowed, it is customary to talk about committees of people, because repeating a person is of course not possible (no cloning allowed!). For example, we might have 13 people, and our goal might be to assign four of them to a committee, where the order within the committee matters.

The answer to our initial question above (namely, how many ordered sets of n objects can be chosen from a given set of N objects, without replacement?) can be obtained quickly with only a slight modification of either of the $N!$ or N^n results in the preceding two sections (as we'll see below). But let's first get a feel for the problem by considering a simple example with small numbers (as is our usual strategy). Let's say that we want to choose a committee of two people from a group of five people. We'll assume that the positions on the committee are distinct. For example, one of the members might be the president. In other words, the order within the pair matters if we're listing the president first. We'll present two ways of counting the number of ordered pairs. The second method is the one that we'll be able to extend to the general case of an ordered set of n objects chosen from a given set of N objects.

Example (Two chosen from five): How many different ordered pairs of people can be chosen from a group of five people?

First solution: Let the five given people be labeled A, B, C, D, E. We'll write down all of the possible ordered pairs of letters, temporarily *including repetitions*, even though we can't actually repeat a person. As we saw in Table 1.3, there are five possibilities for the first entry, and also five possibilities for the second entry, so we end up with the 5 by 5 square of possible pairs shown in Table 1.7.

$$
\begin{array}{lllll}
\mathbf{A\,A} & B\,A & C\,A & D\,A & E\,A \\
A\,B & \mathbf{B\,B} & C\,B & D\,B & E\,B \\
A\,C & B\,C & \mathbf{C\,C} & D\,C & E\,C \\
A\,D & B\,D & C\,D & \mathbf{D\,D} & E\,D \\
A\,E & B\,E & C\,E & D\,E & \mathbf{E\,E}
\end{array}
$$

Table 1.7: Determining the number of ordered pairs chosen from five people. The five pairs in bold aren't allowed.

However, the five pairs with repeated letters (shown in bold along the diagonal of the square) aren't allowed, because the two people on the committee must of course be different. We therefore end up with $5^2 - 5 = 20$ ordered pairs. So that's our answer.

More generally, if we want to pick an ordered pair from a group of N people, we can imagine writing down an N by N square, which yields N^2 pairs, and then subtracting off the N pairs with repeated letters, which leaves us with $N^2 - N$ pairs. Note that this can be written as $N(N - 1)$.

Second solution: This second method is superior to the first one, partly because it is quicker, and partly because it can be generalized easily to larger numbers of people. (For example, we might want to pick an ordered group of four people from a group of 13 people.) Our strategy will be to pick the two committee members one at a time, just as we did at the end of Section 1.2 when we assigned people to seats.

If we have two seats that need to be filled with the two committee members, then there are five possibilities for who goes in the first seat. And then for each of these possibilities, there are four possibilities for who goes in the second seat, because there are only four people left. So there are $5 \cdot 4 = 20$ ways to plop down the two people in the two seats. This is exactly the same reasoning as with the $N!$ ways to assign N people to N seats, except that we're stopping the assignment process after two seats. So we have only the product $5 \cdot 4$ instead of the product $5 \cdot 4 \cdot 3 \cdot 2 \cdot 1$. The number of ordered pairs we can pick from five people is therefore $5 \cdot 4 = 20$, as we found above.

The preceding reasoning generalizes easily to the case where we pick ordered pairs from N people. There are N possibilities for who goes in the first seat, and then for each of these, there are $N - 1$ possibilities for who goes in the second seat. The total number of possible ordered pairs is therefore $N(N - 1)$.

Let's now consider the general case where we pick an ordered set of n objects (without replacement) from a given set of N objects. Equivalently, we're picking a committee of n people from a group N people, where all n positions on the committee are distinct.

The first method in the above example works for any value of N, provided that $n = 2$. However, for larger values of n, it quickly becomes intractable. As in Section 1.3, this is due to the fact that instead of the nice 2-D square we have in Table 1.7, we have a 3-D cube in the $n = 3$ case, and then higher-dimensional objects for larger values of n. Even if you don't want to think about things geometrically, the analogous counting is still difficult, because it is harder to get a handle on the n-tuples with doubly-counted (or triply-counted, etc.) people. In Table 1.7 it was clear that we simply needed to subtract off five pairs from the 25 total (or more generally, N pairs from the N^2 total). But in the $n = 3$ case, it is harder to determine the number of triplets that need to be subtracted off from the naive answer of 5^3. However, see Problem 1.3 if you want to think about how the counting works out.

In contrast with the intractability of the first method above when applied to larger values of n, the second method generalizes quickly. If we imagine assigning people to n ordered seats, there are N ways to assign a person to the first seat. And then for each of these possibilities, there are $N - 1$ ways to assign a person to the second seat (because there are only $N - 1$ people left). So there are $N(N - 1)$ possibilities for the first two seats. And then for each of these possibilities, there are

$N - 2$ ways to assign a person to the third seat (because there are only $N - 2$ people left). So there are $N(N - 1)(N - 2)$ possibilities for the first three seats. And so on, until there are $N(N - 1) \cdots (N - (n - 1))$ possibilities for all n seats. The last factor here is $N - (n - 1)$ because there are only $N - (n - 1)$ people left when choosing the person for the nth seat, since $n - 1$ people have already been chosen. Alternatively, the last factor is $N - (n - 1)$ because that makes there be n factors in the product; this is certainly true in the simple cases of $n = 2$ and $n = 3$.

If we denote by ${}_N P_n$ the number of ordered sets of n objects chosen from N objects (without repetition), then we can write our result as

$$ {}_N P_n = N(N - 1)(N - 2) \cdots (N - (n - 1)). \tag{1.5} $$

If we multiply this by 1 in the form of $(N - n)!/(N - n)!$, we see that the number of ordered sets of n objects chosen from N objects can be written in the concise form,

$$ \boxed{{}_N P_n = \frac{N!}{(N - n)!}} \qquad \text{(ordered subgroups)} \tag{1.6} $$

$\Big($ The ordered sets of n objects chosen from N objects are often called *partial permutations* (because we're permuting a partial set of the N objects) or *k-permutations* (because the letter k is often used in place of the n that we've been using).$\Big)$ Note that ${}_N P_N = N!$ (remember that $0! = 1$) of course, because if $n = N$ then we're forming an ordered list of all N objects. That is, we're forming a permutation of all N objects. So the product in Eq. (1.5) runs from N all the way down to 1.

As mentioned near the beginning of this section, our result for ${}_N P_n$ can be obtained with a quick modification to the reasoning in either of the preceding two sections. From Eq. (1.5) we see that the permutation reasoning in Section 1.2 is modified by simply truncating the product $N(N - 1)(N - 2) \cdots$ after n terms, instead of including all N terms. The modification to Fig. 1.1 is that we stop the branching at the nth level. The reasoning in Section 1.3 (involving ordered sets but with repetitions *allowed*) is modified by simply replacing the N^n product of equal factors N with the $N(N - 1) \cdots (N - (n - 1))$ product of decreasing factors. The factors get smaller because at each stage there is one fewer object/person available, since repetitions aren't allowed. (These decreasing factors lead to the $n \leq N$ restriction, as we noted above.) The modification to Fig. 1.2 is that we decrease the number of branches by one at each stage (with the restriction $n \leq N$).

1.5   Unordered sets, repetitions not allowed

In the preceding section, we considered committees/subgroups in which the order mattered. But what if the order *doesn't* matter? For example, how many ways can we pick a committee of four people from 13 people, where all members of the committee are equivalent? This is the third of the four related scenarios summarized in Tables 1.11 and 1.12. The present scenario is the lower-right one in the table. As usual, let's start off with an example involving small numbers.

Example (Two chosen from five): How many different *unordered* pairs of people can be chosen from a group of five people?

First solution: Let the five given people be labeled A, B, C, D, E. We'll write down all of the possible pairs of letters, *including repetitions* (even though we can't actually repeat a person) and *including different orderings* (even though the order doesn't matter). As in Table 1.7, there are five possibilities for the first entry, and also five possibilities for the second entry, so we end up with the 5 by 5 square of possible pairs shown in Table 1.8.

A A	B A	C A	D A	E A
A B	**B B**	C B	D B	E B
A C	B C	**C C**	D C	E C
A D	B D	C D	**D D**	E D
A E	B E	C E	D E	**E E**

Table 1.8: Determining the number of unordered pairs chosen from five people. The five pairs in bold aren't allowed, and the other pairs are all double counted.

However, as with Table 1.7, the five pairs with repeated letters (shown in bold along the diagonal of the square) aren't allowed, because the two people on the committee must of course be different. Additionally, since we aren't concerned with the order within a given pair, the lower-left triangle of 10 pairs in the table is equivalent to the upper-right triangle of 10 pairs. These two triangles are shown separated in Table 1.9. We see that we have counted every pair twice in Table 1.8. For example, AB represents the same pair as BA, and CE is the same as EC, etc. We therefore have $(5^2 - 5)/2 = 10$ unordered pairs. The subtraction of 5 gets rid of the pairs with repeated letters, and the division by 2 gets rid of the double counting due to the duplicate triangles.

				B A	C A	D A	E A
A B					C B	D B	E B
A C	B C					D C	E C
A D	B D	C D					E D
A E	B E	C E	D E				

Table 1.9: Equivalent sets of unordered pairs of people.

More generally, if we want to pick an unordered pair from a group of N people, we can imagine writing down an N by N square, which yields N^2 pairs, and then subtracting the N pairs with repeated letters. This gives $N^2 - N$ pairs. But we must then divide by 2 to get rid of the double counting; for every pair XY there is an equivalent pair YX. This yields $(N^2 - N)/2$ unordered pairs, which can also be written as $N(N-1)/2$.

Second solution: As in the second solution in the example in Section 1.4, we can imagine picking the committee members one at a time. And as before, this method will generalize quickly to larger numbers of people. If we have two seats that need to be filled with the two committee members, there are five possibilities for who goes in the first seat. And then for each of these possibilities, there are four possibilities for who goes in the second seat, because there are only four people left. So there are $5 \cdot 4 = 20$ ways to plop down the two people in the two seats. However, we double

counted every pair in this reasoning; we counted the pair XY as distinct from the pair
YX. So we need to divide by 2 since we don't care about the order. The number of
unordered pairs we can pick from five people is therefore $(5 \cdot 4)/2 = 10$, as we found
above.

The preceding reasoning generalizes easily to the case where we pick unordered pairs
from N people. There are N possibilities for who goes in the first seat, and then for
each of these, there are $N - 1$ possibilities for who goes in the second seat. This
gives $N(N - 1)$ possibilities. But since we don't care about the order, this reasoning
double counts every pair. We therefore need to divide by 2, yielding the final result of
$N(N - 1)/2$, as we found above.

Let's now consider the general case where we pick an unordered set of n objects
(without replacement) from a given set of N objects. Equivalently, we're picking a
committee of n people from a group N people, where all n positions on the committee are equivalent.

As in Section 1.4, the first method above works for any value of N, provided
that $n = 2$. But for larger values of n, it again quickly becomes intractable. In
contrast, the second method generalizes easily. From Eq. (1.6) we know that there
are $_N P_n = N!/(N - n)!$ ways of assigning people to n *ordered* seats. However,
this expression counts every *unordered* n-tuplet $n!$ times, due to the fact that our
permutation result in Eq. (1.3) tells us that there are $n!$ ways to order any group of
n people. In our $_N P_n$ counting, we counted all of these groups as distinct. Since
they are *not* distinct in the present scenario where the order doesn't matter, we must
divide by $n!$ to get rid of this overcounting. For example, if we're considering
committees of three people, the six triplets XYZ, XZY, YXZ, YZX, ZXY, ZYX are
distinct according to the $_N P_n$ counting. So we must divide by $3! = 6$ to get rid of
this overcounting. We therefore arrive at the general result: The number of sets of
n objects that can be chosen from N objects (where the order doesn't matter, and
where repetitions are not allowed) is

$$\frac{_N P_n}{n!} = \frac{N!}{n!(N - n)!}. \tag{1.7}$$

This result is commonly denoted by the *binomial coefficient* $\binom{N}{n}$, which is read as
"N choose n." We'll have much more to say about binomial coefficients in Section 1.8. Another notation for the above result is $_N C_n$, where the C stands for
"combinations." The result in Eq. (1.7) can therefore be written as

$$\boxed{_N C_n \equiv \binom{N}{n} = \frac{N!}{n!(N - n)!}} \qquad \text{(unordered subgroups)} \tag{1.8}$$

For example, the number of ways to pick an unordered committee of four people
from six people is

$$\binom{6}{4} = \frac{6!}{4!2!} = 15. \tag{1.9}$$

You should check this result by explicitly listing out the 15 groups of four people.

Note that because of our definition of $0! = 1$ in Section 1.1, Eq. (1.8) is valid even in the case of $n = N$, because we have $\binom{N}{N} = N!/N!0! = 1$. And indeed, there is only one way to pick N people from N people. You simply pick them all. Another special case is $n = 0$. This gives $\binom{N}{0} = N!/0!N! = 1$. It's a matter of semantics to say that there is one way to pick zero people from N people; you simply don't pick any of them, and that's the one way. But we'll see later on, especially when dealing with the binomial theorem, that $\binom{N}{0} = 1$ makes perfect sense.

In the end, the only difference between the $\binom{N}{n}$ result in this section (where the order *doesn't* matter) and the $_NP_n$ result in Section 1.4 (where the order *does* matter) is the division by $n!$ to get rid of the overcounting. Remember that neither of these results allows repetitions.

Example (Equal binomial coefficients): We found above that $\binom{6}{4} = 6!/(4!2!) = 15$. But note that $\binom{6}{2} = 6!/(2!4!)$ also equals 15. Both $\binom{6}{4}$ and $\binom{6}{2}$ involve the product of 2! and 4! in the denominator, and since the order doesn't matter in this product, the result is the same. We also have, for example, $\binom{11}{3} = \binom{11}{8}$. Both of these binomial coefficients equal 165. In short, any two n's that add up to N yield the same value of $\binom{N}{n}$.

(a) Demonstrate this fact mathematically.

(b) Explain in words why it is true.

Solution:

(a) Let the two n values be n_1 and n_2. If they add up to N, then they must take the forms of $n_1 = a$ and $n_2 = N - a$, for some value of a. (The above example with $N = 11$ was generated by either $a = 3$ or $a = 8$.) Our goal is to show that $\binom{N}{n_1}$ equals $\binom{N}{n_2}$. And indeed,

$$\binom{N}{n_1} = \frac{N!}{n_1!(N - n_1)!} = \frac{N!}{a!(N - a)!}, \tag{1.10}$$

$$\binom{N}{n_2} = \frac{N!}{n_2!(N - n_2)!} = \frac{N!}{(N - a)!(N - (N - a))!} = \frac{N!}{(N - a)!a!}.$$

The order of the $a!$ and $(N - a)!$ factors in the denominators doesn't matter, so the two results are equal, as desired.

In practice, when calculating $\binom{N}{n}$ by hand or on a calculator, you want to cancel the larger of the factorials in the denominator. For example, you can quickly cancel the 8! in both $\binom{11}{3}$ and $\binom{11}{8}$ and write them as $(11 \cdot 10 \cdot 9)/(3 \cdot 2 \cdot 1) = 165$.

(b) Imagine picking n objects from N objects and then putting them in a box. The number of ways to do this is $\binom{N}{n}$. But note that you generated *two* sets of objects in this process. You generated the n objects in the box, and you also generated the $N - n$ objects *outside* the box. There's nothing special about being inside the box versus being outside, so you can equivalently consider your process as a way of picking the group of $N - n$ objects that remain outside the box. Said in another way, a perfectly reasonable way of picking a committee of n members is to pick the $N - n$ members who are *not* on the committee. There is a one-to-one correspondence between each set of n objects and the complementary (remaining) set of $N - n$ objects. The number of different sets of n objects is therefore equal to the number of different sets of $N - n$ objects, as we wanted to show.

Let's now mix things up a bit and consider an example involving a committee that consists of distinct positions, but with some of the positions being held by more than one person.

Example (Three different titles): From ten people, how many ways can you form a committee of seven people consisting of a president, two (equivalent) vice presidents, and four (equivalent) regular members? We'll give four solutions.

First solution: We can start by picking an *ordered* set of seven people to sit in seven seats in a row. There are $_{10}P_7 = 10 \cdot 9 \cdot 8 \cdot 7 \cdot 6 \cdot 5 \cdot 4$ ways to do this. Let's assume that the president goes in the first seat, the two vice presidents go in the next two, and the four regular members go in the last four. Then the order in which the two vice presidents sit doesn't matter, so $_{10}P_7$ overcounts the number of distinct committees by a factor of 2!. Likewise, the order in which the four regular members sit doesn't matter, so $_{10}P_7$ overcounts the number of distinct committees by an additional factor of 4!. The actual number of distinct committees is therefore

$$\frac{_{10}P_7}{4!2!} = \frac{10 \cdot 9 \cdot 8 \cdot 7 \cdot 6 \cdot 5 \cdot 4}{4! \cdot 2!} = 12,600. \tag{1.11}$$

Second solution: There are 10 (or more precisely, $\binom{10}{1}$) ways to pick the president. And then for each of these possibilities, there are $\binom{9}{2}$ ways to pick the two vice presidents from the remaining nine people; the order doesn't matter between these two people. And then for each scenario of president and vice presidents, there are $\binom{7}{4}$ ways to pick the four regular members from the remaining seven people; again, the order doesn't matter among these four people. The total number of possible committees is therefore

$$\binom{10}{1}\binom{9}{2}\binom{7}{4} = \frac{10}{1!} \cdot \frac{9 \cdot 8}{2!} \cdot \frac{7 \cdot 6 \cdot 5 \cdot 4}{4!} = 12,600. \tag{1.12}$$

We chose not to cancel the factor of 4 in $\binom{7}{4}$ here, so that the agreement with Eq. (1.11) would be clear.

Third solution: There is no reason why the president has to be picked first, so let's instead pick, say, the four regular members first, and then the two vice presidents, and then the president. (Other orders will work perfectly well too.) There are $\binom{10}{4}$ ways to pick the four regular members, and then $\binom{6}{2}$ ways to pick the two vice presidents from the remaining six people, and then $\binom{4}{1}$ ways to pick the president from the remaining four people. The total number of possible committees is therefore

$$\binom{10}{4}\binom{6}{2}\binom{4}{1} = \frac{10 \cdot 9 \cdot 8 \cdot 7}{4!} \cdot \frac{6 \cdot 5}{2!} \cdot \frac{4}{1!} = 12,600. \tag{1.13}$$

We see that the order in which you pick the various subparts of the committee doesn't matter. It had better not matter, of course, because the number of possible committees is a definite number and can't depend on your method of counting it (assuming your method is a valid one!). Mathematically, all of the above solutions yield the same result because all of the calculations have the same product $10 \cdot 9 \cdot 8 \cdot 7 \cdot 6 \cdot 5 \cdot 4$ in the numerator and the same product $1! \cdot 2! \cdot 4!$ in the denominator.

Fourth solution: We can do the counting in yet another way. We can first pick all seven members; there are $\binom{10}{7}$ ways to do this. We can then pick the president from these seven members; there are $\binom{7}{1}$ ways to do this. We can then pick the two vice presidents from the remaining six members; there are $\binom{6}{2}$ ways to do this. We're then stuck with the remaining four members as regular members. The total number of possible committees is therefore

$$\binom{10}{7}\binom{7}{1}\binom{6}{2} = \frac{10 \cdot 9 \cdot 8}{3!} \cdot \frac{7}{1!} \cdot \frac{6 \cdot 5}{2!} = 12,600, \qquad (1.14)$$

If we multiply this expression by 4 over 4, then we have all the same factors in the numerator and denominator as we had in the previous solutions.

Of course, after picking the seven members, we could alternatively then pick, say, the four regular members from these seven, and then pick the two vice presidents from the remaining three. You can verify that this again gives 12,600 possible committees. The moral of all the above solutions is that there are usually many different ways to count things!

For another example, let's do some card counting. A standard deck of cards consists of 52 cards, with four cards (the four suits) for each of the 13 values: 2, 3, ..., 9, 10, J(Jack), Q(Queen), K(King), A(Ace). There is a nearly endless number of subgroup-counting examples relevant to the card game of poker. In the following example, the ordering will matter in some cases but not in others.

Example (Full houses): How many different full-house hands are possible in standard five-card poker? A full house consists of three cards of one value plus two cards of another. An example is 999QQ. (The suits don't matter.)

Solution: Our strategy will be to determine how many hands there are of a given type (999QQ is one type; 88833 is another; etc.) and then multiply this result by the number of different types.

If the hand consists of, say, three 9's and two queens, then there are $\binom{4}{3} = 4$ ways to choose the three 9's from the four 9's (the four suits) in the deck, and likewise $\binom{4}{2} = 6$ ways to choose the two Q's from the four Q's in the deck. So there are $4 \cdot 6 = 24$ possible full houses of the type 999QQ. Note that we used $_4C_3 = \binom{4}{3}$ and $_4C_2 = \binom{4}{2}$ here, instead of $_4P_3$ and $_4P_2$, because the order of the 9's and the order of the Q's in the hand doesn't matter.

How many different AAABB types are there? There are 13 different values of cards in the deck, so there are 13 ways to pick the value that occurs three times. And then there are 12 ways to pick the value that occurs twice, from the remaining 12 values. So there are $13 \cdot 12 = 156$ different types. Note that this result is $_{13}P_2 = 13 \cdot 12$, and *not* $_{13}C_2 = \binom{13}{2} = (13 \cdot 12)/2$, because the order *does* matter. Having three 9's and two Q's is different from having three Q's and two 9's. The total number of possible full-house hands is therefore

$$13 \cdot 12 \cdot \binom{4}{3} \cdot \binom{4}{2} = 156 \cdot 24 = 3,744. \tag{1.15}$$

This should be compared with the total number of possible poker hands, which is much larger: $\binom{52}{5} \approx 2,598,960$. The 3,744 full-house hands account for only about 0.14% of the total number of hands. Many more examples of counting poker hands are given in Problem 1.10.

REMARK: With regard to the $13 \cdot 12 = 156$ number of AAABB types, you can alternatively arrive at this by first noting that there are $\binom{13}{2} = (13 \cdot 12)/2$ possibilities for the two values that appear in the hand, and then realizing that you need to multiply by 2 because each pair of values represents two different types, depending on which of the two values occurs three times. If poker hands instead consisted of only four cards, and if a full house were defined to be a hand of the type AABB, then the number of different types *would* be $\binom{13}{2}$, because the A's and B's are equivalent; each occurs twice. ♣

1.6   What we know so far

In Sections 1.2 through 1.5 we learned how to count various things. Here is a summary of the results:

- Section 1.2: Permutations of N objects:

$$N!$$

- Section 1.3: *Ordered* sets (n objects chosen from N), with repetitions allowed:

$$N^n$$

- Section 1.4: *Ordered* sets (n objects chosen from N), with repetitions *not* allowed:

$${}_N P_n = \frac{N!}{(N-n)!}$$

- Section 1.5: *Unordered* sets (n objects chosen from N), with repetitions *not* allowed:

$${}_N C_n \equiv \binom{N}{n} = \frac{N!}{n!(N-n)!}$$

As we derived these results, we commented along the way on how they relate to each other. It is instructive to pause for a moment and collect all of these relations in one place. They are shown in Fig. 1.3 and summarized as follows.

If we start with the $N!$ result for permutations, we can obtain the ${}_N P_n$ result (for subgroups where the order matters) by simply truncating the product $N(N -$

1)$(N-2)\cdots$ after n terms instead of including all N terms down to 1. The $\binom{N}{n}$ result (for subgroups where the order doesn't matter) is then obtained from $_NP_n$ by dividing by $n!$ to get rid of the overcounting of the equivalent subgroups with different orderings.

If we instead start with the N^n result for a set of n objects chosen from N objects with replacement, we can obtain the $_NP_n$ result (where there is no replacement) by simply changing the product of the n factors $N \cdot N \cdot N \cdots$ to the product of the n factors $N(N-1)(N-2)\cdots$. Each factor decreases by 1 because there is one fewer possibility for each pick, since there is no replacement.

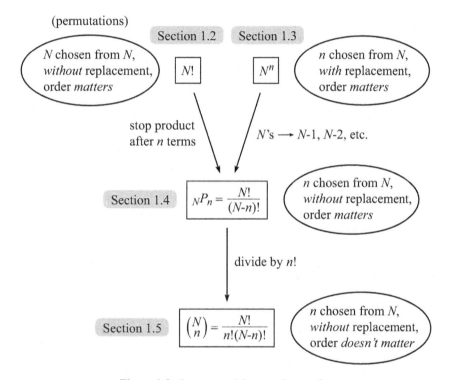

Figure 1.3: Summary of the counting results.

1.7 Unordered sets, repetitions allowed

In Fig. 1.3, the $N!$ result for permutations is somewhat of a different result from the other three (the ones in the righthand column), in that these three involve picking a general number n of objects from N objects. Permutations involve the special case where $n = N$. So let's concentrate on the three results in the righthand column. Since there are two possibilities with regard to replacement (we can replace things or not), and also two possibilities with regard to the order mattering (it matters or it doesn't), there are $2 \cdot 2 = 4$ possible scenarios when picking a general number n of objects from N objects. One scenario is therefore missing in the righthand column. This missing scenario is the one where we pick n objects from N objects, *with* replacement, and with the order *not* mattering. This is indicated in Table 1.10.

	with replacement	*without* replacement
order *matters*	N^n	$\dfrac{N!}{(N-n)!}$
order *doesn't matter*	?	$\dfrac{N!}{n!(N-n)!}$

Table 1.10: The missing result: unordered sets with repetitions allowed.

The missing scenario indicated by the question mark doesn't come up as often as the other three when solving standard problems in combinatorics and probability. And it also doesn't relate to the other three as simply as they relate to each other in Fig. 1.3. But it certainly does have its uses, so let's try to figure it out. We can't just let the question mark sit there, after all! An everyday example of this scenario is the game of Yahtzee™, where five dice are rolled in a group. The order of the dice doesn't matter, so the setup is equivalent to drawing $n = 5$ balls from a box (with replacement, and with the order not mattering), with the balls being labeled with the $N = 6$ numbers 1 through 6.

Before determining what the question mark in Table 1.10 is, let's do a few examples to get a feel for things. These examples will allow us to see a pattern and make a conjecture.

Example 1 ($n = 4$ chosen from $N = 3$): Pick $n = 4$ letters (with replacement, and with the order not mattering) from a hat containing $N = 3$ letters: A, B, C. How many different sets of four letters are possible? It turns out that there are only four different basic types of sets, so let's list them out.

- All four letters can be the same, for example AAAA. There are three sets of this type, because the common letter can be A, B, or C.

- We can have three of one letter and one of another, for example AAAB. (Remember that the order doesn't matter, so AAAB, AABA, ABAA, and BAAA are all equivalent.) There are $3 \cdot 2 = 6$ sets of this type, because there are three choices for the letter that appears three times, and then for each of these choices there are two choices for the letter that appears once.

- We can have two of one letter and two of another, for example AABB. There are three sets of this type, because there are $\binom{3}{2} = 3$ ways to choose the two letters that appear. Note that there are only $\binom{3}{2}$ ways, and not $3 \cdot 2 = 6$ ways as there are for the AAAB type, because two A's and two B's are the same as two B's and two A's.

- We can have two of one letter, one of a second, and one of the third; for example AABC. There are three sets of this type, because there are three ways to choose the letter that appears twice.

We can summarize the above results for the numbers of the different types of sets:

$$\begin{array}{cccc} \text{AAAA} & \text{AAAB} & \text{AABB} & \text{AABC} \\ 3 & 6 & 3 & 3 \end{array}$$

The total number of ways to pick four letters from a set of three letters (with replacement, and with the order not mattering) is therefore $3 + 6 + 3 + 3 = 15$. (Note that 15 can be written as $\binom{6}{2}$.) Be careful not to confuse the actual number of different sets (15 here) with the number of different *types* of sets (4 here).

REMARK: In the above discussion, we counted the *number* of different possible sets. We weren't concerned with the *probability* of actually obtaining a given set. Although we won't tackle probability until Chapter 2, we'll make one comment here.

Consider the set with, say, three C's and one A, which we label as CCCA; remember that the order of the letters doesn't matter. And consider another set with, say, four B's, which we label as BBBB. Each of these sets is one of the 15 sets that we counted above. As far as counting the possible sets goes, these two sets count equally. However, if we're concerned with the *probability* of obtaining a given set, then we must take into account the fact that while four B's can occur in only one way, three C's and one A can occur in four different ways, namely CCCA, CCAC, CACC, and ACCC. Three C's and one A are therefore four times as likely to occur as four B's. (We're assuming that the three letters A, B, C are equally likely to be drawn on each of the four draws.)

Note that when we list out each of the *ordered* sets (such as CCCA, CCAC, CACC, and ACCC) associated with a particular *unordered* set, we are now in the realm of the N^n result in Eq. (1.4). (Each of the four ordered sets just mentioned counts equally as one set in the total number of $N^n = 3^4 = 81$ ordered sets. For more on how this example relates to the N^n result in Eq. (1.4), see the last example in this section.)

But to emphasize, in the present section this difference in probabilities is irrelevant. We are simply counting the number of different unordered sets, paying no attention to the actual probability of each set. We'll have plenty to say about probability in Chapter 2. ♣

Example 2 ($n = 3$ chosen from $N = 4$): Pick $n = 3$ letters (with replacement, and with the order not mattering) from a hat containing $N = 4$ letters: A, B, C, D. There are now only three different basic types of sets. We'll just list them out, along with the number of each:

$$\begin{array}{ccc} \text{AAA} & \text{AAB} & \text{ABC} \\ 4 & 12 & 4 \end{array}$$

You can verify that the three numbers here are correct. For example, there are 12 sets of the AAB type, because there are four ways to choose the letter that appears twice, and then three ways to choose the letter that appears once. Remember that there is a fourth letter D now.

The total number of ways to pick three letters from a set of four letters (with replacement, and with the order not mattering) is therefore $4 + 12 + 4 = 20$. (Note that 20 can be written as $\binom{6}{3}$.)

Example 3 ($n = 5$ chosen from $N = 3$): Pick $n = 5$ letters (with replacement, and with the order not mattering) from a hat containing $N = 3$ letters: A, B, C. There are now five different basic types of sets, and we have:

AAAAA	AAAAB	AAABB	AAABC	AABBC
3	6	6	3	3

Again you can verify that the five numbers here are correct. For example, there are six sets of the AAABB type, because there are three ways to choose the letter that appears three times, and then two ways to choose the letter that appears twice. And there are three sets of the AABBC type, because there are three ways to choose the letter that appears once.

The total number of ways to pick five letters from a set of three letters (with replacement, and with the order not mattering) is therefore $3 + 6 + 6 + 3 + 3 = 21$. (Note that 21 can be written as $\binom{7}{2}$.)

The above results suggest that the answer to our problem (namely, how many ways are there to pick n objects from N objects, *with* replacement, and with the order *not* mattering?) most likely involves binomial coefficients. And a little trial and error shows that the above three results of $\binom{6}{2}$, $\binom{6}{3}$, and $\binom{7}{2}$ are all consistent with the expression $\binom{n+(N-1)}{N-1}$. We'll now explain why this is indeed the general form of the answer.

In the above examples, we concentrated on the different basic types of sets. However, although this gave us enough information to make an educated guess for the general result, this method becomes intractable when dealing with large values of n and N. We therefore need to think about the counting in a different way. This new way of counting is the following.

The different sets (of which there are $\binom{n+(N-1)}{N-1}$, but let's pretend we don't know this yet) are characterized by how many A's, B's, etc., there are. And since the order of the letters doesn't matter, we might as well list out the n letters by putting all the A's first, and then all the B's, and so on. We can therefore imagine putting n objects in a row and labeling them with various letters in alphabetical order. (We'll write these objects as stars, for a reason we'll get to below.) As a concrete example, let's work with $n = 6$ and $N = 3$. So we're picking six letters from A, B, C. Two possible sets are shown in Fig. 1.4. The second set happens to have no A's.

Figure 1.4: Two possible unordered sets of $n = 6$ objects chosen with replacement from $N = 3$ objects.

In writing down an arbitrary set of letters, the decision of how many of each letter to include is equivalent to the decision of where to put the transitions between the letters. If these transitions are represented by vertical dividing lines, then the two sets in Fig. 1.4 can be represented by the two configurations in Fig. 1.5.

Figure 1.5: Separating the different letters in Fig. 1.4.

The number of stars before the first dividing line is the number of A's (which is zero in the second configuration), the number of stars between the two dividing lines is the number of B's, and the number of stars after the second dividing line is the number of C's. Each different placement of the dividing lines produces a different set of letters. So our task reduces to determining how many different ways there are to plop down the dividing lines.

In the present setup with $n = 6$ and $N = 3$, we have six stars and two dividing lines, so we have eight things in all. We can therefore imagine eight spaces lined up that need to be filled, in one way or another, with six stars and two bars.[2] The two configurations in Fig. 1.5 then become the two shown in Fig. 1.6.

Figure 1.6: The stars-and-bars representations of the two sets in Fig. 1.4.

How many different ways can we plop down the stars and bars? The answer is just $\binom{8}{2} = 28$, because we simply need to pick two of the eight spaces as the ones where the bars go. Equivalently, the number of ways is $\binom{8}{6} = 28$, because we need to pick six of the eight spaces as the ones where the stars go. As an exercise, you can verify this result of 28 by explicitly counting the different sets, as we did in the above examples.

We now see where the $\binom{n+(N-1)}{N-1}$ result comes from. If we have N different letters, then we have $N - 1$ bars signifying the transitions between them. If we're picking n letters (with replacement, and with the order not mattering), then we have n stars and $N - 1$ bars that need to be plopped down in a total of $n + (N - 1)$ spaces arranged in a line. There are $\binom{n+(N-1)}{N-1}$, or equivalently $\binom{n+(N-1)}{n}$, ways to do this.

For example, if we're picking $n = 6$ letters (with replacement, and with the order not mattering) from $N = 4$ letters A, B, C, D, then there are $\binom{9}{3} = 84$ ways to do this. The ABBBDD set, for example, is represented by the configuration of six stars and three bars shown in Fig. 1.7.

---

[2]In keeping with the normal convention, we'll refer to the dividing lines as bars. The objects that we're placing down are then stars and bars, and who could possibly dislike a rhyming name like that?

Figure 1.7: The stars-and-bars representation of the ABBBDD set for $n = 6$ and $N = 4$.

If we let $_N U_n$ (with the U standing for "unordered") denote the number of ways to pick n objects from N objects (with replacement, and with the order not mattering), we can write our result as

$$_N U_n = \binom{n + (N - 1)}{N - 1} \qquad \text{(unordered, with repetition)} \qquad (1.16)$$

Remember that it is N (the number of distinct letters, or whatever) that has the 1 subtracted from it in the $\binom{n+(N-1)}{N-1}$ expression, and not n (the number of picks you make). We can now fill in the missing result in Table 1.10, as shown in Table 1.11.

	with replacement	without replacement
order matters	N^n	$\dfrac{N!}{(N-n)!}$
order doesn't matter	$\dfrac{(n+N-1)!}{n!(N-1)!}$	$\dfrac{N!}{n!(N-n)!}$
	$\binom{n+N-1}{n} = \binom{n+N-1}{N-1}$	$\binom{N}{n}$

Table 1.11: Filling in the missing result in Table 1.10.

Let's quickly verify that Eq. (1.16) holds in two simple cases.

- If n is arbitrary and $N = 1$, then Eq. (1.16) gives $_1 U_n = \binom{n}{0} = 1$. This is correct, because if there is only $N = 1$ possible result (call it A) for each of the n picks, then there is only one combined result for all n picks, namely AAAA....

- If $n = 1$ and N is arbitrary, then Eq. (1.16) gives $_N U_1 = \binom{N}{N-1} = N$. This is correct, because if there are N possible results (call them $A_1, A_2, \ldots, A_N$) for each pick, then there are simply N possible results for the $n = 1$ pick, namely A_1 or A_2 or $\ldots A_N$.

See Problems 1.11 and 1.12 for two other simple cases, namely $n = 2$ with arbitrary N, and arbitrary n with $N = 2$. See also Problem 1.13 for an alternative proof of Eq. (1.16) that doesn't use the "stars and bars" reasoning.

Let's now do an example that might seem like a non sequitur at first, but in fact is essentially the same as the "with replacement, order doesn't matter" case that we've been discussing in this section.

Example (Dividing the money): You have ten one-dollar bills that you want to divide among four people. How many different ways can you do this? For example, if the four people are seated in a line, they might receive amounts of $4, 0, 3, 3$, or $1, 6, 2, 1$, or $0, 10, 0, 0$, etc. The dollar bills are identical, but the people are not. So, for example, $4, 0, 3, 3$ is different from $3, 4, 0, 3$.

Solution: This setup is equivalent to a setup where we draw $n = 10$ letters (with replacement, and with the order not mattering) from a box containing $N = 4$ letters A, B, C, D. This equivalence can be seen as follows. Label the four people as A, B, C, D. Reach into the box and pull out a letter, and then give a dollar to the person labeled with that letter. Replace the letter, and then pick a second letter and give a dollar to that person. And so on, a total of ten times. This process will generate a string of letters, for example, CBBDCBDBAD.

Now, since it doesn't matter *when* you give a particular dollar bill to a person, the order of the letters in the string is irrelevant. The above string is therefore equivalent to ABBBBCCDDD, if we arbitrarily list the letters in alphabetical order. The four people A, B, C, D receive, respectively, 1, 4, 2, 3 dollars. There is a one-to-one correspondence between unordered strings of letters and the partitions of the dollar bills. So the desired number of partitions is equal to the number of unordered strings of $n = 10$ letters chosen from $N = 4$ letters, which we know from Eq. (1.16) is equal to

$$\binom{10 + (4 - 1)}{4 - 1} = \binom{13}{3} = 286. \tag{1.17}$$

We have now seen two equivalent scenarios where the stars-and-bars result in Eq. (1.16) applies: unordered strings of letters/objects, and money partitions. Another equivalent scenario is the number of ways to throw n identical balls into N boxes. (This is equivalent to the money-partition scenario, because in that setup we're basically throwing n dollar bills at N people.) Yet another equivalent scenario is the number of ways that N non-negative integers can add up to n. For example, if $N = 4$ and $n = 10$, then one way is $2 + 2 + 4 + 2 = 10$. Another is $3 + 0 + 6 + 1 = 10$. And yet another is $0 + 3 + 6 + 1 = 10$. (We'll assume that the order of the numbers matters.) This is equivalent to the money-partition scenario, because the four numbers in the preceding sums correspond to the amounts of money that the four people get.

The common underlying process in all of these equivalent scenarios is that we're always effectively just throwing down n identical objects onto N spaces. In our original setup with strings of letters, you can imagine throwing n darts at N letters; the number of times a number gets hit is the number of times we write it down. (Picking a letter randomly from a box is equivalent to throwing a dart randomly at the letters in the box.) In the case of N non-negative integers adding up to n, you can imagine throwing n darts at N spaces in a line; the number of times a space gets hit is the number that we write down in that space when forming the sum.

Let's now check that the two entries in the left column in Table 1.11 relate properly, at least in one particular case. The N^n result in the table holds for *ordered* sets (with replacement), while the $\binom{n + (N-1)}{N-1}$ result holds for *unordered* sets (again with replacement). We should be able to extract the former from the latter if we consider how many different ways we can order each of our unordered sets. This is demonstrated in the following example for $n = 4$ and $N = 3$.

Example (Reproducing N^n): The N^n result tells us that there are $3^4 = 81$ different *ordered* sets that are possible when drawing $n = 4$ objects (with replacement) from a hat containing $N = 3$ objects. Let's reproduce this result by considering the number of possible orderings of each of the four different basic types of *unordered* sets (AAAA, AAAB, AABB, AABC) in Example 1 at the beginning of this section. This is clearly a laborious way of producing the simple $3^4 = 81$ result for ordered sets, but it can't hurt to check that it does indeed work.

Solution: From Example 1, we know that the numbers of *unordered* sets of the four types (AAAA, AAAB, AABB, AABC) are, respectively, 3, 6, 3, and 3.

Consider the first type, where all the letters are the same. For a given unordered set of this type, there is only one way to order the letters, since they are all the same. So the total number of *ordered* sets associated with the three *unordered* sets of the AAAA type is simply $3 \cdot 1 = 3$.

Now consider the second type of set, with three of one letter and one of another. For a given unordered set of this type, there are four ways to order the letters (for example, AAAB, AABA, ABAA, BAAA). So the total number of *ordered* sets associated with the six *unordered* sets of the AAAB type is $6 \cdot 4 = 24$.

Now consider the third type of set, with two of one letter and two of another. For a given unordered set of this type, there are six ways to order the letters (for example, AABB, ABAB, ABBA, BAAB, BABA, BBAA). So the total number of *ordered* sets associated with the three *unordered* sets of the AABB type is $3 \cdot 6 = 18$.

Finally, consider the fourth type of set, with two of one letter, one of another, and one of the third. For a given unordered set of this type, there are 12 ways to order the letters. (We won't list them out, but for AABC there are four places to put the B, and then three places to put the C. Alternatively, there are $\binom{4}{2} = 6$ ways to assign the A's, and then two possible ways to order B and C.) So the total number of *ordered* sets associated with the three *unordered* sets of the AABC type is $3 \cdot 12 = 36$.

The complete total of the number of *ordered* sets involving $n = 4$ letters chosen from $N = 3$ letters (with replacement) is therefore $3 + 24 + 18 + 36 = 81$, as desired.

As mentioned at the beginning of this section, the case of unordered sets with replacement doesn't come up as often in standard probability setups as the other three cases in Table 1.11. One reason for this is that the sets in each of the other three cases are all equally likely (assuming that all of the objects in a box at a given time are equally likely to be drawn), whereas the sets in the case of unordered draws with replacement are *not* all equally likely (as we noted in the remark in the first example in this section). Counting sets in the latter case is therefore not as useful in probability as it is in the other three cases.

As we noted at the beginning of Section 1.3, the phrase "with replacement" (or "with repetition") means the same thing as "identical trials." Examples include rolling dice, flipping coins, and drawing balls from a box with replacement. In contrast, "without replacement" means the same thing as "depleting trials," that is, trials where the number of possible outcomes decreases by 1 after each trial. Examples include picking committees of people (because a given person can't be repeated, of course), assigning people to seats, and drawing balls from a box without replacement.

As far as the "order matters" and "order doesn't matter" descriptors in Table 1.11 go, you can simply imagine the ordered objects appearing in a line, and the unordered objects appearing in an amorphous group, or blob. We can therefore describe the four possibilities in Table 1.11 with the four phrases given in Table 1.12. A standard example of each case is given in Table 1.13.

	with replacement	*without* replacement
order *matters*	identical trials in a line	depleting trials in a line
order *doesn't matter*	identical trials in a blob	depleting trials in a blob

Table 1.12: Descriptions of the four possibilities with regard to ordering and replacement.

	with replacement	*without* replacement
order *matters*	ordered dice rolls	committees with distinct assignments
order *doesn't matter*	unordered dice rolls	committees with equivalent members

Table 1.13: Examples of the four possibilities with regard to ordering and replacement.

1.8 Binomial coefficients

1.8.1 Coins and Pascal's triangle

Let's look at the coin-flipping example at the end of Section 1.3 in more detail. We found that with four coins there are six different ways to obtain exactly two Heads. How many ways are there to obtain other numbers of Heads? From Table 1.6, we see that the numbers of ways to obtain exactly zero, one, two, three, or four Heads are, respectively, 1, 4, 6, 4, 1. (These same numbers are relevant for Tails too, of course.) The sum of these numbers correctly equals the total number of possibilities, which is $2^4 = 16$ from Eq. (1.4) with $N = 2$ and $n = 4$.

If we instead consider only three coins, the $2^3 = 8$ possibilities are obtained by taking either column in Table 1.6 and removing the first letter. We quickly see that the numbers of ways to obtain exactly zero, one, two, or three Heads are 1, 3, 3, 1. With two coins, the numbers for zero, one, or two Heads are 1, 2, 1. And for one coin, the numbers for zero or one Heads are just 1, 1. Also, for zero coins, you can only obtain zero Heads, and there's just one way to do this; you simply don't list anything down, and that's that. This is somewhat a matter of semantics, but if we use a "1" for this case, it will fit in nicely with the rest of the results below.

Note that for three coins, $1 + 3 + 3 + 1 = 2^3$. And for two coins, $1 + 2 + 1 = 2^2$. And for one coin, $1 + 1 = 2^1$. So in each case the total number of possibilities for n flips ends up being 2^n, consistent with Eq. (1.4).

We can collect the above results and arrange them as shown in Table 1.14. Each row lists the number of different ways we can obtain the various possible numbers of Heads. These numbers range from 0 to n.

$n = 0$:				1			
$n = 1$:			1		1		
$n = 2$:		1		2		1	
$n = 3$:	1		3		3		1
$n = 4$:	1	4		6		4	1

Table 1.14: Pascal's triangle up to $n = 4$.

The arrangement in Table 1.14 is known as *Pascal's triangle* (for $n = 4$). Do these numbers look familiar? A couple more rows might help. If you figure things out for the $n = 5$ and $n = 6$ coin-flipping cases by explicitly listing out the possibilities, you will arrive at Table 1.15.

$n = 0$:						1					
$n = 1$:					1		1				
$n = 2$:				1		2		1			
$n = 3$:			1		3		3		1		
$n = 4$:		1		4		6		4		1	
$n = 5$:	1		5		10		10		5		1
$n = 6$:	1	6		15		20		15		6	1

Table 1.15: Pascal's triangle up to $n = 6$.

At this point, you might be getting a feeling of deja vu with the 10's and 15's, since we've seen them before at various times in this chapter. You might then make the (correct) guess that the entries in Table 1.15 are nothing other than the binomial coefficients! We defined these coefficients in Section 1.5 as the number of ways of picking unordered subgroups; see Eq. (1.8). Written out explicitly in terms of the binomial coefficients, Table 1.15 becomes Table 1.16.

$n = 0$: $\binom{0}{0}$

$n = 1$: $\binom{1}{0}$ $\binom{1}{1}$

$n = 2$: $\binom{2}{0}$ $\binom{2}{1}$ $\binom{2}{2}$

$n = 3$: $\binom{3}{0}$ $\binom{3}{1}$ $\binom{3}{2}$ $\binom{3}{3}$

$n = 4$: $\binom{4}{0}$ $\binom{4}{1}$ $\binom{4}{2}$ $\binom{4}{3}$ $\binom{4}{4}$

$n = 5$: $\binom{5}{0}$ $\binom{5}{1}$ $\binom{5}{2}$ $\binom{5}{3}$ $\binom{5}{4}$ $\binom{5}{5}$

$n = 6$: $\binom{6}{0}$ $\binom{6}{1}$ $\binom{6}{2}$ $\binom{6}{3}$ $\binom{6}{4}$ $\binom{6}{5}$ $\binom{6}{6}$

Table 1.16: Binomial coefficients up to $n = 6$.

Now, observing a pattern and guessing the correct rule is most of the battle. But is there a way to prove that the entries in Table 1.16 (which are the numbers of ways of obtaining the various numbers of Heads) are in fact equal to the binomial coefficients (which we defined as the numbers of ways of picking unordered subgroups)? For example, can we demonstrate that the number of ways of obtaining two Heads in six coin flips is $\binom{6}{2}$? Indeed we can. It's actually almost a matter of definition, as the following reasoning shows.

If we flip six coins, we can imagine having six blank spaces on the paper that need to be filled with H's and T's. If we're considering the scenarios where two Heads come up, then we need to fill two of the blanks with H's and four of them with T's. So the question reduces to: How many different ways can we plop down two H's in six possible spots? But this is *exactly* the same question as: How many different (unordered) committees of two people can we form from six people? The equivalence of these two questions is made clear if we imagine six people sitting in a row, and we plop down an H on two of them, with the understanding that the two people who get tagged with an H are the two people on the committee.

In general, the $\binom{n}{k}$ ways that k Heads can come up in n flips of a coin correspond exactly to the $\binom{n}{k}$ committees of k people that can be chosen from n people. Each coin flip corresponds to a person, and that person is declared to be on the committee if the result of the coin flip is Heads.

1.8.2 $(a + b)^n$ and Pascal's triangle

A quick examination of Pascal's triangle in Table 1.16 shows (as we observed above) that the sum of the numbers in a given row equals 2^n. For example,

$$\binom{4}{0} + \binom{4}{1} + \binom{4}{2} + \binom{4}{3} + \binom{4}{4} = 2^4, \tag{1.18}$$

or more generally,

$$\binom{n}{0} + \binom{n}{1} + \binom{n}{2} + \cdots + \binom{n}{n-1} + \binom{n}{n} = 2^n \qquad (1.19)$$

We know that this relation must be true, because both sides represent the total number of possible outcomes for n coin flips (with the lefthand side enumerated according to how many Heads appear). But is there a way to demonstrate this equality
without invoking the fact that both sides are relevant to coin flips? Indeed there is.
We'll give the proof in Section 1.8.3, but first we need some background.

Consider the quantity $(a+b)^n$. You can quickly show that $(a+b)^2 = a^2 + 2ab + b^2$. And then you can multiply this by $(a+b)$ to arrive at $(a+b)^3 = a^3 + 3a^2b + 3ab^2 + b^3$. And then you can multiply by $(a+b)$ again to obtain the expression for
$(a+b)^4$, and so on. The results are shown in Table 1.17.

$$(a+b)^1 = \qquad\qquad a+b$$
$$(a+b)^2 = \qquad\qquad a^2 + 2ab + b^2$$
$$(a+b)^3 = \qquad\qquad a^3 + 3a^2b + 3ab^2 + b^3$$
$$(a+b)^4 = \qquad\qquad a^4 + 4a^3b + 6a^2b^2 + 4ab^3 + b^4$$
$$(a+b)^5 = \qquad a^5 + 5a^4b + 10a^3b^2 + 10a^2b^3 + 5ab^4 + b^5$$
$$(a+b)^6 = \quad a^6 + 6a^5b + 15a^4b^2 + 20a^3b^3 + 15a^2b^4 + 6ab^5 + b^6$$

Table 1.17: Binomial expansion up to $n = 6$.

The coefficients here are exactly the numbers in Table 1.15! And there is a very
good reason for this. Consider, for example, $(a+b)^5$. This is shorthand for

$$(a+b)(a+b)(a+b)(a+b)(a+b). \qquad (1.20)$$

In multiplying this out, we obtain a number of terms; 32 of them in fact, although
many take the same form. There are 32 terms because in multiplying out the five
factors of $(a+b)$, every term in the result will involve either the a or the b from
the first $(a+b)$ factor, and similarly either the a or the b from the second $(a+b)$
factor, and so on with the third, fourth, and fifth $(a+b)$ factors. Since there are two
possibilities (the a or the b) for each factor, we end up with $2^5 = 32$ different terms.

However, many of the terms are equivalent. For example, if we pick the a from
the first and third terms, and the b from the second, fourth, and fifth terms, then we
obtain $ababb$, which equals a^2b^3. Alternatively, we can pick, say, the a from the
second and fifth terms, and the b from the first, third, and fourth terms, which gives
$babba$, which also equals a^2b^3.

How many ways can we obtain an a^2b^3 product? Well, we have five choices
(the five $(a+b)$ factors) of where to pick the three b's from (or equivalently five
choices of where to pick the two a's from). So the number of ways to obtain an
a^2b^3 product is $\binom{5}{3} = 10$ (or equivalently $\binom{5}{2} = 10$), in agreement with Table 1.17.
This reasoning makes it clear why the coefficients of the terms in the expansion of

$(a + b)^n$ take the general form of $\binom{n}{k}$, where k is the power of b in a given term. (It also works if k is the power of a.)

In general, just as with the coin flips in Section 1.8.1, the $\binom{n}{k}$ ways that k b's can be chosen from the n factors of $(a+b)$ correspond exactly to the $\binom{n}{k}$ committees of k people that can be chosen from n people. Each factor of $(a + b)$ corresponds to a person, and that person is declared to be on the committee if the b is chosen from that factor.

To sum up, we have encountered three situations (committees, coins, and $(a + b)^n$) that involve binomial coefficients. And they all involve binomial coefficients for the same reason: they all deal with the number of ways that k things can be chosen from n things (unordered, and without repetition). The answer to all three of the following questions is the binomial coefficient $\binom{n}{k}$.

- How many different (unordered) committees of k people can be chosen from n people?

- Flip a coin n times. How many different outcomes involve exactly k Heads?

- Expand $(a + b)^n$. What is the coefficient of $a^{n-k} b^k$?

In each case, a binary choice is made n times, with k choices having the same result: k of the n people are given a "yes" to be on the committee, or k of the n coin flips are Heads, or k of the n factors of $(a + b)$ have a b chosen from them. Note that, as we have observed on various occasions, the three bullet points above still have the answer of $\binom{n}{k}$ if we make the complementary substitution of $k \rightarrow n - k$. This substitution is equivalent to picking k people to *not* be on the committee, or replacing Heads with Tails, or replacing $a^{n-k} b^k$ with $a^k b^{n-k}$.

A word on the order of our logic in this section. We originally *defined* the binomial coefficients in Section 1.5 as the number of ways of picking unordered subgroups (see Eq. (1.8)), and then we *showed* here that the binomial coefficients are also the coefficients that arise in the binomial expansion. This might seem a little backwards, because the name "binomial coefficients" suggests that they should be defined via the binomial expansion. But since we encountered unordered subgroups first in this chapter, we chose to take the "backwards" route. In the end, it doesn't matter, of course, because both results are equal to $N!/n!(N - n)!$, and it's just semantics what name we use for this quantity.

See Problem 1.8 for a generalization of the binomial coefficient, called the *multinomial coefficient*.

1.8.3 Properties of Pascal's triangle

Having established that the coefficients of the terms in the expansion of $(a + b)^n$ take the form of $\binom{n}{k}$, we can now quickly explain why the relation in Eq. (1.19) holds, without invoking anything about coin flips. The general form of the results in Table 1.17 is

$$(a+b)^n = \binom{n}{0}a^n + \binom{n}{1}a^{n-1}b + \binom{n}{2}a^{n-2}b^2 + \cdots + \binom{n}{n-1}ab^{n-1} + \binom{n}{n}b^n$$

$$= \sum_{k=0}^{n} \binom{n}{k}a^{n-k}b^k. \tag{1.21}$$

This is known as the *binomial expansion,* or *binomial theorem,* or *binomial formula.* It holds for any values of a and b. Therefore, since we are free to pick a and b to be whatever we want, let's pick them both to be 1. Multiplication by 1 doesn't affect anything, so we can just erase all of the a's and b's on the righthand side of Eq. (1.21). We then see that the righthand side is equal to the lefthand side of Eq. (1.19). But the lefthand side of Eq. (1.21) is $(1+1)^n$, which is simply 2^n, which is equal to the righthand side of Eq. (1.19). We have therefore proved Eq. (1.19).

Another nice property of Pascal's triangle, which you can verify by looking at Table 1.15, is that each number is the sum of the two numbers above it (or just the "1" above it, if it occurs at the end of a line). For example, in the $n = 6$ line, 20 is the sum of the two 10's above it (that is, $\binom{6}{3} = \binom{5}{2} + \binom{5}{3}$). And the first 15 is the sum of the 5 and 10 above it (that is, $\binom{6}{2} = \binom{5}{1} + \binom{5}{2}$). Likewise for the second 15 and all the other numbers. Written out explicitly, the general property is

$$\boxed{\binom{n}{k} = \binom{n-1}{k-1} + \binom{n-1}{k}} \tag{1.22}$$

The task of Problem 1.15 is to give a mathematical proof of this relation, using the explicit form of the binomial coefficients. But let's demonstrate it here in a more intuitive way by taking advantage of what the binomial coefficients mean in terms of choosing committees. Relations among binomial coefficients often have intuitive proofs like this which involve no (or very little) math.

In words, Eq. (1.22) says that the number of ways to pick k people from n people equals the number of ways to pick $k-1$ people from $n-1$ people, plus the number of ways to pick k people from $n-1$ people. Does this make sense? Yes indeed, due to the following reasoning.

Let's single out one of the n people, whom we will call Alice. There are two types of committees of k people: those that contain Alice, and those that don't. How many committees of each type are there? If the committee *does* contain Alice, then the other $k-1$ members must be chosen from the remaining $n-1$ people. There are $\binom{n-1}{k-1}$ ways to do this. If the committee *does not* contain Alice, then all k of the members must be chosen from the remaining $n-1$ people. There are $\binom{n-1}{k}$ ways to do this. Since each of the total $\binom{n}{k}$ number of committees falls into one or the other of these two categories, we therefore arrive at Eq. (1.22).

The task of Problem 1.16 is to reproduce the reasoning in the preceding paragraph to demonstrate Eq. (1.22), but instead in the language of coin flips or the $(a+b)^n$ binomial expansion.

1.9 Summary

In this chapter we learned how to count things. In particular, we learned:

- $N!$ ("N factorial") is defined to be the product of the first N integers:

$$N! = 1 \cdot 2 \cdot 3 \cdots (N-2) \cdot (N-1) \cdot N. \qquad (1.23)$$

- The number of different permutations of N objects (that is, the number of different ways of ordering them) is $N!$.

- Consider a process for which there are N possible results each time it is performed. If it is performed n times, then the total number of possible combined outcomes, where the order *does* matter, equals

$$N^n \qquad \text{(ordered, with repetition)} \qquad (1.24)$$

Examples include rolling an N-sided die n times, or drawing one of N balls from a box n times, with replacement each time (so that all of the draws are equivalent).

- Given N people, the number of different ways to choose an n-person committee where the order *does* matter (for example, where there are n distinct positions) equals

$$_N P_n = \frac{N!}{(N-n)!} \qquad \text{(ordered, without repetition)} \qquad (1.25)$$

- Given N people, the number of different ways to choose an n-person committee where the order *doesn't* matter is denoted by $\binom{N}{n}$, and it equals

$$_N C_n \equiv \binom{N}{n} = \frac{N!}{n!(N-n)!} \qquad \text{(unordered, without repetition)} \qquad (1.26)$$

- Consider a process for which there are N possible results each time it is performed. If it is performed n times, then the total number of possible combined outcomes, where the order *doesn't* matter, equals

$$_N U_n = \binom{n+(N-1)}{N-1} \qquad \text{(unordered, with repetition)} \qquad (1.27)$$

- The binomial coefficients $\binom{n}{k}$, which appear in Pascal's triangle, are relevant in three situations we have discussed: (1) choosing committees, (2) flipping coins, and (3) expanding $(a+b)^n$. All three of these situations involve counting the number of ways that k things can be chosen from n things (unordered, and without repetition).

1.10 Exercises

See **www.people.fas.harvard.edu/~djmorin/book.html** for a supply of problems without included solutions.

1.11   Problems

Section 1.2: Permutations

1.1. **Assigning seats** ∗

Six girls and four boys are to be assigned to ten seats in a row, with the
stipulations that a girl sits in the third seat and a boy sits in the eighth seat.
How many arrangements are possible?

Section 1.3: Ordered sets, repetitions allowed

1.2. **Number of outcomes** ∗

One person rolls(two)(six-sided) dice, and another person flips(six(two-sided)
coins. Which setup has the larger number of possible outcomes, assuming
that the order matters?

Section 1.4: Ordered sets, repetitions not allowed

1.3. **Subtracting the repeats** ∗∗

(a) From Eq. (1.6) we know that the number of ordered sets of three people
chosen from five people is $5 \cdot 4 \cdot 3 = 60$. Reproduce this result by
starting with the naive answer of $5^3 = 125$ ordered sets where repetitions
are allowed, and then subtracting off the number of triplets that have
repeated people.

(b) It's actually not much more difficult to solve this problem in the general
case where triplets are chosen from N people, instead of five. Repeat
part (a) for a general N.

1.4. **Subtracting the repeats, again** ∗∗

Repeat the task of Problem 1.3(a), but now in the case where you pick quadruplets (instead of triplets) from five people.

Section 1.5: Unordered sets, repetitions not allowed

1.5. **Sum from 1 to N** ∗

In Table 1.9 we saw that if we pick two (unordered) people from a group of
five people, the $\binom{5}{2} = 10$ possibilities can be listed as shown in Table 1.18.

$$
\begin{array}{llll}
A B & & & \\
A C & B C & & \\
A D & B D & C D & \\
A E & B E & C E & D E
\end{array}
$$

Table 1.18: Unordered pairs chosen from five people.

If we look at the number of pairs in each row, we see that we can write 10
as $1 + 2 + 3 + 4$. If we add on a sixth person, we'll need to add on a fifth
row (AF, BF, CF, DF, EF), so we see that the number of possibilities, namely

$\binom{6}{2}$ = 15, can be written as $1 + 2 + 3 + 4 + 5$. This pattern continues, and we find that the number of possible (unordered) pairs that we can pick from N people equals the sum of 1 through $N - 1$. But we already know that the number of pairs equals $\binom{N}{2} = N(N-1)/2$. So it must be the case that the sum of 1 through $N - 1$ equals $N(N - 1)/2$. Equivalently, if we replace $N - 1$ by N here, it must be the case that the sum of 1 through N equals $(N + 1)N/2$, which people usually write as $N(N + 1)/2$. Demonstrate this result in two other ways:

(a) Write down the numbers 1 through N in increasing order in a horizontal line. And then below this string of numbers, write them down again but in decreasing order. Then add each number to the one above/below it, and take it from there.

(b) First, quickly verify that the result holds for $N = 1$. Second, demonstrate mathematically that *if* the result holds for the sum of 1 through N, *then* it also holds for the sum of 1 through $N + 1$. Since the latter sum is simply $N + 1$ larger than the former, this amounts to demonstrating that $N(N + 1)/2 + (N + 1) = (N + 1)(N + 2)/2$. (The righthand side here is the proposed result, with N replaced by $N + 1$.) Third, explain why the preceding two facts imply that the result is valid for all N. The technique here is called *mathematical induction*. (This problem is an exercise more in mathematical induction than in combinatorics. But it's included here because the induction technique is something that everyone should see at least once!)

1.6. **Many ways to count** *

How many different orderings are there of the six letters: A, A, A, B, B, C? How many different ways can you think of to answer this question?

1.7. **Committees with a president** **

choose the committee
from the group

Two students are given the following problem: From N people, how many ways are there to choose a committee of n people, with one person chosen as the president? One student gives an answer of $n\binom{N}{n}$, while the other student gives an answer of $N\binom{N-1}{n-1}$. *the other committee members from remainder of group* *choose the president from the committee* *president from group*

(a) By writing out the binomial coefficients, show that the two answers are equal.

(b) Explain the (valid) reasonings that lead to these two (correct) answers.

1.8. **Multinomial coefficients** **

(a) A group of ten people are divided into three committees. Three people are on committee A, two are on committee B, and five are on committee C. How many different ways are there to divide up the people?

(b) A group of N people are divided into k committees. n_1 people are on committee 1, n_2 people are on committee 2, ..., and n_k people are on committee k, with $n_1 + n_2 + \ldots + n_k = N$. How many different ways are there to divide up the people?

1.9. **One heart and one 7** **

How many different five-card poker hands contain exactly one heart and exactly one 7? (If the hand contains the 7 of hearts, then this one card satisfies both requirements.) *Do this card first. (+ 4 other cards in hand)*
Then separate heart + 7 into 2 cards

1.10. **Poker hands** ***

In a standard 52-card deck, how many different five-card poker hands are there of each of the following types? For each type, it is understood that we don't count hands that also fall into a higher category. For example, when counting the three-of-a-kind hands, we *don't* count the full-house or four-of-a-kind hands, even though they technically contain three cards of the same kind.

   (a) Full house (three cards of one value, two of another). We already solved this in the last example in Section 1.5, but we're listing it again here so that all of the results for the various hands are contained in one place.

   (b) Straight flush (five consecutive values, all of the same suit). In the spirit of being realistic, assume that aces can be either high (above kings) or low (below 2's).

   (c) Flush (five cards of the same suit), excluding straight flushes.

   (d) Straight (five consecutive values), excluding straight flushes.

   (e) One pair.

   (f) Two pairs.

   (g) Three of a kind.

   (h) Four of a kind.

   (i) None of the above.

Section 1.7: Unordered sets, repetitions allowed

1.11. **Rolling two dice** *

   (a) Two standard 6-sided dice are rolled. Find the total number of *unordered* outcomes by looking at Table 1.5.

   (b) Find the total number of unordered outcomes by using Eq. (1.16).

   (c) By taking the lead from Table 1.5, find the total number of unordered outcomes for two N-sided dice, and then verify that your result agrees with Eq. (1.16).

1.12. **Unordered coins** *

If you flip n coins and write down the *unordered* list of Heads and Tails that you obtain, what does Eq. (1.16) give for the number of possible outcomes?

The simplicity of the result you just obtained suggests that there is alternative way of deriving it. Give an intuitive explanation of your answer that doesn't rely on Eq. (1.16).

1.13. **Proof without stars and bars** ∗∗∗

This problem gives a (longer) proof of Eq. (1.16) that doesn't rely on the stars-and-bars reasoning that we used in Section 1.7.

(a) When explicitly counting (that is, without using Eq. (1.16)) the number of unordered outcomes for n identical trials, each with N possible outcomes, we saw in Problem 1.12 that it is helpful to list the outcomes according to how many times a given individual result (such as Heads) appears. Use this strategy to count the number of possible outcomes for the $N = 3$ case (with arbitrary n). You may assume that you already know the result for the $N = 2$ case in Problem 1.12. You will need to use the result from Problem 1.5.

(b) The way in which the $N = 3$ result (with arbitrary n) follows from the $N = 2$ result (with arbitrary n) suggests an inductive proof of Eq. (1.16) for general N. By again listing (or imagining listing) the outcomes according to how many times a given individual result appears, and by making use of Problem 1.17 below (so you should look at that problem before solving this one), show inductively that if Eq. (1.16) holds for $N - 1$, then it also holds for N. (See Problem 1.5 for an explanation of mathematical induction.)

1.14. **Yahtzee** ∗∗∗

In the game of Yahtzee™, five dice are rolled in a group, with the order not mattering.

(a) Using Eq. (1.16), how many unordered rolls (sets) are possible?

(b) In the spirit of the examples at the beginning of Section 1.7, reproduce the result in part (a) by determining how many unordered rolls there are of each general type (for example, three of one number and two of another, etc.).

(c) In the spirit of the example at the end of Section 1.7, show that the total number of *ordered* Yahtzee rolls is $6^5 = 7776$.

Section 1.8: Binomial coefficients

1.15. **Pascal sum 1** ∗

Using $\binom{n}{k} = n!/k!(n-k)!$, show that

$$\binom{n}{k} = \binom{n-1}{k-1} + \binom{n-1}{k}. \tag{1.28}$$

1.16. **Pascal sum 2** **

At the end of Section 1.8.3, we demonstrated the relation $\binom{n}{k} = \binom{n-1}{k-1} + \binom{n-1}{k}$ by using an argument involving committees. Repeat this reasoning, but now in terms of:

(a) coin flips,

(b) the $(a + b)^n$ binomial expansion.

1.17. **Pascal diagonal sum** **

(a) If we pick an unordered committee of three people from five people (A, B, C, D, E), we can list the $\binom{5}{3} = 10$ possibilities as show in Table 1.19. We have grouped them according to which letter comes first. (The order of letters doesn't matter, so we've written each triplet in increasing alphabetical order.) The columns in the table tell us that we can think of 10 as equaling $6 + 3 + 1$. Explain why it makes sense to write this sum as $\binom{4}{2} + \binom{3}{2} + \binom{2}{2}$.

$$
\begin{array}{llll}
ABC & & & \\
ABD & & & \\
ABE & & & \\
ACD & BCD & & \\
ACE & BCE & & \\
ADE & BDE & CDE &
\end{array}
$$

Table 1.19: Unordered triplets chosen from five people.

(b) You can also see from Tables 1.15 and 1.16 that, for example, $\binom{6}{3} = \binom{5}{2} + \binom{4}{2} + \binom{3}{2} + \binom{2}{2}$. More generally,

$$\binom{n}{k} = \binom{n-1}{k-1} + \binom{n-2}{k-1} + \binom{n-3}{k-1} + \cdots + \binom{k}{k-1} + \binom{k-1}{k-1}. \quad (1.29)$$

In words: A given number (for example, $\binom{6}{3}$) in Pascal's triangle equals the sum of the numbers in the diagonal string that starts with the number that is above and to the left of the given number ($\binom{5}{2}$ in this case) and then proceeds upward to the right. So the string contains $\binom{5}{2}$, $\binom{4}{2}$, $\binom{3}{2}$, and $\binom{2}{2}$ in this case.

Prove Eq. (1.29) by making repeated use of Eq. (1.22), which says that each number in Pascal's triangle is the sum of the two numbers above it (or just the "1" above it, if it occurs at the end of a line). *Hint:* No math needed! You just need to draw a few pictures of Pascal's triangle after successive applications of Eq. (1.22).

1.12 Solutions

1.1. **Assigning seats**

There are six ways to pick the girl who sits in the third seat, and then for each of these choices there are four ways to pick the boy who sits in the eighth seat. For each of these $6 \cdot 4 = 24$ combinations, there are $8! = 40,320$ permutations of the remaining eight people in the remaining eight seats. The total number of possible arrangements with the given stipulations is therefore $24 \cdot 40,320 = 967,680$. This is smaller than the answer of $10!$ in the case with no stipulations, by a factor of $(6 \cdot 4 \cdot 8!)/10! = (6 \cdot 4)/(10 \cdot 9) \approx 0.27$.

1.2. **Number of outcomes**

In the case of the two six-sided dice, using $N = 6$ and $n = 2$ in Eq. (1.4) gives $6^2 = 36$ possible outcomes. In the case of the six two-sided coins, using $N = 2$ and $n = 6$ in Eq. (1.4) gives $2^6 = 64$ possible outcomes. The latter setup therefore has the larger number of possible outcomes.

If we replace the number 6 in this problem with, say, 20 (for example, we can roll the icosahedral die on the cover of this book), and if we keep the 2 the same, then the above two results become, respectively, $20^2 = 400$ and $2^{20} = 1,048,576$. The latter result is larger than the former by a factor of about 2600, whereas in the original problem the factor was only about 1.8. The two results are equal if we replace the 6 with 4 (which corresponds to a tetrahedral die).

1.3. **Subtracting the repeats**

(a) If repetitions are allowed, there are two general types of ordered triplets that contain repeated people: all three people can be the same (such as AAA), or two people can be the same, with the third being different (such as AAB). Since we are choosing from five people, there are five triplets of the first type (AAA through EEE).

How many triplets are there of the second type? There are five ways to pick the letter that appears twice, and then four ways to pick the letter that appears once from the remaining four letters. And then for each of these $5 \cdot 4 = 20$ combinations, there are three ways to order the letters (AAB, ABA, BAA). So there are $20 \cdot 3 = 60$ ordered triplets of the general type AAB.

The total number of ordered triplets that contain repeated people is therefore $5 + 60 = 65$. Subtracting this from the $5^3 = 125$ total number of ordered triplets (*with* repetitions allowed) gives $125 - 65 = 60$ ordered triplets *without* repetitions, as desired.

(b) Again, if repetitions are allowed, there are two general types of ordered triplets that contain repeated people: AAA and AAB. Since we are choosing from N people, there are now N possible letters, so there are N triplets of the first type.

How many triplets are there of the second type? There are N ways to pick the letter that appears twice, and then $N - 1$ ways to pick the letter that appears once from the remaining $N - 1$ letters. And then for each of these $N(N - 1)$ combinations, there are three ways to order the letters (AAB, ABA, BAA). So there are $N(N - 1) \cdot 3$ ordered triplets of the general type AAB.

The total number of ordered triplets that contain repeated people is therefore $N + 3N(N-1) = 3N^2 - 2N$. Our goal is to show that when this is subtracted from the N^3 total number of ordered triplets (*with* repetitions allowed), we obtain the

$N(N-1)(N-2)$ result in Eq. (1.6) for triplets *without* repetitions. So we want to show that

$$N^3 - (3N^2 - 2N) = N(N-1)(N-2). \qquad (1.30)$$

If you multiply out the righthand side, you will quickly see that the desired equality holds.

1.4. Subtracting the repeats, again

Our goal is to show that when the number of ordered quadruplets with repeated people is subtracted from the $5^4 = 625$ total number of ordered quadruplets (with repetitions allowed), we obtain the correct number $5 \cdot 4 \cdot 3 \cdot 2 = 120$ of ordered quadruplets without repetitions. If repetitions are allowed, there are four general types of ordered quadruplets that contain repeated people: AAAA, AAAB, AABB, and AABC. Let's look at each of these in turn.

- FIRST TYPE: Since we are choosing from five people, there are five quadruplets of this type (AAAA through EEEE).

- SECOND TYPE: There are five ways to pick the letter that appears three times, and then four ways to pick the letter that appears once from the remaining four letters. And then for each of these $5 \cdot 4 = 20$ combinations, there are four ways to order the letters (AAAB, AABA, ABAA, BAAA). So there are $20 \cdot 4 = 80$ ordered quadruplets of the general type AAAB.

- THIRD TYPE: There are $\binom{5}{2} = 10$ ways to pick the two letters that appear. And then for each of these combinations, there are $\binom{4}{2} = 6$ ways to order the letters (AABB, ABAB, ABBA, BBAA, BABA, BAAB). So there are $10 \cdot 6 = 60$ ordered quadruplets of the general type AABB.

- FOURTH TYPE: There are five ways to pick the letter that appears twice, and then $\binom{4}{2} = 6$ ways to pick the other two letters from the remaining four letters. And then for each of these $5 \cdot 6 = 30$ combinations, there are 12 ways to order the letters (four ways to pick the location of one of the single letters, and then three for the other). So there are $30 \cdot 12 = 360$ ordered quadruplets of the general type AABC.

The total number of ordered quadruplets that contain repeated people is therefore $5 + 80 + 60 + 360 = 505$. Subtracting this from the $5^4 = 625$ total number of ordered quadruplets (*with* repetitions allowed) gives $625 - 505 = 120$ ordered quadruplets *without* repetitions, as desired.

In the same manner as in Problem 1.3(b), you can solve this problem in the general case where quadruplets are chosen from N people, instead of five. The math gets a little messy, but in the end it comes down to replacing every 5 in the above solution with an N, and replacing the appropriate 4's with $(N-1)$'s.

1.5. Sum from 1 to N

(a) Our instructions are to write down the following two horizontal strings of numbers:

$$
\begin{array}{ccccccc}
1 & 2 & 3 & \cdots & N-2 & N-1 & N \\
N & N-1 & N-2 & \cdots & 3 & 2 & 1
\end{array}
$$

Note that every column of two numbers has the same sum, namely $N+1$. And since there are N columns, the total sum of the two rows (viewed as N columns) is $N(N+1)$. We have counted every number twice, so the sum of the numbers 1 through N is half of $N(N + 1)$, that is, $N(N + 1)/2$. As we've seen many times throughout this chapter, and as we'll see many more times, things become much clearer if you group objects in certain ways!

REMARK: One day when he was in grade school (or so the story goes), the German mathematician Carl Friedrich Gauss (1777-1855) encountered the above problem. His teacher was trying to quiet the students by giving them the task of adding up the numbers 1 through 100, thinking that it would occupy them for a while. But to the teacher's amazement, Gauss quickly came up with the correct answer, 5050, by cleverly thinking of the above method on the spot. ♣

(b) Our first task is easy. If $N = 1$ then the sum of the numbers 1 through $N = 1$ is simply 1, which equals $N(N + 1)/2$ when $N = 1$.

For our second task, if we assume that the sum of 1 through N equals $N(N + 1)/2$, then the sum of 1 through $N + 1$ is $N + 1$ more than that, so it equals

$$
\begin{aligned}
1 + 2 + 3 + \cdots + N + (N + 1) &= \left(1 + 2 + 3 + \cdots + N\right) + (N + 1) \\
&= \frac{N(N + 1)}{2} + (N + 1) \\
&= (N + 1)\left(\frac{N}{2} + 1\right) \\
&= (N + 1)\left(\frac{N + 2}{2}\right) \\
&= \frac{(N + 1)(N + 2)}{2},
\end{aligned}
\tag{1.31}
$$

which is the proposed result with N replaced by $N + 1$, as desired.

Now for the third task. We have demonstrated two facts: First, we have shown that the result (that the sum of 1 through N equals $N(N+1)/2$) holds for $N = 1$. And second, we have shown that *if* the result holds for N, *then* it also holds for $N + 1$. (This second fact is called the *inductive step* in the proof.) The combination of these two facts implies that the result holds for all N, by the following reasoning. Since the result holds for $N = 1$, the second fact implies that it also holds for $N = 2$. And then since it holds for $N = 2$, the second fact implies that it also holds for $N = 3$. And then since it holds for $N = 3$, the second fact implies that it also holds for $N = 4$. And so on. The result therefore holds for all N (positive integers).

REMARKS: This method of proof (mathematical induction) requires that you already have a guess for what the answer is. The induction reasoning then lets you rigorously prove that your guess is correct. If you don't already know the answer (which is $N(N + 1)/2$ in the present case), then mathematical induction doesn't help you. In short, with mathematical induction, you can *prove* a result, but you can't *derive* it.

Note that although it was trivial to demonstrate, the first of the above two facts (that the result holds for $N = 1$) is critical in an inductive proof. The second fact alone isn't sufficient for the proof. As an example of why this is true, let's say that someone proposes that the sum $1 + 2 + 3 + \cdots + N$ equals $N(N + 1)/2 + 73$. (Any other additive constant would serve the purpose here just as well.) This expression is obviously incorrect, even though it *does* satisfy the inductive step. This can be seen by tacking a 73 on to the $N(N + 1)/2$ term in the second line of Eq. (1.31). So our new (incorrect) guess does indeed satisfy the statement, "If it holds for N, then it also holds for $N + 1$." The problem, however, is that the "if" part of this statement is never satisfied. The guess doesn't hold for $N = 1$ (or any other value of N), so there is no number at which we can start the inductive chain of reasoning. ♣

1.6. Many ways to count

We'll present four solutions:

FIRST SOLUTION: There are $\binom{6}{3} = 20$ ways to choose where the three A's go in the six possible places. For each of these 20 ways, there are $\binom{3}{2} = 3$ ways to choose where the two B's go in the remaining three places (or equivalently $\binom{3}{1} = 3$ ways to choose where the one C goes). The total number of orderings is therefore $20 \cdot 3 = 60$.

SECOND SOLUTION: There are $\binom{6}{2} = 15$ ways to choose where the two B's go in the six possible places. For each of these 15 ways, there are $\binom{4}{3} = 4$ ways to choose where the three A's go in the remaining four places (or equivalently $\binom{4}{1} = 4$ ways to choose where the one C goes). The total number of orderings is therefore $15 \cdot 4 = 60$.

THIRD SOLUTION: There are $\binom{6}{1} = 6$ ways to choose where the C goes in the six possible places. For each of these 6 ways, there are $\binom{5}{3} = 10$ ways to choose where the three A's go in the remaining five places (or equivalently $\binom{5}{2} = 10$ ways to choose where the two B's go). The total number of orderings is therefore $6 \cdot 10 = 60$.

FOURTH SOLUTION: Let's forget for a moment that the three A's, along with the two B's, are equivalent. If we treat all six letters as distinguishable, then there are $6! = 720$ ways to order them. However, since the three A's are in fact indistinguishable, we have overcounted the number of orderings by a factor of $3!$, because that is the number of ways to order the three A's. Similarly, the two B's are indistinguishable, so we have also overcounted by $2!$. The actual number of different orderings is therefore $6!/(3!2!) = 720/(6 \cdot 2) = 60$.

1.7. Committees with a president

(a) If we write out the binomial coefficients, the equality to be demonstrated is

$$n\binom{N}{n} = N\binom{N-1}{n-1}$$

$$\Longleftrightarrow \quad n\frac{N!}{n!(N-n)!} = N\frac{(N-1)!}{(n-1)!(N-n)!}$$

$$\Longleftrightarrow \quad \frac{N!}{(n-1)!(N-n)!} = \frac{N!}{(n-1)!(N-n)!}, \tag{1.32}$$

which is indeed true.

(b) FIRST STUDENT'S REASONING: Imagine *first* picking the n committee members (there are $\binom{N}{n}$ ways to do this), and *then* picking the president from these n people (there are n ways to do this). The total number of ways to form a committee with a president is therefore $n\binom{N}{n}$.

SECOND STUDENT'S REASONING: Imagine *first* picking the president from the complete set of N people (there are N ways to do this), and *then* picking the other $n - 1$ committee members from the remaining $N - 1$ people (there are $\binom{N-1}{n-1}$ ways to do this). The total number of ways to form a committee with a president is therefore $N\binom{N-1}{n-1}$.

1.8. **Multinomial coefficients**

(a) FIRST SOLUTION: There are $\binom{10}{3}$ ways to choose the three members of committee A. And then from the remaining seven people, there are $\binom{7}{2}$ ways to choose the two members of committee B. The five remaining people are then on committee C. The total number of ways to choose the committees is therefore

$$\binom{10}{3}\binom{7}{2} = \frac{10!}{3!7!}\frac{7!}{2!5!} = \frac{10!}{3!2!5!} = 2,520. \tag{1.33}$$

Alternatively, we can use the above reasoning but consider the committees in a different order. For example, we can first pick the two members of committee B, and then the five members of committee C. This yields an answer of

$$\binom{10}{2}\binom{8}{5} = \frac{10!}{2!8!}\frac{8!}{5!3!} = \frac{10!}{2!5!3!} = 2,520. \tag{1.34}$$

Considering the committees in any other order will give the same answer, as you can check. One of the factorials will always cancel, and you will be left with the product $3!2!5!$ in the denominator.

SECOND SOLUTION: Since the numbers of people on the committees are 3, 2, and 5, the appearance of the product $3!2!5!$ in the denominator suggests that there is a more streamlined way of obtaining the answer. And indeed, imagine lining up ten seats, with the first three labeled A, the next two labeled B, and the last five labeled C. There are 10! different ways to assign the ten people to the ten seats. But the 3! possible permutations of the first three people don't change the committee A assignments, because we don't care about the order of people within a committee. So the 10! figure overcounts the number of committee assignments by 3!. We therefore need to divide 10! by 3!. Likewise, the 2! permutations of the people in the B seats and the 5! permutations of the people in the C seats don't change the committee assignments. So we also need to divide by 2! and 5!. The correct number of different committee assignments is therefore $10!/(3!2!5!)$.

(b) The reasoning in the second solution above immediately extends to the general case, so the answer is

$$\frac{N!}{n_1!n_2!\cdots n_k!}. \tag{1.35}$$

In short, there are $N!$ ways to assign N people to N seats in a row. But the $n_i!$ permutations of the people within each committee don't change the committee assignments. So $N!$ overcounts the true number of assignments by the product $n_1!n_2!\cdots n_k!$. We must therefore divide $N!$ by this product.

Alternatively, we can use the reasoning in the first solution above. There are $\binom{N}{n_1}$ ways to choose the n_1 members of committee 1. And then from the remaining $N - n_1$ people, there are $\binom{N-n_1}{n_2}$ ways to choose the n_2 members of committee 2. And so on. The total number of ways to choose the committees is therefore

$$\binom{N}{n_1}\binom{N - n_1}{n_2}\binom{N - n_1 - n_2}{n_3}\cdots$$
$$= \frac{N!}{n_1!(N - n_1)!}\cdot\frac{(N - n_1)!}{n_2!(N - n_1 - n_2)!}\cdot\frac{(N - n_1 - n_2)!}{n_3!(N - n_1 - n_2 - n_3)!}\cdots$$
$$= \frac{N!}{n_1!n_2!\cdots n_k!}. \tag{1.36}$$

Most of the factorials cancel in pairs. The last factorial in the denominator, namely $(N - n_1 - n_2 - \cdots - n_k)!$, equals $0! = 1$, because the sum of the n_i equals N.

The above result can be extended quickly to the case where only a subset of the N people are assigned to be on the committees, that is, where $\sum n_i < N$. In this case, we can simply pretend that the leftover people are on one additional committee. So we now have $k + 1$ committees, where $\sum n_i = N$. For example, if the task of this problem were instead to pick the three committees (with 3, 2, and 5 people) from a set of 16 people, then the number of possible ways would be $16!/(3!2!5!6!)$, which is about 20 million.

REMARK: The expression in Eq. (1.35) is called a *multinomial coefficient* (analogous to the binomial coefficient) and is denoted by

$$\binom{N}{n_1, n_2, \ldots, n_k} \equiv \frac{N!}{n_1! n_2! \cdots n_k!}, \tag{1.37}$$

where it is understood that $n_1 + n_2 + \cdots + n_k = N$. In the multinomial-coefficient notation, the standard binomial coefficient $\binom{N}{n}$ is written as $\binom{N}{n, N-n}$. But in this $k = 2$ case, people always just write $\binom{N}{n}$. However, for all other k, the convention is to list all k numbers in the lower entry of the coefficient.

The multinomial coefficients appear the expansion,

$$(x_1 + x_2 + \cdots + x_k)^N = \sum_{\sum n_i = N} \binom{N}{n_1, n_2, \ldots, n_k} x_1^{n_1} x_2^{n_2} \cdots x_k^{n_k}. \tag{1.38}$$

The multinomial coefficients appear here for exactly the same reason they appear in the above solution involving the number of committees. If we look at a particular $x_1^{n_1} x_2^{n_2} \cdots x_k^{n_k}$ term on the righthand side of Eq. (1.38), the n_1 factors of x_1 can come from any n_1 of the N factors of $(x_1 + x_2 + \cdots + x_k)$ on the lefthand side. Picking these n_1 factors is equivalent to picking a specific set of n_1 people to be on committee 1. Likewise for the $x_2^{n_2}$ factor and the n_2 people on committee 2. And so on. The number of ways to pick a particular $x_1^{n_1} x_2^{n_2} \cdots x_k^{n_k}$ product is therefore equal to the number of ways to pick committees of n_1, n_2, ..., n_k people. That is, the coefficient in the sum in Eq. (1.38) equals the expression in Eq. (1.35). The reasoning we used here is basically the same as the reasoning we used in Section 1.8.2 for the case of binomial coefficients. ♣

1.9. One heart and one 7

It is easiest to deal with the 7 of hearts separately. If the hand contains this card, then none of the other four cards in the hand can be a heart or a 7. There are 12 other hearts and three other 7's. So including the 7 of hearts, 16 cards are ruled out, which leaves 36. The number of ways to choose four cards from 36 is $\binom{36}{4} = 58,905$. This is therefore the number of desired hands that contain the 7 of hearts.

Now consider the hands that don't contain the 7 of hearts. There are 12 other hearts and three other 7's to choose from. So there are $12 \cdot 3 = 36$ ways to choose the two cards of the required type. For the remaining three cards, there are again 36 cards to choose from, yielding $\binom{36}{3} = 7,140$ possibilities. The total number of desired hands that lack the 7 of hearts is then $36 \cdot 7,140 = 257,040$.

The total number of desired hands (with or without the 7 of hearts) is therefore $58,905 + 257,040 = 315,945$. This is about 12% of the $\binom{52}{5} = 2,598,960$ total number of poker hands.

1.10. **Poker hands**

(a) FULL HOUSE: There are 13 ways to choose the value that appears three times, and $\binom{4}{3} = 4$ ways to choose the specific three cards from the four (the four suits) that have this value. And then there are 12 ways to choose the value that appears twice from the remaining 12 values, and $\binom{4}{2} = 6$ ways to choose the specific two cards from the four that have this value. The total number of full-house hands is therefore

$$13 \cdot \binom{4}{3} \cdot 12 \cdot \binom{4}{2} = 3{,}744. \tag{1.39}$$

(b) STRAIGHT FLUSH: The five consecutive values can be A, 2, 3, 4, 5, or 2, 3, 4, 5, 6, and so on until 10, J, Q, K, A. There are 10 of these sequences; remember that aces can be high or low. Each sequence can occur in four possible suits, so the total number of straight-flush hands is

$$4 \cdot 10 = 40. \tag{1.40}$$

Of these 40 hands, four of them are the Royal flushes, consisting of 10, J, Q, K, A (one for each suit).

(c) FLUSH: The number of ways to pick five cards from the 13 cards of a given suit is $\binom{13}{5}$. Since there are four suits, the total number of flush hands is $4 \cdot \binom{13}{5} = 5{,}148$. However, 40 of these were already counted in the straight-flush category above, so that leaves

$$4 \cdot \binom{13}{5} - 40 = 5{,}108 \tag{1.41}$$

hands that are "regular" flushes.

(d) STRAIGHT: The 10 sequences listed in part (b) are relevant here. But now there are four possible choices (the four suits) for each of the five cards. The total number of straight hands is therefore $10 \cdot 4^5 = 10{,}240$. However, 40 of these were already counted in the straight-flush category above, so that leaves

$$10 \cdot 4^5 - 40 = 10{,}200 \tag{1.42}$$

hands that are "regular" straights.

(e) ONE PAIR: There are 13 ways to pick the value that appears twice, and $\binom{4}{2} = 6$ ways to choose the specific two cards from the four that have this value. The other three values must all be different, and they must be chosen from the remaining 12 values. There are $\binom{12}{3}$ ways to do this. And then there are four possible choices (the four suits) for each of these three values, which brings in a factor of 4^3. The total number of pair hands is therefore

$$13 \cdot \binom{4}{2} \cdot \binom{12}{3} \cdot 4^3 = 1{,}098{,}240. \tag{1.43}$$

Alternatively, you can count this as $13 \cdot \binom{4}{2} \cdot 48 \cdot 44 \cdot 40/3! = 1{,}098{,}240$, because after picking the pair, there are 48 choices for the third card (because one value is off limits), then 44 choices for the fourth card (because two values are off limits), and then 40 choices for the fifth card (because three values are off limits). But we have counted the 3! possible permutations of a given set of third/fourth/fifth cards as distinct. Since the order doesn't matter, we must correct for this by dividing by 3!, which gives the above result.

Note that when counting the pair hands, we don't need to worry about double counting any flushes, because the two cards in the pair necessarily have different suits. Likewise, we don't need to worry about double counting any straights, because the two cards in the pair have the same value, by definition.

(f) TWO PAIRS: There are $\binom{13}{2}$ ways to choose the two values for the two pairs. For each pair, there are $\binom{4}{2} = 6$ ways to choose the specific two cards from the four that have that value. This brings in a factor of 6^2. And then there are 44 choices for the fifth card, since two values are off limits. The total number of two-pair hands is therefore

$$\binom{13}{2} \cdot \binom{4}{2}^2 \cdot 44 = 123{,}552. \tag{1.44}$$

(g) THREE OF A KIND: There are 13 ways to pick the value that appears three times, and $\binom{4}{3} = 4$ ways to choose the specific three cards from the four that have this value. The other two values must be different, and they must be chosen from the remaining 12 values. There are $\binom{12}{2}$ to do this. And then there are four possible choices for each of these two values, which brings in a factor of 4^2. The total number of three-of-a-kind hands is therefore

$$13 \cdot \binom{4}{3} \cdot \binom{12}{2} \cdot 4^2 = 54{,}912. \tag{1.45}$$

Alternatively, as in part (e), you can think of this as $13 \cdot \binom{4}{3} \cdot 48 \cdot 44/2! = 54{,}912$.

(h) FOUR OF A KIND: There are 13 ways to pick the value that appears four times, and then only $\binom{4}{4} = 1$ way to choose the specific four cards from the four that have this value. There are 48 choices for the fifth card, so the total number of four-of-a-kind hands is

$$13 \cdot \binom{4}{4} \cdot 48 = 624. \tag{1.46}$$

(i) NONE OF THE ABOVE: The easy way to calculate this number is to subtract the sum of the results in parts (a) through (h) from the total number of possible poker hands, namely $\binom{52}{5} = 2{,}598{,}960$. But let's do it the hard way.

We'll start by considering only the values of the cards and ignoring the suits. Since we don't want any pairs, we're concerned with hands where all five values are different (for example, $3, 4, 7, J, K$). There are $\binom{13}{5}$ ways to pick these five values. However, we also don't want any straights (such as $3, 4, 5, 6, 7$), so we must exclude these. As in parts (b) and (d), there are 10 different sequences of straights (remembering that aces can be high or low). So the number of possible none-of-the-above sets of values is $\binom{13}{5} - 10$.

We must now account for the possibility of different suits. For each of the $\binom{13}{5} - 10$ sets of values, each value has four options for its suit, so this brings in a factor of 4^5. However, we don't want any flushes, so we must exclude these. There are four possible flushes (one for each suit) for each set of values, so the number of possible none-of-the-above suit combinations for each of the $\binom{13}{5} - 10$ sets of values is $4^5 - 4$. The total number of none-of-the-above hands is therefore

$$\left(\binom{13}{5} - 10\right) \cdot (4^5 - 4) = 1{,}302{,}540. \tag{1.47}$$

These none-of-the-above hands are commonly known as "high card" hands, because the hand's rank is determined by the highest card it contains (or the second highest if there is a tie, etc.).

Let's now check that all of our results correctly add up to the total number of possible hands, $\binom{52}{5} = 2,598,960$. The various results (along with their percentages) are listed in Table 1.20 in order of increasing frequency. We see that they do indeed add up correctly. Note that one-pair and none-of-the-above hands account for 92% of the total number of hands.

Royal flush =	4	0.00015%
Straight flush (not Royal) =	36	0.0014%
Four of a kind =	624	0.024%
Full house =	3,744	0.14%
Flush (not straight flush) =	5,108	0.20%
Straight (not straight flush) =	10,200	0.39%
Three of a kind =	54,912	2.1%
Two pairs =	123,552	4.8%
One pair =	1,098,240	42.3%
None of the above =	1,302,540	50.1%
Total =	2,598,960	

Table 1.20: The numbers of different poker hands.

1.11. Rolling two dice

(a) Table 1.5 lists all $6^2 = 36$ *ordered* outcomes of two rolls. Since we aren't concerned with the order here, we are interested only in the upper-right, or the lower-left, triangle of the square (with non-repeated numbers), along with the diagonal (with repeated numbers). The upper-right, or the lower-left, triangle has $\binom{6}{2} = 15$ entries. And the diagonal has six entries. So the total number of unordered outcomes is $15 + 6 = 21$.

Alternatively, if we ignore the duplicate lower-left triangle, there are six entries in the top row, five in the second, four in the third, etc. So the total number of unordered outcomes is the sum $6 + 5 + 4 + 3 + 2 + 1 = 21$.

(b) This setup is the $N = 6$ and $n = 2$ case of Eq. (1.16), because there are $N = 6$ possible results for each of the $n = 2$ rolls. So Eq. (1.16) gives the total number of unordered outcomes of two rolls as

$$\binom{2 + (6 - 1)}{6 - 1} = \binom{7}{5} = 21. \tag{1.48}$$

(c) If we generalize Table 1.5 to an N by N square, then the upper-right, or the lower-left, triangle has $\binom{N}{2} = N(N-1)/2$ entries. And the diagonal has N entries. So the total number of unordered outcomes is $N(N-1)/2 + N = N(N+1)/2$.

Alternatively, as in part (a), if we ignore the duplicate lower-left triangle, there are N entries in the top row, $N - 1$ in the second, $N - 2$ in the third, etc. So the total number of unordered outcomes is the sum of 1 through N, which equals $N(N+1)/2$ from Problem 1.5.

This result agrees with Eq. (1.16) when $n = 2$ (with general N), because that equation gives

$$_N U_2 = \binom{2 + (N - 1)}{N - 1} = \binom{N + 1}{N - 1} = \frac{(N + 1)N}{2}. \tag{1.49}$$

1.12. **Unordered coins**

This is the $N = 2$ case (with arbitrary n) of Eq. (1.16), because there are $N = 2$ possible results for each of the n coin flips. So Eq. (1.16) gives the total number of unordered outcomes for n flips as

$$_2U_n = \binom{n + (2 - 1)}{2 - 1} = \binom{n + 1}{1} = n + 1. \tag{1.50}$$

To see why this result makes sense, consider the concrete case with, say, $n = 5$ flips. The possible outcomes are (if we arbitrarily list the Heads first in each string, since the order doesn't matter):

$$\text{HHHHH} \quad \text{HHHHT} \quad \text{HHHTT} \quad \text{HHTTT} \quad \text{HTTTT} \quad \text{TTTTT}$$

If we label each of these outcomes by the number of Tails, then we can write them as 0, 1, 2, 3, 4, and 5. There are six possibilities here. More generally, if we have n flips, the number of Tails can range from 0 to n. There are $n + 1$ possibilities here, so this is the number of unordered outcomes.

1.13. **Proof without stars and bars**

(a) With the notation $_N U_n$ for the result in Eq. (1.16), our goal is to determine $_3U_n$. Let the $N = 3$ individual results be labeled A, B, and C. We can categorize the unordered outcomes of the n trials according to the number of A's that appear. Let's do this for the concrete case of $n = 4$, to get a feel for what's going on. We'll then consider general n. We'll need to list *all* of the unordered outcomes here, as opposed to just one of each general type (as we did in the examples at the beginning of Section 1.7). The possible unordered outcomes are shown in Table 1.21.

BBBB	ABBB	AABB	AAAB	AAAA
BBBC	ABBC	AABC	AAAC	
BBCC	ABCC	AACC		
BCCC	ACCC			
CCCC				

Table 1.21: Unordered lists of $n = 4$ letters chosen from $N = 3$ letters, with replacement. The lists are grouped in columns according to how many A's appear.

This table is consistent with the results in the first example in Section 1.7, where we found that there are three sets of the AAAA type, six of the AAAB type, three of the AABB type, and three of the AABC type.

Look at each column in the table. The first column has no A's, so we're forming sets of $n = 4$ letters from the $N = 2$ other letters, B and C. The first column therefore has $_2U_4$ entries, which we see equals 5 (consistent with Problem 1.12). The second column has one A, so we're forming sets of $n = 4 - 1 = 3$ letters from the $N = 2$ letters B and C. The second column therefore has $_2U_3$ entries, which we see equals 4. Similarly, the third column has three entries, the fourth has two, and the fifth has one.

Note that even if we don't know what all the various $_2U_n$ values are, the reasoning in the preceding paragraph still tells us that if we group the sets according to the number of A's that appear, we can write down the relation,

$$_3U_4 = {}_2U_4 + {}_2U_3 + {}_2U_2 + {}_2U_1 + {}_2U_0. \tag{1.51}$$

If we then invoke the $_2U_n = n + 1$ result from Problem 1.12, the righthand side of Eq. (1.51) equals $5 + 4 + 3 + 2 + 1 = 15$. This agrees with the $\binom{6}{2}$ result in Eq. (1.16) for $n = 4$ and $N = 3$.

Now consider a general value of n instead of the specific $n = 4$ value we used above (but still with $N = 3$). The list of unordered outcomes has the same general form as in Table 1.21, except that there are now $n + 1$ columns instead of five. In the first column (with no A's), we're forming sets of n letters from the $N = 2$ other letters, B and C. In the second column (with one A), we're forming sets of $n - 1$ letters from the $N = 2$ letters B and C. And so on, until the last column has one set with n A's. For example, the possible outcomes for $n = 6$ are shown in Table 1.22.

```
BBBBBB  ABBBBB  AABBBB  AAABBB  AAAABB  AAAAAB  AAAAAA
BBBBBC  ABBBBC  AABBBC  AAABBC  AAAABC  AAAAAC
BBBBCC  ABBBCC  AABBCC  AAABCC  AAAACC
BBBCCC  ABBCCC  AABCCC  AAACCC
BBCCCC  ABCCCC  AACCCC
BCCCCC  ACCCCC
CCCCCC
```

Table 1.22: Unordered lists with $n = 6$ and $N = 3$.

The same reasoning that led to Eq. (1.51) carries through here, and we end up with

$$_3U_n = {}_2U_n + {}_2U_{n-1} + {}_2U_{n-2} + \cdots + {}_2U_1 + {}_2U_0. \qquad (1.52)$$

If we then invoke the $_2U_n = n + 1$ result from Problem 1.12, we obtain

$$_3U_n = (n + 1) + (n) + (n - 1) + \cdots + 2 + 1$$
$$= \frac{(n + 1)(n + 2)}{2}, \qquad (1.53)$$

in agreement with the $\binom{n+2}{2}$ result in Eq. (1.16) for $N = 3$. We have used the result from Problem 1.5 that the sum of the first k integers equals $k(k + 1)/2$, with $k = n + 1$ here.

(b) In the case of general N (and n), we can again group the sets of letters according to how many times a given individual letter (call it A) appears. If A doesn't appear, then we're forming sets of n letters from the $N - 1$ other letters, B, C, ... If A appears once, then we're forming sets of $n - 1$ letters from the $N - 1$ other letters. If A appears twice, then we're forming sets of $n-2$ letters from the $N - 1$ other letters. And so on, until A appears all n times, and we're forming sets of zero letters from the $N-1$ other letters. (There's only one way to do that; simply don't pick any letters.) If we add up all of these possibilities, we obtain

$$_NU_n = {}_{N-1}U_n + {}_{N-1}U_{n-1} + {}_{N-1}U_{n-2} + \cdots + {}_{N-1}U_1 + {}_{N-1}U_0. \qquad (1.54)$$

If we then invoke the inductive hypothesis that $_{N-1}U_n$ equals $\binom{n+N-2}{N-2}$ for any n, from Eq. (1.16), we can rewrite Eq. (1.54) as

$$_NU_n = \binom{n + N - 2}{N - 2} + \binom{n + N - 3}{N - 2} + \binom{n + N - 4}{N - 2} + \cdots + \binom{N - 1}{N - 2} + \binom{N - 2}{N - 2}. \qquad (1.55)$$

But this sum takes exactly the same form as the sum in the Eq. (1.29) result in Problem 1.17, which we'll copy here:

$$\binom{n}{k} = \binom{n-1}{k-1} + \binom{n-2}{k-1} + \binom{n-3}{k-1} + \cdots + \binom{k}{k-1} + \binom{k-1}{k-1}. \qquad (1.56)$$

When applying this equation, all we have to do is observe that the two entries in the binomial coefficient on the lefthand side are each 1 more than the corresponding entries in the first binomial coefficient on the righthand side. Applying this result to Eq. (1.55) yields

$$_N U_n = \binom{n+N-1}{N-1}, \qquad (1.57)$$

in agreement with Eq. (1.16), as desired.

Note that if $N = 1$ (with arbitrary n), then Eq. (1.16) gives $_1U_n = \binom{n}{0} = 1$. This is correct, because if there is only $N = 1$ possible outcome (call it A) for each of the n trials, then there is only one possible combined outcome for all n trials, namely AAAA....

We have therefore shown two things: (1) Eq. (1.16) holds for $N = 1$ (and all n), and (2) if Eq. (1.16) holds for $N - 1$ (and all n) then it also holds for N (and all n). It therefore follows inductively that Eq. (1.16) holds for all N (and n), as desired.

1.14. Yahtzee

(a) As mentioned near the beginning of Section 1.7, a roll of five dice is equivalent to drawing $n = 5$ balls in succession from a box (with replacement, and with the order not mattering), with the balls being labeled with the $N = 6$ numbers 1 through 6. So Eq. (1.16) does indeed apply. With $n = 5$ and $N = 6$, we obtain $\binom{10}{5} = 252$ possible rolls.

(b) There are seven different basic types of unordered rolls (sets):

1. All five numbers are the same, for example 11111: There are six sets of this type, because the common number can be 1, 2, 3, 4, 5, or 6.

2. Four of one number and one of another, for example 11112: (Remember that the order doesn't matter, so 11112, 11121, etc. are all equivalent.) There are $6 \cdot 5 = 30$ sets of this type, because there are six choices for the number that appears four times, and then for each of these choices there are five choices for the number that appears once.

3. Three of one number and two of another, for example 11122: There are again $6 \cdot 5 = 30$ sets of this type, because there are six choices for the number that appears three times, and then five choices for the number that appears twice.

4. Three of one number, one of a second, and one of a third, for example 11123: There are $6 \cdot 10 = 60$ sets of this type, because there are six choices for the number that appears three times, and then $\binom{5}{2} = 10$ ways to choose the other two numbers from the remaining five.

5. Two of one number, two of a second, and one of a third, for example 11223: There are again $6 \cdot 10 = 60$ sets of this type, because there are six choices for the number that appears once, and then $\binom{5}{2} = 10$ ways to choose the other two (repeated) numbers from the remaining five.

6. Two of one number and one each of three other numbers, for example 11234: There are $6 \cdot 10 = 60$ sets of this type, because there are six choices for the number that appears twice, and then $\binom{5}{3} = 10$ ways to choose the other three numbers from the remaining five.

7. One each of five numbers, for example 12345: There are six sets of this type, because there are six ways to choose the number that doesn't appear.

Let's summarize the above results for the numbers of each of the different types of unordered sets:

11111	11112	11122	11123
6	30	30	60

11223	11234	12345
60	60	6

The total number of (unordered) 5-dice Yahtzee rolls is therefore

$$6 + 30 + 30 + 60 + 60 + 60 + 6 = 252, \tag{1.58}$$

in agreement with the result in part (a).

(c) We'll now determine the number of *ordered* sets associated with each of the above seven types of *unordered* sets.

1. All five numbers are the same: For a given unordered set of this type, there is only one way to order the numbers, because they are all the same. So the total number of *ordered* sets associated with the six *unordered* sets of the 11111 type is simply $6 \cdot 1 = 6$.

2. Four of one number and one of another: For a given unordered set of this type, there are five ways to order the numbers, because there are five places to put the single number. So the total number of *ordered* sets associated with the 30 *unordered* sets of the 11112 type is $30 \cdot 5 = 150$.

3. Three of one number and two of another: For a given unordered set of this type, there are $\binom{5}{2} = 10$ ways to order the numbers, because there are $\binom{5}{2}$ places to put the two common numbers. So the total number of *ordered* sets associated with the 30 *unordered* sets of the 11122 type is $30 \cdot 10 = 300$.

4. Three of one number, one of a second, and one of a third: For a given unordered set of this type, there are 20 ways to order the numbers, because there are five places to put one of the single numbers, and then four places to put the other. So the total number of *ordered* sets associated with the 60 *unordered* sets of the 11123 type is $60 \cdot 20 = 1200$.

5. Two of one number, two of a second, and one of a third: For a given unordered set of this type, there are 30 ways to order the numbers, because there are five places to put the single number, and then $\binom{4}{2} = 6$ ways to place one of the pairs. So the total number of *ordered* sets associated with the 60 *unordered* sets of the 11223 type is $60 \cdot 30 = 1800$.

6. Two of one number and one each of three other numbers: For a given unordered set of this type, there are 60 ways to order the numbers, because there are five places to put one of the single numbers, four places to put the second, and three places to put the third. So the total number of *ordered* sets associated with the 60 *unordered* sets of the 11234 type is $60 \cdot 60 = 3600$.

7. One each of five numbers: For a given unordered set of this type, there are 120 ways to order the numbers, because there are 5! permutations of the five numbers. So the total number of *ordered* sets associated with the 6 *unordered* sets of the 12345 type is $6 \cdot 120 = 720$.

These results are summarized in Table 1.23. The entries in the "Unordered" row are the results from part (b) for the number of unordered sets of each type. Each entry in the "Ordered" row is the number of ordered sets for *each* of the unordered sets. For example, there are 5 ordered sets for *each* of the 30 un-ordered sets of the 11112 type. Each entry in the "Total" row is the total number of ordered sets of a certain type; this is the product of the entries in the "Un-ordered" and "Ordered" rows. The complete total number of *ordered* sets (rolls) involving $n = 5$ dice, each with $N = 6$ sides, is therefore

$$6 + 150 + 300 + 1200 + 1800 + 3600 + 720 = 7776, \qquad (1.59)$$

which equals 6^5, as desired.

Type	11111	11112	11122	11123	11223	11234	12345
Unordered	6	30	30	60	60	60	6
Ordered	1	5	10	20	30	60	120
Total	6	150	300	1200	1800	3600	720

Table 1.23: Verifying that the total number of ordered rolls of five dice is $6^5 = 7776$.

1.15. **Pascal sum 1**

Using $\binom{n}{k} = n!/k!(n-k)!$, the righthand side of Eq. (1.28) can be written as

$$\binom{n-1}{k-1} + \binom{n-1}{k} = \frac{(n-1)!}{(k-1)!(n-k)!} + \frac{(n-1)!}{k!(n-k-1)!}. \qquad (1.60)$$

Let's get a common denominator in these fractions, so that we can add them. The common denominator is $k!(n-k)!$, so multiplying the first fraction by k/k and the second by $(n-k)/(n-k)$ gives

$$\binom{n-1}{k-1} + \binom{n-1}{k} = \frac{k(n-1)!}{k!(n-k)!} + \frac{(n-k)(n-1)!}{k!(n-k)!}. \qquad (1.61)$$

If we cancel the $\pm k(n-1)!$ terms in the numerators, we obtain

$$\binom{n-1}{k-1} + \binom{n-1}{k} = \frac{n(n-1)!}{k!(n-k)!}$$

$$= \frac{n!}{k!(n-k)!}$$

$$= \binom{n}{k}, \qquad (1.62)$$

as desired.

1.16. **Pascal sum 2**

(a) The binomial coefficients give the number of ways of obtaining k Heads in n coin flips. So to demonstrate the given relation, we want to show that the number of ways of obtaining k Heads in n coin flips equals the number of ways of obtaining $k-1$ Heads in $n-1$ coin flips, plus the number of ways of obtaining k Heads in $n-1$ coin flips. This is true due to the following reasoning.

If we single out the first coin flip, we see that there are two basic ways to obtain k Heads: either we obtain a Heads on the first flip, or we don't. How many possibilities are there for each of these two ways? If the first flip *is* a Heads, then the other $k-1$ Heads must come from the remaining $n-1$ flips. There are $\binom{n-1}{k-1}$ ways for this to happen. If the first flip *isn't* a Heads, then all k Heads must come from the remaining $n-1$ flips. There are $\binom{n-1}{k}$ ways for this to happen. Since each of the total $\binom{n}{k}$ number of ways of obtaining k Heads falls into one or the other of these two categories, we therefore arrive at Eq. (1.22).

(b) The binomial coefficients are the coefficients of the terms in the binomial expansion of $(a+b)^n$. So to demonstrate the given equation, we want to show that the coefficient of the term involving b^k in $(a+b)^n$ equals the coefficient of the term involving b^{k-1} in $(a+b)^{n-1}$, plus the coefficient of the term involving b^k in $(a+b)^{n-1}$. This is true due to the following reasoning.

Let's write $(a+b)^n$ in the form of $(a+b) \cdot (a+b)^{n-1}$, and imagine multiplying out the $(a+b)^{n-1}$ part. The result contains many terms, but the two relevant ones are $\binom{n-1}{k-1}a^{n-k}b^{k-1}$ and $\binom{n-1}{k}a^{n-k-1}b^k$. So we have

$$(a+b)^n = (a+b)\left(\cdots + \binom{n-1}{k-1}a^{n-k}b^{k-1} + \binom{n-1}{k}a^{n-k-1}b^k + \cdots \right).$$
$$(1.63)$$

There are two ways to obtain a b^k term on the righthand side. Either the b in the first factor gets multiplied by the $\binom{n-1}{k-1}a^{n-k}b^{k-1}$ term in the second factor, or the a in the first factor gets multiplied by the $\binom{n-1}{k}a^{n-k-1}b^k$ term in the second factor. The net coefficient of the b^k term on the righthand side is therefore $\binom{n-1}{k-1} + \binom{n-1}{k}$. But the coefficient of the b^k term on the lefthand side is $\binom{n}{k}$, so we have demonstrated Eq. (1.22).

1.17. **Pascal diagonal sum**

(a) The $\binom{4}{2}$ comes from the fact that once we've chosen the first letter to be A, there are $\binom{4}{2} = 6$ ways to pick the other two letters from B, C, D, E. This yields the first column in the table. Likewise, the second column has $\binom{3}{2} = 3$ triplets starting with B and involving two letters from C, D, E. (We've already listed all the groups with A.) And the third column has $\binom{2}{2} = 1$ triplet starting with C and involving the two letters D, E. (We've already listed all the groups with A and B.)

(b) Consider an arbitrary number in Pascal's triangle, such as the one represented by the circled dot in the first triangle in Fig. 1.8. The number happens to be $\binom{5}{2}$, but the actual value isn't important here. By Eq. (1.22) this number equals the sum of the two numbers above it, as shown in the second triangle. At every stage from here on, we will replace the *righthand* of the two numbers (that were just circled) with the two numbers above it. This doesn't affect the sum, due to

Eq. (1.22). The number that just got replaced will be shown with a dotted circle. The end result is the four circled numbers in the fifth triangle in the figure; this is the desired diagonal string of numbers. Since the sum is unaffected by the replacements at each stage, the sum of the numbers in the diagonal string equals the original number in the first triangle. In this specific case, we have shown that $\binom{5}{2} = \binom{4}{1} + \binom{3}{1} + \binom{2}{1} + \binom{1}{1}$. But the result holds for any starting point.

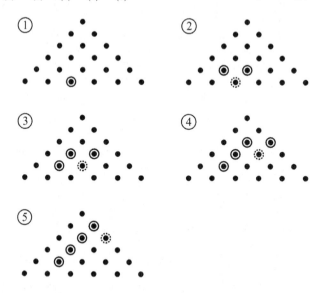

Figure 1.8: Illustration of $\binom{5}{2} = \binom{4}{1} + \binom{3}{1} + \binom{2}{1} + \binom{1}{1}$. Each number (dot) can be replaced with the two numbers above it.

REMARK: We can give another proof of Eq. (1.29) by generalizing what we observed about Table 1.19 in part (a). Let's imagine picking a committee of k people from n people, and let's label the people as 1, 2, 3, etc. When we list out the $\binom{n}{k}$ possible committees, we can arrange them in groups according to what the lowest number in the committee is. For example, some committees have a 1; other committees don't have a 1 but have a 2; other committees don't have a 1 or a 2 but have a 3; and so on. How many committees are there of each of these types?

If the lowest number is a 1, then the other $k - 1$ people on the committee must be chosen from the $n - 1$ people who are 2 or higher. There are $\binom{n-1}{k-1}$ ways to do this. Similarly, if the lowest number is a 2, then the other $k - 1$ people must be chosen from the $n - 2$ people who are 3 or higher. There are $\binom{n-2}{k-1}$ ways to do this. Likewise, if the lowest number is a 3, then the other $k - 1$ people must be chosen from the $n - 3$ people who are 4 or higher. There are $\binom{n-3}{k-1}$ ways to do this. This method of counting continues until we reach the stage where there are only $k - 1$ numbers higher than the lowest one (which occurs when the lowest number equals $n - (k - 1)$), in which case there is just $\binom{k-1}{k-1} = 1$ way to choose the other $k - 1$ people. Since the total number of possible committees is $\binom{n}{k}$, we therefore arrive at Eq. (1.29). ♣

Chapter 2

Probability

Having learned in Chapter 1 how to count things, we can now talk about probability. We will find that in many situations it is a trivial matter to generate probabilities from our counting results. So we will be justly rewarded for the time and effort we spent in Chapter 1.

The outline of this chapter is as follows. In Section 2.1 we give the definition of probability. Although this definition is fairly easy to apply in most cases, there are a number of subtleties that come up. These are discussed in Appendix A. In Section 2.2 we present the various rules of probability. We show how these can be applied in a few simple examples, and then we work through a number of more substantial examples in Section 2.3. In Section 2.4 we present four classic probability problems that many people find counterintuitive. Section 2.5 is devoted to *Bayes' theorem*, which is a relation between certain conditional probabilities. Finally, in Section 2.6 we discuss *Stirling's formula*, which gives an approximation to the ubiquitous factorial, $n!$.

2.1 Definition of probability

Probability gives a measure of how likely it is for something to happen. It can be defined as follows:

> DEFINITION OF PROBABILITY: Consider a very large number of identical trials of a certain process; for example, flipping a coin, rolling a die, picking a ball from a box (with replacement), etc. If the probability of a particular event occurring (for example, getting a Heads, rolling a 5, or picking a blue ball) is p, then the event will occur in a fraction p of the trials, on average.

Some examples are:

- The probability of getting a Heads on a coin flip is $1/2$ (or equivalently 50%). This is true because the probabilities of getting a Heads or a Tails are equal, which means that these two outcomes must each occur half of the time, on average.

- The probability of rolling a 5 on a standard 6-sided die is 1/6. This is true because the probabilities of rolling a 1, 2, 3, 4, 5, or 6 are all equal, which means that these six outcomes must each happen one sixth of the time, on average.

- If there are three red balls and seven blue balls in a box, then the probabilities of picking a red ball or a blue ball are, respectively, 3/10 and 7/10. This follows from the fact that the probabilities of picking each of the ten balls are all equal (or at least let's assume they are), which means that each ball will be picked one tenth of the time, on average. Since there are three red balls, a red ball will therefore be picked 3/10 of the time, on average. And since there are seven blue balls, a blue ball will be picked 7/10 of the time, on average.

Note the inclusion of the words "on average" in the above definition and examples. We'll discuss this in detail in the subsection below.

Many probabilistic situations have the property that they involve a number of different possible outcomes, *all of which are equally likely*. For example, Heads and Tails on a coin are equally likely to be tossed, the numbers 1 through 6 on a die are equally likely to be rolled, and the ten balls in the above box are all equally likely to be picked. In such a situation, the probability of a certain scenario happening is given by

$$p = \frac{\text{number of desired outcomes}}{\text{total number of possible outcomes}} \qquad \text{(for equally likely outcomes)} \quad (2.1)$$

Calculating a probability then simply reduces to a matter of counting the number of desired outcomes, along with the total number of outcomes. For example, the probability of rolling an even number on a die is 1/2, because there are three desired outcomes (2, 4, and 6) and six total possible outcomes (the six numbers). And the probability of picking a red ball in the above example is 3/10, as we already noted, because there are three desired outcomes (picking any of the three red balls) and ten total possible outcomes (the ten balls). These two examples involved trivial counting, but we'll encounter many examples where it is more involved. This is why we did all of that counting in Chapter 1!

It should be stressed that Eq. (2.1) holds only under the assumption that all of the possible outcomes are equally likely. But this usually isn't much of a restriction, because this assumption will generally be valid in the setups we'll be dealing with in this book. In particular, it holds in setups dealing with permutations and subgroups, both of which we studied in detail in Chapter 1. Our ability to count these sorts of things will allow us to easily calculate probabilities via Eq. (2.1). Many examples are given in Section 2.3 below.

There are three words that people often use interchangeably: "probability," "chance," and "odds." The first two of these mean the same thing. That is, the statement, "There is a 40% chance that the bus will be late," is equivalent to the statement, "There is a 40% probability that the bus will be late." However, the word "odds" has a different meaning; see Problem 2.1 for a discussion of this.

The importance of the words "on average"

The above definition of probability includes the words "on average." These words are critical, because the definition wouldn't make any sense if we omitted them and instead went with something like: "If the probability of a particular event occurring is *p*, then the event will occur in *exactly* a fraction *p* of the trials." This can't be a valid definition of probability, for the following reason. Consider the roll of one die, for which the probability of each number occurring is 1/6. This definition would imply that on one roll of a die, we will get 1/6 of a 1, and 1/6 of a 2, and so on. But this is nonsense; you can't roll 1/6 of a 1. The number of times a 1 appears on one roll must of course be either zero or one. And in general for many rolls, the number must be an integer, 0, 1, 2, 3, ...

There is a second problem with this definition, in addition to the problem of non integers. What if we roll a die six times? This definition would imply that we will get exactly $(1/6) \cdot 6 = 1$ of each number. This prediction is a little better, in that at least the proposed numbers are integers. But it still can't be correct, because if you actually do the experiment and roll a die six times, you will find that you are certainly *not* guaranteed to get each of the six numbers exactly once. This scenario *might* happen (we'll calculate the probability in Section 2.3.4 below), but it is more likely that some numbers will appear more than once, while other numbers won't appear at all.

Basically, for a small number of trials (such as six), the fractions of the time that the various events occur will most likely not look much like the various probabilities. This is where the words "very large number" in our original definition come in. The point is that if you roll a die a huge number of times, then the fractions of the time that each of the six numbers appears will be *approximately* equal to 1/6. And the larger the number of rolls, the closer the fractions will generally be to 1/6.

In Chapter 5 we'll explain why the fractions are expected to get closer and closer to the actual probabilities, as the number of trials gets larger and larger. For now, just take it on faith that if you flip a coin 100 times, the probability of obtaining either 49, 50, or 51 Heads isn't so large. It happens to be about 24%, which tells you that there is a decent chance that the fraction of Heads will deviate moderately from 1/2. However, if you flip a coin 100,000 times, the probability of obtaining Heads between 49% and 51% of the time is 99.999999975%, which tells you that there is virtually no chance that the fraction of Heads will deviate much from 1/2. If you increase the number of flips to 10^9 (a billion), this result is even more pronounced; the probability of obtaining Heads in the narrow range between 49.99% and 50.01% of the time is 99.999999975% (the same percentage as above). We'll discuss such matters in detail in Section 5.2. For more commentary on the words "on average," see the last section in Appendix A.

2.2 The rules of probability

So far we've talked only about the probabilities of single events, for example, rolling an even number on a die, getting a Heads on a coin toss, or picking a blue ball from a box. We'll now consider two (or more) events. Reasonable questions we can ask are: What is the probability that both of the events occur? What is the probability that either of the events occurs? The rules presented below will answer these questions. We'll provide a few simple examples for each rule, and then we'll work through some longer examples in Section 2.3.

2.2.1   AND: The "intersection" probability, $P(A \text{ and } B)$

Let A and B be two events. For example, if we roll two dice, we can let $A = \{$rolling a 2 on the left die$\}$ and $B = \{$rolling a 5 on the right die$\}$. Or we might have $A = \{$picking a red ball from a box$\}$ and $B = \{$picking a blue ball without replacement after the first pick$\}$. What is the probability that A and B both occur? In answering this question, we must consider two cases: (1) A and B are independent events, or (2) A and B are dependent events. Let's look at each of these in turn. In each case, the probability that A and B both occur is known as the *joint probability*.

Independent events

Two events are said to be *independent* if they don't affect each other, or more precisely, if the occurrence of one doesn't affect the probability that the other occurs. An example is the first setup mentioned above – rolling two dice, with $A = \{$rolling a 2 on the left die$\}$ and $B = \{$rolling a 5 on the right die$\}$. The probability of obtaining a 5 on the right die is 1/6, independent of what happens with the left die. And similarly the probability of obtaining a 2 on the left die is 1/6, independent of what happens with the right die. Independence requires that neither event affects the other. The events in the second setup mentioned above with the balls in the box are *not* independent; we'll talk about this below.

Another example of independent events is picking *one* card from a deck, with $A = \{$the card is a king$\}$ and $B = \{$the (same) card is a heart$\}$. The probability of the card being a heart is 1/4, independent of whether or not it is a king. And the probability of the card being a king is 1/13, independent of whether or not it is a heart. Note that it is possible to have two different events even if we have only one card. This card has two qualities (its suit and its value), and we can associate an event with each of these qualities.

REMARK: A note on terminology: The words "event" and "outcome" sometimes mean the same thing in practice, but there is technically a difference. An *outcome* is the result of an experiment. If we draw a card from a deck, then there are 52 possible outcomes; for example, the 4 of clubs, the jack of diamonds, etc. An *event* is a set of outcomes. For example, an event might be "drawing a heart." This event contains 13 outcomes, namely the 13 cards that are hearts. A given card may belong to many events. For example, in addition to belonging to the A and B events in the preceding paragraph, the king of hearts belongs to the events $C = \{$the card is red$\}$, $D = \{$the card's value is higher than 8$\}$, $E = \{$the card is the king of hearts$\}$, and so on. As indicated by the event E, an event may consist of a single outcome. An event may also be the empty set (which occurs with probability 0), or the entire set of all possible outcomes (which occurs with probability 1), which is known as the *sample space*. ♣

The "And" rule for independent events is:

- *If events A and B are independent, then the probability that they both occur equals the product of their individual probabilities:*

$$P(A \text{ and } B) = P(A) \cdot P(B) \tag{2.2}$$

We can quickly apply this rule to the two examples mentioned above. The probability of rolling a 2 on the left die and a 5 on the right die is

$$P(2 \text{ and } 5) = P(2) \cdot P(5) = \frac{1}{6} \cdot \frac{1}{6} = \frac{1}{36} . \tag{2.3}$$

This agrees with the fact that one out of the 36 pairs of (ordered) numbers in Table 1.5 is "2, 5." Similarly, the probability that a card is both a king and a heart is

$$P(\text{king and heart}) = P(\text{king}) \cdot P(\text{heart}) = \frac{1}{13} \cdot \frac{1}{4} = \frac{1}{52} . \tag{2.4}$$

This makes sense, because one of the 52 cards in a deck is the king of hearts.

The logic behind Eq. (2.2) is the following. Consider N trials of a given process, where N is very large. In the case of the two dice, a trial consists of rolling both dice. The outcome of such a trial takes the form of an ordered pair of numbers. The first number is the result of the left roll, and the second number is the result of the right roll. On average, the fraction of the outcomes that have a 2 as the first number is $(1/6) \cdot N$.

Let's now consider only this "2-first" group of outcomes and ignore the rest. Then on average, a fraction $1/6$ *of these outcomes* have a 5 as the second number. This is where we are invoking the independence of the events. As far as the second roll is concerned, the set of $(1/6) \cdot N$ trials that have a 2 as the first roll is no different from any other set of $(1/6) \cdot N$ trials, so the probability of obtaining a 5 on the second roll is simply $1/6$. Putting it all together, the average number of trials that have both a 2 as the first number *and* a 5 as the second number is $1/6$ of $(1/6) \cdot N$, which equals $(1/6) \cdot (1/6) \cdot N$.

In the case of general probabilities $P(A)$ and $P(B)$, it is easy to see that the two $(1/6)$'s in the above result get replaced by $P(A)$ and $P(B)$. So the average number of outcomes where A and B both occur is $P(A) \cdot P(B) \cdot N$. And since we performed N trials, the fraction of outcomes where A and B both occur is $P(A) \cdot P(B)$, on average. From the definition of probability in Section 2.1, this fraction is the probability that A and B both occur, in agreement with Eq. (2.2).

If you want to think about the rule in Eq. (2.2) in terms of a picture, then consider Fig. 2.1. Without worrying about specifics, let's assume that different points within the overall square represent different outcomes. And let's assume that they're all equally likely, which means that the area of a region gives the probability that an outcome located in that region occurs (assuming that the area of the whole region is 1). The figure corresponds to $P(A) = 0.2$ and $P(B) = 0.4$. Outcomes to the left of the vertical line are ones where A occurs, and outcomes to the right of the vertical line are ones where A doesn't occur. Likewise for B and outcomes above and below the horizontal line.

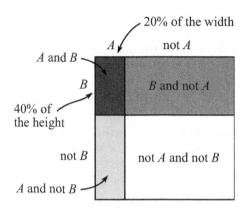

Figure 2.1: A probability square for independent events.

From the figure, we see that not only is 40% of the entire square above the vertical line, but also that 40% of the left vertical strip (where A occurs) is above the vertical line, and likewise for the right vertical strip (where A doesn't occur). In other words, B occurs 40% of the time, independent of whether or not A occurs. Basically, B couldn't care less what happens with A. Similar statements hold with A and B interchanged. So this type of figure, with a square divided by horizontal and vertical lines, does indeed represent independent events.

The darkly shaded "A and B" region is the intersection of the region to the left of the vertical line (where A occurs) and the region above the horizontal line (where B occurs). Hence the word "intersection" in the title of this section. The area of the darkly shaded region is 20% of 40% (or 40% of 20%) of the total area, that is, $(0.2)(0.4) = 0.08$ of the total area. The total area corresponds to a probability of 1, so the darkly shaded region corresponds to a probability of 0.08. Since we obtained this probability by multiplying $P(A)$ by $P(B)$, we have therefore given a pictorial proof of Eq. (2.2).

Dependent events

Two events are said to be *dependent* if they *do* affect each other, or more precisely, if the occurrence of one *does* affect the probability that the other occurs. An example is picking two balls in succession from a box containing two red balls and three blue balls (see Fig. 2.2), with $A = \{$choosing a red ball on the first pick$\}$ and $B = \{$choosing a blue ball on the second pick, *without replacement* after the first pick$\}$. If you pick a red ball first, then the probability of picking a blue ball second is 3/4, because there are three blue balls and one red ball left. On the other hand, if you *don't* pick a red ball first (that is, if you pick a blue ball first), then the probability of picking a blue ball second is 2/4, because there are two red balls and two blue balls left. So the occurrence of A certainly affects the probability of B.

Another example might be something like: $A = \{$it rains at 6:00$\}$ and $B = \{$you walk to the store at 6:00$\}$. People are generally less likely to go for a walk when it's raining outside, so (at least for most people) the occurrence of A affects the probability of B.

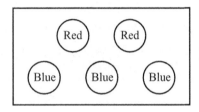

Figure 2.2: A box with two red balls and three blue balls.

The "And" rule for dependent events is:

- *If events A and B are dependent, then the probability that they both occur equals*

$$P(A \text{ and } B) = P(A) \cdot P(B|A) \qquad (2.5)$$

where $P(B|A)$ stands for the probability that B occurs, given that A occurs. It is called a "conditional probability," because we are assuming a given condition, namely that A occurs. It is read as "the probability of B, given A."

There is actually no need for the "dependent" qualifier in the first line of this rule, as we'll see in the second remark near the end of this section.

The logic behind Eq. (2.5) is the following. Consider N trials of a given process, where N is very large. In the above setup with the balls in a box, a "trial" consists of picking two balls in succession, without replacement. On average, the fraction of the outcomes in which a red ball is drawn on the first pick is $P(A) \cdot N$. Let's now consider only these outcomes and ignore the rest. Then a fraction $P(B|A)$ of *these outcomes* have a blue ball drawn second, by the definition of $P(B|A)$. So the number of outcomes where A and B both occur is $P(B|A) \cdot P(A) \cdot N$. And since we performed N trials, the fraction of outcomes where A and B both occur is $P(A) \cdot P(B|A)$, on average. This fraction is the probability that A and B both occur, in agreement with the rule in Eq. (2.5).

The reasoning in the previous paragraph is equivalent to the mathematical identity,

$$\frac{n_{A \text{ and } B}}{N} = \frac{n_A}{N} \cdot \frac{n_{A \text{ and } B}}{n_A}, \qquad (2.6)$$

where n_A is the number of trials where A occurs, etc. By definition, the lefthand side of this equation equals $P(A \text{ and } B)$, the first term on the righthand side equals $P(A)$, and the second term on the righthand side equals $P(B|A)$. So Eq. (2.6) is equivalent to the relation,

$$P(A \text{ and } B) = P(A) \cdot P(B|A), \qquad (2.7)$$

which is Eq. (2.5). In terms of the Venn-diagram type of picture in Fig. 2.3, Eq. (2.6) is the statement that the darkly shaded area (which represents $P(A \text{ and } B)$) equals the area of the A region (which represents $P(A)$) multiplied by the fraction of the A region that is taken up by the darkly shaded region. This fraction is $P(B|A)$, by definition.

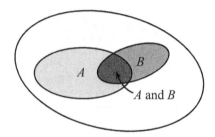

Figure 2.3: Venn diagram for probabilities of dependent events.

As in Fig. 2.1, we're assuming in Fig. 2.3 that different points within the over-all boundary represent different outcomes, and that they're all equally likely. This means that the area of a region gives the probability that an outcome located in that region occurs (assuming that the area of the whole region is 1). We're using Fig. 2.3 for its qualitative features only, so we're drawing the various regions as general blobs, as opposed to the specific rectangles in Fig. 2.1, which we used for a quantitative calculation.

Because the "*A* and *B*" region in Fig. 2.3 is the intersection of the *A* and *B* regions, and because the intersection of two sets is usually denoted by $A \cap B$, you will often see the $P(A \text{ and } B)$ probability written as $P(A \cap B)$. That is,

$$P(A \cap B) \equiv P(A \text{ and } B). \tag{2.8}$$

But we'll stick with the $P(A \text{ and } B)$ notation in this book.

There is nothing special about the order of *A* and *B* in Eq. (2.5). We could just as well interchange the letters and write $P(B \text{ and } A) = P(B) \cdot P(A|B)$. However, we know that $P(B \text{ and } A) = P(A \text{ and } B)$, because it doesn't matter which event you say first when you say that two events both occur. So we can also write $P(A \text{ and } B) = P(B) \cdot P(A|B)$. Combining this with Eq. (2.5), we see that we can write $P(A \text{ and } B)$ in two different ways:

$$P(A \text{ and } B) = P(A) \cdot P(B|A)$$
$$= P(B) \cdot P(A|B). \tag{2.9}$$

The fact that $P(A \text{ and } B)$ can be written in these two ways will be critical when we discuss Bayes' theorem in Section 2.5.

Example (Balls in a box): Let's apply Eq. (2.5) to the setup with the balls in the box in Fig. 2.2 above. Let $A = \{$choosing a red ball on the first pick$\}$ and $B = \{$choosing a blue ball on the second pick, *without replacement* after the first pick$\}$. For shorthand, we'll denote these events by Red_1 and Blue_2, where the subscript refers to the first or second pick. We noted above that $P(\text{Blue}_2|\text{Red}_1) = 3/4$. And we also know that $P(\text{Red}_1)$ is simply $2/5$, because there are initially two red balls and three blue balls. So Eq. (2.5) gives the probability of picking a red ball first and a blue ball second (without replacement after the first pick) as

$$P(\text{Red}_1 \text{ and } \text{Blue}_2) = P(\text{Red}_1) \cdot P(\text{Blue}_2|\text{Red}_1) = \frac{2}{5} \cdot \frac{3}{4} = \frac{3}{10}. \tag{2.10}$$

We can verify that this is correct by listing out all of the possible pairs of balls that can be picked. If we label the balls as 1, 2, 3, 4, 5, and if we let 1, 2 be the red balls, and 3, 4, 5 be the blue balls, then the possible outcomes are shown in Table 2.1. The first number stands for the first ball picked, and the second number stands for the second ball picked.

	Red first		Blue first		
Red second	—	2 1	3 1	4 1	5 1
	1 2	—	3 2	4 2	5 2
Blue second	1 3	2 3	—	4 3	5 3
	1 4	2 4	3 4	—	5 4
	1 5	2 5	3 5	4 5	—

Table 2.1: Ways to pick two balls from the box in Fig. 2.2, without replacement.

The "—" entries stand for the outcomes that aren't allowed; we can't pick two of the same ball, because we're not replacing the ball after the first pick. The dividing lines are drawn for clarity. The internal vertical line separates the outcomes where a red or blue ball is drawn on the first pick, and the internal horizontal line separates the outcomes where a red or blue ball is drawn on the second pick. The six pairs in the lower left corner are the outcomes where a red ball (numbered 1 and 2) is drawn first and a blue ball (numbered 3, 4, and 5) is drawn second. Since there are 20 possible outcomes in all, the desired probability is $6/20 = 3/10$, in agreement with Eq. (2.10).

Table 2.1 also gives a verification of the $P(\text{Red}_1)$ and $P(\text{Blue}_2|\text{Red}_1)$ probabilities we wrote down in Eq. (2.10). $P(\text{Red}_1)$ equals $2/5$ because eight of the 20 entries are to the left of the vertical line. And $P(\text{Blue}_2|\text{Red}_1)$ equals $3/4$ because six of these eight entries are below the horizontal line.

The task of Problem 2.4 is to verify that the second expression in Eq. (2.9) also gives the correct result for $P(\text{Red}_1 \text{ and Blue}_2)$ in this setup.

We can think about the rule in Eq. (2.5) in terms of a picture analogous to Fig. 2.1. If we consider the above example with the red and blue balls, then the first thing we need to do is recast Table 2.1 in a form where equal areas yield equal probabilities. If we get rid of the "—" entries in Table 2.1, then all entries have equal probabilities, and we end up with Table 2.2.

1 2	2 1	3 1	4 1	5 1
1 3	2 3	3 2	4 2	5 2
1 4	2 4	3 4	4 3	5 3
1 5	2 5	3 5	4 5	5 4

Table 2.2: Rewriting Table 2.1.

In the spirit of Fig. 2.1, this table becomes the square shown in Fig. 2.4. The upper left region corresponds to red balls on both picks. The lower left region corresponds to a red ball and then a blue ball. The upper right region corresponds to a blue ball and then a red ball. And the lower right region corresponds to blue balls on both picks. This figure makes it clear why we formed the product $(2/5) \cdot (3/4)$ in Eq. (2.10). The $2/5$ gives the fraction of the outcomes that lie to the left of the vertical line (these are the ones that have a red ball first), and the $3/4$ gives the fraction *of these outcomes* that lie below the horizontal line (these are the ones that have a blue ball second). The product of these fractions gives the overall fraction (namely $3/10$) of the outcomes that lie in the lower left region.

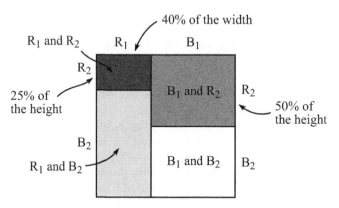

Figure 2.4: Pictorial representation of Table 2.2.

The main difference between Fig. 2.4 and Fig. 2.1 is that the one horizontal line in Fig. 2.1 is now two different horizontal lines in Fig. 2.4. The heights of the horizontal lines in Fig. 2.4 depend on which vertical strip we're dealing with. This is the visual manifestation of the fact that the red/blue probabilities on the second pick depend on what happens on the first pick.

REMARKS:

1. The method of explicitly counting the possible outcomes in Table 2.1 shows that you don't *have* to use the rule in Eq. (2.5), or similarly the rule in Eq. (2.2), to calculate probabilities. You can often instead just count up the various outcomes and solve the problem from scratch. However, the rules in Eqs. (2.2) and (2.5) allow you to take a shortcut that avoids listing out all the outcomes, which might be rather difficult if you're dealing with large numbers.

2. The rule in Eq. (2.2) for independent events is a special case of the rule in Eq. (2.5) for dependent events. This is true because if A and B are independent, then $P(B|A)$ is simply equal to $P(B)$, because the probability of B occurring is just $P(B)$, independent of whether or not A occurs. Eq. (2.5) then reduces to Eq. (2.2) when $P(B|A) = P(B)$. Therefore, there was technically no need to introduce Eq. (2.2) first. We could have started with Eq. (2.5), which covers all possible scenarios, and then showed that it reduces to Eq. (2.2) when the events are independent. But pedagogically, it is often better to start with a special case and then work up to the more general case.

3. In the above "balls in a box" example, we encountered the conditional probability $P(\text{Blue}_2|\text{Red}_1)$. We can also talk about the "reversed" conditional probability, $P(\text{Red}_1|\text{Blue}_2)$. However, since the second pick happens *after* the first pick, you might wonder how much sense it makes to talk about the probability of the Red$_1$

event, *given* the Blue$_2$ event. Does the second pick somehow influence the first pick, even though the second pick hasn't happened yet? When you make the first pick, are you being affected by a mysterious influence that travels backward in time?

No, and no. When we talk about $P(\text{Red}_1|\text{Blue}_2)$, or about any other conditional probability in the example, everything we might want to know can be read off from Table 2.1. Once the table has been created, we can forget about the temporal order of the events. By looking at the Blue$_2$ pairs (below the horizontal line), we see that $P(\text{Red}_1|\text{Blue}_2) = 6/12 = 1/2$. This should be contrasted with $P(\text{Red}_1|\text{Red}_2)$, which is obtained by looking at the Red$_2$ pairs (above the horizontal line); we find that $P(\text{Red}_1|\text{Red}_2) = 2/8 = 1/4$. Therefore, the probability that your first pick is red *does* depend on whether your second pick is blue or red. But this doesn't mean that there is a backward influence in time. All it says is that if you perform a large number of trials of the given process (drawing two balls, without replacement), and if you look at all of the cases where your second pick is blue (or conversely, red), then you will find that your first pick is red in $1/2$ (or conversely, $1/4$) of these cases, on average. In short, the second pick has no causal influence on the first pick, but the after-the-fact *knowledge* of the second pick affects the *probability* of what the first pick *was*.

4. A trivial yet extreme example of dependent events is the two events: A, and "not A." The occurrence of A highly affects the probability of "not A" occurring. If A occurs, then "not A" occurs with probability zero. And if A doesn't occur, then "not A" occurs with probability 1. ♣

In the second remark above, we noted that if A and B are independent (that is, if the occurrence of one doesn't affect the probability that the other occurs), then $P(B|A) = P(B)$. Similarly, we also have $P(A|B) = P(A)$. Let's prove that one of these relations implies the other. Assume that $P(B|A) = P(B)$. Then if we equate the two righthand sides of Eq. (2.9) and use $P(B|A) = P(B)$ to replace $P(B|A)$ with $P(B)$, we obtain

$$P(A) \cdot P(B|A) = P(B) \cdot P(A|B)$$
$$\implies P(A) \cdot P(B) = P(B) \cdot P(A|B)$$
$$\implies P(A) = P(A|B). \tag{2.11}$$

So $P(B|A) = P(B)$ implies $P(A|B) = P(A)$, as desired. In other words, if B is independent of A, then A is also independent of B. We can therefore talk about two events being independent, without worrying about the direction of the independence. The condition for independence is therefore either of the relations,

$$\boxed{P(B|A) = P(B) \quad \text{or} \quad P(A|B) = P(A)} \quad \text{(independence)} \tag{2.12}$$

Alternatively, the condition for independence may be expressed by Eq. (2.2),

$$\boxed{P(A \text{ and } B) = P(A) \cdot P(B)} \quad \text{(independence)} \tag{2.13}$$

because this equation implies (by comparing it with Eq. (2.5), which is valid in any case) that $P(B|A) = P(B)$.

2.2.2   OR: The "union" probability, $P(A$ or $B)$

Let A and B be two events. For example, let $A = \{$rolling a 2 on a die$\}$ and $B = \{$rolling a 5 on the *same* die$\}$. Or we might have $A = \{$rolling an even number (that is, 2, 4, or 6) on a die$\}$ and $B = \{$rolling a multiple of 3 (that is, 3 or 6) on the *same* die$\}$. A third example is $A = \{$rolling a 1 on one die$\}$ and $B = \{$rolling a 6 on *another* die$\}$. What is the probability that either A or B (or both) occurs? In answering this question, we must consider two cases: (1) A and B are exclusive events, or (2) A and B are nonexclusive events. Let's look at each of these in turn.

Exclusive events

Two events are said to be *exclusive* if one precludes the other. That is, they can't both happen. An example is rolling one die, with $A = \{$rolling a 2 on the die$\}$ and $B = \{$rolling a 5 on the *same* die$\}$. These events are exclusive because it is impossible for one number to be both a 2 and a 5. (The events in the second and third scenarios mentioned above are *not* exclusive; we'll talk about this below.) Another example is picking one card from a deck, with $A = \{$the card is a diamond$\}$ and $B = \{$the card is a heart$\}$. These events are exclusive because it is impossible for one card to be both a diamond and a heart.

The "Or" rule for exclusive events is:

- *If events A and B are exclusive, then the probability that either of them occurs equals the sum of their individual probabilities:*

$$\boxed{P(A \text{ or } B) = P(A) + P(B)} \qquad (2.14)$$

The logic behind this rule boils down to Fig. 2.5. The key feature of this figure is that there is *no overlap* between the two regions, because we are assuming that A and B are exclusive. If there were a region that was contained in both A and B, then the outcomes in that region would be ones for which A and B both occur, which would violate the assumption that A and B are exclusive. The rule in Eq. (2.14) is simply the statement that the area of the union (hence the word "union" in the title of this section) of regions A and B equals the sum of their areas. There is nothing fancy going on here. This statement is no deeper than the statement that if you have two separate bowls, the total number of apples in the two bowls equals the number of apples in one bowl plus the number of apples in the other bowl.

We can quickly apply this rule to the two examples mentioned above. In the example with the die, the probability of rolling a 2 or a 5 on one die is

$$P(2 \text{ or } 5) = P(2) + P(5) = \frac{1}{6} + \frac{1}{6} = \frac{1}{3}. \qquad (2.15)$$

This makes sense, because two of the six numbers on a die are the 2 and the 5. In the card example, the probability of a card being either a diamond or a heart is

$$P(\text{diamond or heart}) = P(\text{diamond}) + P(\text{heart}) = \frac{1}{4} + \frac{1}{4} = \frac{1}{2}. \qquad (2.16)$$

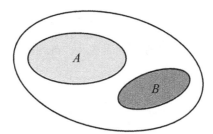

Figure 2.5: Venn diagram for the probabilities of exclusive events.

This makes sense, because half of the 52 cards in a deck are diamonds or hearts.

A special case of Eq. (2.14) is the "Not" rule, which follows from letting $B =$ "not A."

$$P(A \text{ or } (\text{not } A)) = P(A) + P(\text{not } A)$$
$$\implies 1 = P(A) + P(\text{not } A)$$
$$\implies P(\text{not } A) = 1 - P(A). \tag{2.17}$$

The first equality here follows from Eq. (2.14), because A and "not A" are certainly exclusive events; you can't both have something and not have it. To obtain the second line in Eq. (2.17), we have used $P(A \text{ or } (\text{not } A)) = 1$, which holds because every possible outcome belongs to either A or "not A."

Nonexclusive events

Two events are said to be *nonexclusive* if it is possible for both to happen. An example is rolling one die, with $A =$ {rolling an even number (that is, 2, 4, or 6)} and $B =$ {rolling a multiple of 3 (that is, 3 or 6) on the *same* die}. If you roll a 6, then A and B both occur. Another example is picking one card from a deck, with $A =$ {the card is a king} and $B =$ {the card is a heart}. If you pick the king of hearts, then A and B both occur.

The "Or" rule for nonexclusive events is:

- *If events A and B are nonexclusive, then the probability that either (or both) of them occurs equals*

$$\boxed{P(A \text{ or } B) = P(A) + P(B) - P(A \text{ and } B)} \tag{2.18}$$

The "or" here is the so-called "inclusive or," in the sense that we say "A or B occurs" if either *or both* of the events occur. As with the "dependent" qualifier in the "And" rule in Eq. (2.5), there is actually no need for the "nonexclusive" qualifier in the "Or" rule here, as we'll see in the third remark below.

The logic behind Eq. (2.18) boils down to Fig. 2.6. The rule in Eq. (2.18) is the statement that the area of the union of regions A and B equals the sum of their areas *minus the area of the overlap*. This subtraction is necessary so that we don't double count the region that belongs to both A and B. This region isn't "doubly good" just because it belongs to both A and B. As far as the "A or B" condition goes, the overlap region is just the same as any other part of the union of A and B.

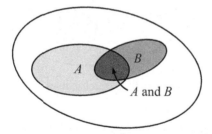

Figure 2.6: Venn diagram for the probabilities of nonexclusive events.

In terms of a physical example, the rule in Eq. (2.18) is equivalent to the statement that if you have two bird cages that have a region of overlap, then the total number of birds in the cages equals the number of birds in one cage, plus the number in the other cage, minus the number in the overlap region. In the situation shown in Fig. 2.7, we have $7 + 5 - 2 = 10$ birds (which oddly all happen to be flying at the given moment).

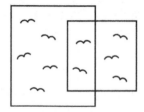

Figure 2.7: Birds in overlapping cages.

Things get more complicated if you have three or more events and you want to calculate probabilities like $P(A$ or B or $C)$. But in the end, the main task is to keep track of the overlaps of the various regions; see Problem 2.2.

Because the "A or B" region in Fig. 2.6 is the union of the A and B regions, and because the union of two sets is usually denoted by $A \cup B$, you will often see the $P(A$ or $B)$ probability written as $P(A \cup B)$. That is,

$$P(A \cup B) \equiv P(A \text{ or } B). \tag{2.19}$$

But we'll stick with the $P(A$ or $B)$ notation in this book.

We can quickly apply Eq. (2.18) to the two examples mentioned above. In the example with the die, the only way to roll an even number *and* a multiple of 3 on a single die is to roll a 6, which happens with probability $1/6$. So Eq. (2.18) gives the probability of rolling an even number or a multiple of 3 as

$$P(\text{even or mult of 3}) = P(\text{even}) + P(\text{mult of 3}) - P(\text{even and mult of 3})$$

$$= \frac{1}{2} + \frac{1}{3} - \frac{1}{6} = \frac{4}{6} = \frac{2}{3}. \tag{2.20}$$

This makes sense, because four of the six numbers on a die are even numbers or multiples of 3, namely 2, 3, 4, and 6. (Remember that whenever we use "or," it means the "inclusive or.") We subtracted off the 1/6 in Eq. (2.20) so that we didn't double count the roll of a 6.

In the card example, the only way to pick a king *and* a heart with a single card is to pick the king of hearts, which happens with probability 1/52. So Eq. (2.18) gives the probability that a card is a king or a heart as

$$P(\text{king or heart}) = P(\text{king}) + P(\text{heart}) - P(\text{king and heart})$$

$$= \frac{1}{13} + \frac{1}{4} - \frac{1}{52} = \frac{16}{52} = \frac{4}{13}. \qquad (2.21)$$

This makes sense, because 16 of the 52 cards in a deck are kings or hearts, namely the 13 hearts, plus the kings of diamonds, spades, and clubs; we already counted the king of hearts. As in the previous example with the die, we subtracted off the 1/52 here so that we didn't double count the king of hearts.

REMARKS:

1. If you want, you can think of the area of the union of A and B in Fig. 2.6 as the area of *only* A, plus the area of *only* B, plus the area of "A and B." (Equivalently, the number of birds in the cages in Fig. 2.7 is $5 + 3 + 2 = 10$.) This is easily visualizable, because these three areas are the ones you see in the figure. However, the probabilities of *only* A and of *only* B are often a pain to deal with, so it's generally easier to think of the area of the union of A and B as the area of A, plus the area of B, minus the area of the overlap. This way of thinking corresponds to Eq. (2.18).

2. As we mentioned in the first remark on page 66, you don't *have* to use the above rules of probability to calculate things. You can often instead just count up the various outcomes and solve the problem from scratch. In many cases you're doing basically the same thing with the two methods, as we saw in the above examples with the die and the cards.

3. As with Eqs. (2.2) and (2.5), the rule in Eq. (2.14) for exclusive events is a special case of the rule in Eq. (2.18) for nonexclusive events. This is true because if A and B are exclusive, then $P(A \text{ and } B) = 0$, by definition. Eq. (2.18) then reduces to Eq. (2.14) when $P(A \text{ and } B) = 0$. Likewise, Fig. 2.5 is a special case of Fig. 2.6 when the regions have zero overlap. There was therefore technically no need to introduce Eq. (2.14) first. We could have started with Eq. (2.18), which covers all possible scenarios, and then showed that it reduces to Eq. (2.14) when the events are exclusive. But as in Section 2.2.1, it is often better to start with a special case and then work up to the more general case. ♣

2.2.3 (In)dependence and (non)exclusiveness

Two events are either independent or dependent, and they are also either exclusive or nonexclusive. There are therefore $2 \cdot 2 = 4$ combinations of these characteristics. Let's see which combinations are possible. You'll need to read this section *very* slowly if you want to keep everything straight. This discussion is given for curiosity's sake only, in case you were wondering how the dependent/independent characteristic relates to the exclusive/nonexclusive characteristic. There is no need to memorize the results below. Instead, you should think about each situation individually and determine its properties from scratch.

- EXCLUSIVE AND INDEPENDENT: This combination isn't possible. If two events are independent, then their probabilities are independent of each other, which means that there is a nonzero probability (namely, the product of the individual probabilities) that both events happens. Therefore, they cannot be exclusive.

  Said in another way, if two events A and B are exclusive, then the probability of B given A is zero. But if they are also independent, then the probability of B is independent of what happens with A. So the probability of B must be zero, period. Such a B is a very uninteresting event, because it never happens.

- EXCLUSIVE AND DEPENDENT: This combination is possible. An example consists of the events

$$A = \{\text{rolling a 2 on a die}\},$$
$$B = \{\text{rolling a 5 on the } same \text{ die}\}. \tag{2.22}$$

  Another example consists of A as one event and $B = \{\text{not } A\}$ as the other. In both of these examples the events are exclusive, because they can't both happen. Furthermore, the occurrence of one event certainly affects the probability of the other occurring, in that the probability $P(B|A)$ takes the extreme value of zero, due to the exclusive nature of the events. The events are therefore quite dependent (in a negative sort of way). In short, *if two events are exclusive, then they are necessarily also dependent.*

- NONEXCLUSIVE AND INDEPENDENT: This combination is possible. An example consists of the events

$$A = \{\text{rolling a 2 on a die}\},$$
$$B = \{\text{rolling a 5 on } another \text{ die}\}. \tag{2.23}$$

  Another example consists of the events $A = \{\text{getting a Heads on a coin flip}\}$ and $B = \{\text{getting a Heads on another coin flip}\}$. In both of these examples the events are clearly independent, because they involve different dice or coins. And the events *can* both happen (a fact that is guaranteed by their independence, as mentioned in the "Exclusive and Independent" case above), so they are nonexclusive. In short, *if two events are independent, then they are necessarily also nonexclusive.* This statement is the logical "contrapositive" of the corresponding statement in the "Exclusive and Dependent" case above.

- NONEXCLUSIVE AND DEPENDENT: This combination is possible. An example consists of the events

$$A = \{\text{rolling a 2 on a die}\},$$
$$B = \{\text{rolling an even number on the } same \text{ die}\}. \tag{2.24}$$

Another example consists of picking balls *without replacement* from a box with two red balls and three blue balls, with the events being A = {picking a red ball on the first pick} and B = {picking a blue ball on the second pick}. In both of these examples the events are dependent, because the occurrence of A affects the probability of B. (In the die example, $P(B|A)$ takes on the extreme value of 1, which isn't equal to $P(B) = 1/2$. Also, $P(A|B) = 1/3$, which isn't equal to $P(A) = 1/6$. Likewise for the box example.) And the events can both happen, so they are nonexclusive.

To sum up, we see that all exclusive events must be dependent, but nonexclusive events can be either independent or dependent. Similarly, all independent events must be nonexclusive, but dependent events can be either exclusive or nonexclusive. These facts are summarized in Table 2.3, which indicates which combinations are possible.

	Independent	Dependent
Exclusive	NO	YES
Nonexclusive	YES	YES

Table 2.3: Relations between (in)dependence and (non)exclusiveness.

2.2.4 Conditional probability

In Eq. (2.5) we introduced the concept of conditional probability, with $P(B|A)$ denoting the probability that B occurs, given that A occurs. In this section we'll talk more about conditional probabilities. In particular, we'll show that two probabilities that you might naively think are equal are in fact not equal. Consider the following example.

Fig. 2.8 gives a pictorial representation of the probability that a random person's height is greater than 6′3″ (6 feet, 3 inches) or less than 6′3″, along with the probability that a random person's last name begins with Z or not Z. We haven't tried to mimic the exact numbers, but we have indicated that the vast majority of people are under 6′3″ (this case takes up most of the vertical span of the square), and also that the vast majority of people have a last name that doesn't begin with Z (this case takes up most of the horizontal span of the square). We'll assume that the probabilities involving heights and last-name letters are independent. This independence manifests itself in the fact that the horizontal and vertical dividers of the square are straight lines (as opposed to, for example, the shifted lines in Fig. 2.4). This independence makes things a little easier to visualize, but it isn't critical in the following discussion.

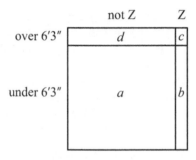

Figure 2.8: Probability square for independent events (height, and first letter of last name).

Let's now look at some conditional probabilities. Let the areas of the four rectangles in Fig. 2.8 be a, b, c, d, as indicated. The area of a region represents the
probability that a given person is in that region. Let Z stand for "having a last name
that begins with Z," and let U stand for "being under 6'3" in height."

Consider the conditional probabilities $P(Z|U)$ and $P(U|Z)$. $P(Z|U)$ deals with
the subset of cases where we know that U occurs. These cases are associated with
the area below the horizontal dividing line in the figure. So $P(Z|U)$ equals the
fraction of the area *below the horizontal line* (which is $a + b$) that is also *to the right
of the vertical line* (which is b). This fraction $b/(b + a)$ is very small.

In contrast, $P(U|Z)$ deals with the subset of cases where we know that Z occurs.
These cases are associated with the area to the right of the vertical dividing line in
the figure. So $P(U|Z)$ equals the fraction of the area *to the right of the vertical line*
(which is $b + c$) that is also *below the horizontal line* (which is b). This fraction
$b/(b + c)$ is very close to 1. To sum up, we have

$$P(Z|U) = \frac{b}{b + a} \approx 0,$$

$$P(U|Z) = \frac{b}{b + c} \approx 1. \tag{2.25}$$

We see that $P(Z|U)$ is *not* equal to $P(U|Z)$. If we were dealing with a situation
where $a = c$, then these conditional probabilities *would* be equal. But that is an
exception. In general, the two probabilities are not equal.

If you're too hasty in your thinking, you might say something like, "Since U
and Z are independent, one doesn't affect the other, so the conditional probabilities should be the same." This conclusion is incorrect. The correct statement is,
"Since U and Z are independent, one doesn't affect the other, so the conditional
probabilities are equal to the corresponding unconditional probabilities." That is,
$P(Z|U) = P(Z)$ and $P(U|Z) = P(U)$. But $P(Z)$ and $P(U)$ are vastly different, with
the former being approximately zero, and the latter being approximately 1.

In order to make it obvious that the two conditional probabilities $P(A|B)$ and
$P(B|A)$ aren't equal in general, we picked an example where the various probabilities were all either close to zero or close to 1. We did this solely for pedagogical
purposes; the non-equality of the conditional probabilities holds in general (except
in the $a = c$ case). Another extreme example that makes it clear that the two conditional probabilities are different is: The probability that a living thing is human,

given that it has a brain, is very small; but the probability that a living thing has a brain, given that it is human, is 1.

The takeaway lesson here is that when thinking about the conditional probability $P(A|B)$, the order of A and B is critical. Great confusion can arise if one forgets this fact. The classic example of this confusion is the "Prosecutor's fallacy," discussed below in Section 2.4.3. That example should convince you that a lack of basic knowledge of probability can have significant and possibly tragic consequences in real life.

2.3 Examples

Let's now do some examples. Introductory probability problems generally fall into a few main categories, so we've divided the examples into the various subsections below. There is no better way to learn how to solve probability problems (or any kind of problem, for that matter) than to just sit down and do a bunch of them, so we've presented quite a few.

If the statement of a given problem lists out the specific probabilities of the possible outcomes, then the rules in Section 2.2 are often called for. However, in many problems you encounter, you'll be calculating probabilities from scratch (by counting things), so the rules in Section 2.2 generally don't come into play. You simply have to do lots of counting. This will become clear in the examples below. For all of these, *be sure to try the problem for a few minutes on your own before looking at the solution.*

In virtually all of these examples, we'll be dealing with situations in which the various possible outcomes are equally likely. For example, we'll be tossing coins, picking cards, forming committees, forming permutations, etc. We will therefore be making copious use of Eq. (2.1),

$$p = \frac{\text{number of desired outcomes}}{\text{total number of possible outcomes}} \qquad \text{(for equally likely outcomes)} \qquad (2.26)$$

We won't, however, bother to specifically state each time that the different outcomes are all equally likely. Just remember that they are, and that this fact is necessary for Eq. (2.1) to be valid.

Before getting into the examples, let's start off with a problem-solving strategy that comes in very handy in certain situations.

2.3.1 The art of "not"

There are many setups in which the easiest way to calculate the probability of a given event A is not to calculate it directly, but rather to calculate the probability of "not A" and then subtract the result from 1. This yields $P(A)$ because we know from Eq. (2.17) that $P(A) = 1 - P(\text{not } A)$. The event "not A" is called the *complement* of the event A.

The most common situation of this type involves a question along the lines of, "What is the probability of obtaining at least one of such-and-such?" The "at least" part appears to make things difficult, because it could mean one, or two, or three, etc.

It would be at best rather messy, and at worst completely intractable, to calculate the individual probabilities of all the different numbers and then add them up to obtain the answer. The "at least one" question is very different from the "exactly one" question.

The key point that simplifies things is that the only way to *not* get at least one of something is to get *exactly zero* of it. This means that we can just calculate the probability of getting zero, and then subtract the result from 1. We therefore need to calculate only *one* probability, instead of a potentially large number of probabilities.

Example (At least one 6): Three dice are rolled. What is the probability of obtaining at least one 6?

Solution: We'll find the probability of obtaining zero 6's and then subtract the result from 1. In order to obtain zero 6's, we must obtain something other than a 6 on the first die (which happens with 5/6 probability), and likewise on the second die (5/6 probability again), and likewise on the third die (5/6 probability again). These are independent events, so the probability of obtaining zero 6's equals $(5/6)^3 = 125/216$. The probability of obtaining at least one 6 is therefore $1 - (5/6)^3 = 91/216$, which is about 42%.

If you want to solve this problem the long way, you can add up the probabilities of obtaining exactly one, two, or three 6's. This is the task of Problem 2.11.

REMARK: Beware of the following incorrect reasoning for this problem: There is a 1/6 chance of obtaining a 6 on each of the three rolls. The total probability of obtaining at least one 6 therefore seems like it should be $3 \cdot (1/6) = 1/2$. This is incorrect because we're trying to find the probability of "a 6 on the first roll" *or* "a 6 on the second roll" *or* "a 6 on the third roll." (This "or" combination is equivalent to obtaining at least one 6. Remember that when we write "or," we mean the "inclusive or.") But from Eq. (2.14) (or its simple extension to three events) it is appropriate to add up the individual probabilities *only if the events are exclusive.* For nonexclusive events, we must subtract off the "overlap" probabilities, as we did in Eq. (2.18); see Problem 2.2(d) for the case of three events. The above three events (rolling 6's) are clearly nonexclusive, because it is possible to obtain a 6 on, say, both the first roll *and* the second roll. We have therefore double (or triple) counted many of the outcomes, and this is why the incorrect answer of 1/2 is larger than the correct answer of 91/216. The task of Problem 2.12 is to solve this problem by using the result in Problem 2.2(d) to keep track of all the double (and triple) counting.

Another way of seeing why the "$3 \cdot (1/6) = 1/2$" reasoning can't be correct is that it would imply that if we had, say, 12 dice, then the probability of obtaining at least one 6 would be $12 \cdot (1/6) = 2$. But probabilities larger than 1 are nonsensical. ♣

2.3.2 Picking seats

Situations often come up where we need to assign various things to various spots. We'll generally talk about assigning people to seats. There are two common ways to solve problems of this sort: (1) You can count up the number of desired outcomes, along with the total number of outcomes, and then take their ratio via Eq. (2.1), or (2) you can imagine assigning the seats one at a time, finding the probability of success at each stage, and using the rules in Section 2.2, or their extensions to more than two events. It's personal preference which method you use. But it never hurts to solve a problem both ways, of course, because that allows you to double check your answer.

Example 1 (Middle in the middle): Three chairs are arranged in a line, and three people randomly take seats. What is the probability that the person with the middle height ends up in the middle seat?

First solution: Let the people be labeled from tallest to shortest as 1, 2, and 3. Then the $3! = 6$ possible orderings are

$$1\,2\,3 \qquad 1\,3\,2 \qquad 2\,1\,3 \qquad 2\,3\,1 \qquad 3\,1\,2 \qquad 3\,2\,1 \qquad (2.27)$$

We see that two of these ($1\,2\,3$ and $3\,2\,1$) have the middle-height person in the middle seat. So the probability is $2/6 = 1/3$.

Second solution: Imagine assigning the people randomly to the seats, and let's assign the middle-height person first, which we are free to do. There is a $1/3$ chance that this person ends up in the middle seat (or any other seat, for that matter). So $1/3$ is the desired answer. Nothing fancy going on here.

Third solution: If you want to assign the tallest person first, then there is a $1/3$ chance that she ends up in the middle seat, in which case there is zero chance that the middle-height person ends up there. There is a $2/3$ chance that the tallest person *doesn't* end up in the middle seat, in which case there is a $1/2$ chance that the middle-height person ends up there (because there are two seats remaining, and one yields success). So the total probability that the middle-height person ends up in the middle seat is

$$\frac{1}{3} \cdot 0 + \frac{2}{3} \cdot \frac{1}{2} = \frac{1}{3}. \qquad (2.28)$$

REMARK: The preceding equation technically comes from one application of Eq. (2.14) and two applications of Eq. (2.5). If we let T stand for tallest and M stand for middle-height, and if we use the notation T_{mid} to mean that the tallest person is in the middle seat, etc., then we can write

$$P(M_{mid}) = P(T_{mid} \text{ and } M_{mid}) + P(T_{not\ mid} \text{ and } M_{mid})$$
$$= P(T_{mid}) \cdot P(M_{mid}|T_{mid}) + P(T_{not\ mid}) \cdot P(M_{mid}|T_{not\ mid})$$
$$= \frac{1}{3} \cdot 0 + \frac{2}{3} \cdot \frac{1}{2} = \frac{1}{3}. \qquad (2.29)$$

Eq. (2.14) is relevant in the first line because the two events "T_{mid} and M_{mid}" and "$T_{not\ mid}$ and M_{mid}" are exclusive events, since T can't be both in the middle seat and not in the middle seat.

However, when solving problems of this kind, although it is sometimes helpful to explicitly write down the application of Eqs. (2.14) and (2.5) as we just did, this often isn't necessary. It is usually quicker to imagine a large number of trials and then calculate the number of these trials that yield success. For example, if we do 600 trials of the present setup, then $(1/3) \cdot 600 = 200$ of them (on average) have T in the middle seat, in which case failure is guaranteed. Of the other $(2/3) \cdot 600 = 400$ trials where T isn't in the middle seat, half of them (which is $(1/2) \cdot 400 = 200$) have M in the middle seat. So the desired probability is $200/600 = 1/3$. In addition to being more intuitive, this method is safer than just plugging things into formulas (although it's really the same reasoning in the end). ♣

Example 2 (Order of height in a line): Five chairs are arranged in a line, and five people randomly take seats. What is the probability that they end up in order of decreasing height, from left to right?

First solution: There are $5! = 120$ possible arrangements of the five people in the seats. But there is only one arrangement where they end up in order of decreasing height. So the probability is $1/120$.

Second solution: If we randomly assign the tallest person to a seat, there is a $1/5$ chance that she ends up in the leftmost seat. Assuming that she ends up there, there is a $1/4$ chance that the second tallest person ends up in the second leftmost seat (because there are only four seats left). Likewise, the chances that the other people end up where we want them are $1/3$, then $1/2$, and then $1/1$. (If the first four people end up in the desired seats, then the shortest person is guaranteed to end up in the rightmost seat.) So the probability is $1/5 \cdot 1/4 \cdot 1/3 \cdot 1/2 \cdot 1/1 = 1/120$.

The product of these five probabilities comes from the extension of Eq. (2.5) to five events (see Problem 2.2(b) for the three-event case), which takes the form,

$$P(A \text{ and } B \text{ and } C \text{ and } D \text{ and } E) = P(A) \cdot P(B|A) \cdot P(C|A \text{ and } B)$$
$$\cdot P(D|A \text{ and } B \text{ and } C) \qquad (2.30)$$
$$\cdot P(E|A \text{ and } B \text{ and } C \text{ and } D).$$

We will use similar extensions repeatedly in the examples below.

Alternatively, instead of assigning people to seats, we can assign seats to people. That is, we can assign the first seat to one of the five people, and then the second seat to one of the remaining four people, and so on. Multiplying the probabilities of success at each stage gives the same product as above, $1/5 \cdot 1/4 \cdot 1/3 \cdot 1/2 \cdot 1/1 = 1/120$.

Example 3 (Order of height in a circle): Five chairs are arranged in a circle, and five people randomly take seats. What is the probability that they end up in order of decreasing height, going clockwise? The decreasing sequence of people can start anywhere in the circle. That is, it doesn't matter which seat has the tallest person.

First solution: As in the previous example, there are $5! = 120$ possible arrangements of the five people in the seats. But now there are *five* arrangements where they end up in order of decreasing height. This is true because the tallest person can take five possible seats, and once her seat is picked, the positions of the other people are uniquely determined if they are to end up in order of decreasing height. The probability is therefore $5/120 = 1/24$.

Second solution: If we randomly assign the tallest person to a seat, it doesn't matter where she ends up, because all five seats in the circle are equivalent. But given that she ends up in a certain seat, the second tallest person needs to end up in the seat next to her in the clockwise direction. This happens with probability 1/4. Likewise, the third tallest person has a 1/3 chance of ending up in the next seat in the clockwise direction. And then 1/2 for the fourth tallest person, and 1/1 for the shortest person. The probability is therefore $1/4 \cdot 1/3 \cdot 1/2 \cdot 1/1 = 1/24$.

If you want, you can preface this product with a "5/5" for the tallest person, because there are five possible seats she can take (this is the denominator), and there are also five successful seats she can take (this is the numerator) because it doesn't matter where she ends up.

Example 4 (Three girls and three boys): Six chairs are arranged in a line, and three girls and three boys randomly pick seats. What is the probability that the three girls end up in the three leftmost seats?

First solution: The total number of possible seat arrangements is $6! = 720$. There are $3! = 6$ different ways that the three girls can be arranged in the three leftmost seats, and $3! = 6$ different ways that the three boys can be arranged in the other three (the rightmost) seats. So the total number of successful arrangements is $3! \cdot 3! = 36$. The desired probability is therefore $3!3!/6! = 36/720 = 1/20$.

Second solution: Let's assume that the girls pick their seats first, one at a time. The first girl has a 3/6 chance of picking one of the three leftmost seats. Then, given that she is successful, the second girl has a 2/5 chance of success, because only two of the remaining five seats are among the left three. And finally, given that she too is successful, the third girl has a 1/4 chance of success, because only one of the remaining four seats is among the left three. If all three girls are successful, then all three boys are guaranteed to end up in the three rightmost seats. The desired probability is therefore $3/6 \cdot 2/5 \cdot 1/4 = 1/20$.

Third solution: The $3!3!/6!$ result in the first solution looks suspiciously like the inverse of the binomial coefficient $\binom{6}{3} = 6!/3!3!$. This suggests that there is another way to solve the problem. And indeed, imagine randomly choosing three of the six seats for the girls. There are $\binom{6}{3}$ ways to do this, all equally likely. Only one of these is the successful choice of the three leftmost seats, so the desired probability is $1/\binom{6}{3} = 3!3!/6! = 1/20$.

2.3.3 Socks in a drawer

Picking colored socks from a drawer is a classic probabilistic setup. As usual, if you want to deal with such setups by counting things, then subgroups and binomial coefficients will come into play. If, however, you want to imagine picking the socks in succession, then you'll end up multiplying various probabilities and using the rules in Section 2.2.

Example 1 (Two blue and two red): A drawer contains two blue socks and two red socks. If you randomly pick two socks, what is the probability that you obtain a matching pair?

First solution: There are $\binom{4}{2} = 6$ possible pairs you can pick. Of these, two are matching pairs (one blue pair, one red pair). So the probability is $2/6 = 1/3$. If you want to list out all the pairs, they are (with 1 and 2 being the blue socks, and 3 and 4 being the red socks):

$$\mathbf{1,2} \quad 1,3 \quad 1,4 \quad 2,3 \quad 2,4 \quad \mathbf{3,4} \tag{2.31}$$

The pairs in bold are the matching pairs.

Second solution: After you pick the first sock, there is one sock of that color (whatever it may be) left in the drawer, and two of the other color. So of the three socks left, one gives you a matching pair, and two don't. The desired probability is therefore $1/3$. See Problem 2.9 for a generalization of this example.

Example 2 (Four blue and two red): A drawer contains four blue socks and two red socks, as shown in Fig. 2.9. If you randomly pick two socks, what is the probability that you obtain a matching pair?

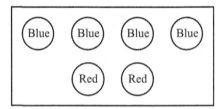

Figure 2.9: A box with four blue socks and two red socks.

First solution: There are $\binom{6}{2} = 15$ possible pairs you can pick. Of these, there are $\binom{4}{2} = 6$ blue pairs and $\binom{2}{2} = 1$ red pair. The desired probability is therefore

$$\frac{\binom{4}{2} + \binom{2}{2}}{\binom{6}{2}} = \frac{7}{15}. \tag{2.32}$$

Second solution: There is a $4/6$ chance that the first sock you pick is blue. If this happens, there is a $3/5$ chance that the second sock you pick is also blue (because there are three blue and two red socks left in the drawer). Similarly, there is a $2/6$ chance that the first sock you pick is red. If this happens, there is a $1/5$ chance that the second sock you pick is also red (because there are one red and four blue socks left in the drawer). The probability that the socks match is therefore

$$\frac{4}{6} \cdot \frac{3}{5} + \frac{2}{6} \cdot \frac{1}{5} = \frac{14}{30} = \frac{7}{15}. \tag{2.33}$$

If you want to explicitly justify the sum on the lefthand side here, it comes from the sum on the righthand side of the following relation (with B_1 standing for a blue sock on the first pick, etc.):

$$P(B_1 \text{ and } B_2) + P(R_1 \text{ and } R_2) = P(B_1) \cdot P(B_2|B_1) + P(R_1) \cdot P(R_2|R_1). \quad (2.34)$$

However, equations like this can be a bit intimidating, so it's often better to think in terms of a large set of trials, as mentioned in the remark in the first example in Section 2.3.2.

2.3.4 Coins and dice

There is never a shortage of probability examples involving dice rolls or coin flips.

Example 1 (One of each number): Six dice are rolled. What is the probability of obtaining exactly one of each of the numbers 1 through 6?

First solution: The total number of possible (ordered) outcomes for what all six dice show is 6^6, because there are six possibilities for each die. How many outcomes are there that have each number appearing once? This is simply the question of how many permutations there are of six numbers, because we need all six numbers to appear, but it doesn't matter in what order. There are 6! permutations, so the desired probability is

$$\frac{6!}{6^6} = \frac{5}{324} \approx 1.5\%. \quad (2.35)$$

Second solution: Let's imagine rolling six dice in succession, with the goal of having each number appear once. On the first roll, we get what we get, and there's no way to fail. So the probability of success on the first roll is 1. However, on the second roll, we don't want to get a repeat of the number that appeared on the first roll (whatever that number happened to be). Since there are five "good" options left, the probability of success on the second roll is 5/6. On the third roll, we don't want to get a repeat of either of the numbers that appeared on the first and second rolls, so the probability of success on the third roll (given success on the first two rolls) is 4/6. Likewise, the fourth roll has a 3/6 chance of success, the fifth has 2/6, and the sixth has 1/6. The probability of complete success all the way through is therefore

$$1 \cdot \frac{5}{6} \cdot \frac{4}{6} \cdot \frac{3}{6} \cdot \frac{2}{6} \cdot \frac{1}{6} = \frac{5}{324}, \quad (2.36)$$

in agreement with the first solution. Note that if we write the initial 1 here as 6/6, then this expression becomes $6!/6^6$, which is the fraction that appears in Eq. (2.35).

Example 2 (Three pairs): Six dice are rolled. What is the probability of getting three pairs, that is, three different numbers that each appear twice?

Solution: We'll count the total number of (ordered) ways to get three pairs, and then we'll divide that by the total number of possible (ordered) outcomes for the six rolls, which is 6^6.

There are two steps in the counting. First, how many different ways can we pick the three different numbers that show up? We need to pick three numbers from six, so the number of ways is $\binom{6}{3} = 20$.

Second, given the three numbers that show up, how many different (ordered) ways can two of each appear on the dice? Let's says the numbers are 1, 2, and 3. We can imagine plopping two of each of these numbers down on six blank spots (which represent the six dice) on a piece of paper. There are $\binom{6}{2} = 15$ ways to pick where the two 1's go. And then there are $\binom{4}{2} = 6$ ways to pick where the two 2's go in the four remaining spots. And then finally there is $\binom{2}{2} = 1$ way to pick where the two 3's go in the two remaining spots.

The total number of ways to get three pairs is therefore $\binom{6}{3} \cdot \binom{6}{2} \cdot \binom{4}{2} \cdot \binom{2}{2}$. So the probability of getting three pairs is

$$p = \frac{\binom{6}{3} \cdot \binom{6}{2} \cdot \binom{4}{2} \cdot \binom{2}{2}}{6^6} = \frac{20 \cdot 15 \cdot 6 \cdot 1}{6^6} = \frac{25}{648} \approx 3.9\%. \tag{2.37}$$

If you try to solve this problem in a manner analogous to the second solution in the previous example (that is, by multiplying probabilities for the successive rolls), then things get a bit messy because there are many different scenarios that lead to three pairs.

Example 3 (Five coin flips): A coin is flipped five times. Calculate the probabilities of getting the various possible numbers of Heads (0 through 5).

Solution: We'll count the number of (ordered) ways to get the different numbers of Heads, and then we'll divide that by the total number of possible (ordered) outcomes for the five flips, which is 2^5.

There is only $\binom{5}{0} = 1$ way to get zero Heads, namely TTTTT. There are $\binom{5}{1} = 5$ ways to get one Heads (such as HTTTT), because there are $\binom{5}{1}$ ways to choose the one coin that shows Heads. There are $\binom{5}{2} = 10$ ways to get two Heads, because there are $\binom{5}{2}$ ways to choose the two coins that show Heads. And so on. The various probabilities are therefore

$$P(0) = \frac{\binom{5}{0}}{2^5}, \quad P(1) = \frac{\binom{5}{1}}{2^5}, \quad P(2) = \frac{\binom{5}{2}}{2^5},$$

$$P(3) = \frac{\binom{5}{3}}{2^5}, \quad P(4) = \frac{\binom{5}{4}}{2^5}, \quad P(5) = \frac{\binom{5}{5}}{2^5}. \tag{2.38}$$

Plugging in the values of the binomial coefficients gives

$$P(0) = \frac{1}{32}, \quad P(1) = \frac{5}{32}, \quad P(2) = \frac{10}{32},$$

$$P(3) = \frac{10}{32}, \quad P(4) = \frac{5}{32}, \quad P(5) = \frac{1}{32}. \tag{2.39}$$

The sum of all these probabilities correctly equals 1. The physical reason for this is that the number of Heads must be *something*, which means that the sum of all the probabilities must be 1. (This holds for any number of flips, of course, not just 5.) The mathematical reason is that the sum of the binomial coefficients (the numerators in the above fractions) equals 2^5 (which is the denominator). See Section 1.8.3 for the explanation of this.

2.3.5 Cards

We already did a lot of card counting in Chapter 1 (particularly in Problem 1.10), and some of those results will be applicable here. As we have mentioned a number of times, exercises in probability are often just exercises in counting. There is effectively an endless number of probability questions we can ask about cards. In the following examples, we will always assume a standard 52-card deck.

Example 1 (Royal flush from seven cards): A few variations of poker involve being dealt seven cards (in one way or another) and forming the best five-card hand that can be made from these seven cards. What is the probability of being able to form a Royal flush in this setup? A Royal flush consists of $10, J, Q, K, A$, all from the same suit.

Solution: The total number of possible seven-card hands is $\binom{52}{7} = 133,784,560$. The number of seven-card hands that contain a Royal flush is $4 \cdot \binom{47}{2} = 4,324$, because there are four ways to choose the five Royal flush cards (the four suits), and then $\binom{47}{2}$ ways to choose the other two cards from the remaining $52 - 5 = 47$ cards in the deck. The probability is therefore

$$\frac{4 \cdot \binom{47}{2}}{\binom{52}{7}} = \frac{4,324}{133,784,560} \approx 0.0032\%. \tag{2.40}$$

This is larger than the result for five-card hands. In that case, only four of the $\binom{52}{5} = 2,598,960$ hands are Royal flushes, so the probability is $4/2,598,960 \approx 0.00015\%$, which is about 20 times smaller than 0.0032%. As an exercise, you can show that the ratio happens to be exactly 21.

Example 2 (Suit full house): In a five-card poker hand, what is the probability of getting a "full house" of suits, that is, three cards of one suit and two of another? (This isn't an actual poker hand worth anything, but that won't stop us from calculating the probability!) How does your answer compare with the probability of getting an actual full house, that is, three cards of one value and two of another? Feel free to use the result from part (a) of Problem 1.10.

Solution: There are four ways to choose the suit that appears three times, and $\binom{13}{3} = 286$ ways to choose the specific three cards from the 13 of this suit. And then there are three ways to choose the suit that appears twice from the remaining three suits, and $\binom{13}{2} = 78$ ways to choose the specific two cards from the 13 of this suit. The total number of suit-full-house hands is therefore $4 \cdot \binom{13}{3} \cdot 3 \cdot \binom{13}{2} = 267,696$. Since there is a total of $\binom{52}{5}$ possible hands, the desired probability is

$$\frac{4 \cdot \binom{13}{3} \cdot 3 \cdot \binom{13}{2}}{\binom{52}{5}} = \frac{267,696}{2,598,960} \approx 10.3\%. \qquad (2.41)$$

From part (a) of Problem 1.10, the total number of actual full-house hands is 3,744, which yields a probability of $3,744/2,598,960 \approx 0.14\%$. It is therefore much more likely (by a factor of about 70) to get a full house of suits than an actual full house of values. (You can show that the exact ratio is 71.5.) This makes intuitive sense; there are more values than suits (13 compared with four), so it is harder to have all five cards involve only two values as opposed to only two suits.

Example 3 (Only two suits):   In a five-card poker hand, what is the probability of having all of the cards be members of at most two suits? (A single suit falls into this category.) The suit full house in the previous example is a special case of "at most two suits." This problem is a little tricky, at least if you solve it a certain way; be careful about double counting some of the hands!

First solution: If two suits appear, then there are $\binom{4}{2} = 6$ ways to pick them. For a given choice of two suits, there are $\binom{26}{5}$ ways to pick the five cards from the $2 \cdot 13 = 26$ cards of these two suits. It therefore seems like there should be $\binom{4}{2} \cdot \binom{26}{5} = 394,680$ different hands that consist of cards from at most two suits.

However, this isn't correct, because we double (or actually triple) counted the hands that involve only one suit (the flushes). For example, if all five cards are hearts, then we counted such a hand in the heart/diamond set of $\binom{26}{5}$ hands, and also in the heart/spade set, and also in the heart/club set. We counted it three times when we should have counted it only once. Since there are $\binom{13}{5}$ hands that are heart flushes, we have included an extra $2 \cdot \binom{13}{5}$ hands, so we need to subtract these from our total. Likewise for the diamond, spade, and club flushes. The total number of hands that involve at most two suits is therefore

$$\binom{4}{2}\binom{26}{5} - 4 \cdot 2 \cdot \binom{13}{5} = 394,680 - 10,296 = 384,384. \qquad (2.42)$$

The desired probability is then

$$\frac{\binom{4}{2}\binom{26}{5} - 8 \cdot \binom{13}{5}}{\binom{52}{5}} = \frac{384,384}{2,598,960} \approx 14.8\%. \qquad (2.43)$$

This is larger than the result in Eq. (2.41), as it should be, because suit full houses are a subset of the hands that involve at most two suits.

Second solution: There are three general ways that we can have at most two suits: (1) all five cards can be of the same suit (a flush), (2) four cards can be of one suit, and one card of another, or (3) three cards can be of one suit, and two cards of another; this is the suit full house from the previous example. We will denote these types of hands by $(5,0)$, $(4,1)$, and $(3,2)$, respectively. How many hands of each type are there?

There are $4 \cdot \binom{13}{5} = 5,148$ hands of the $(5,0)$ type, because there are $\binom{13}{5}$ ways to pick five cards from the 13 cards of a given suit, and there are four suits. From the previous example, there are $4 \cdot \binom{13}{3} \cdot 3 \cdot \binom{13}{2} = 267,696$ hands of the $(3,2)$ type. To figure out the number of hands of the $(4,1)$ type, we can use exactly the same kind of reasoning as in the previous example. This gives $4 \cdot \binom{13}{4} \cdot 3 \cdot \binom{13}{1} = 111,540$ hands. Adding up these three results gives the total number of "at most two suits" hands as

$$4 \cdot \binom{13}{5} + 4 \cdot \binom{13}{4} \cdot 3 \cdot \binom{13}{1} + 4 \cdot \binom{13}{3} \cdot 3 \cdot \binom{13}{2}$$

$$= 5,148 + 111,540 + 267,696$$

$$= 384,384, \tag{2.44}$$

in agreement with the first solution. (The repetition of the "384" here is due in part to the factors of 13 and 11 in all of the terms in the first line of Eq. (2.44). These numbers are factors of 1001.) The hands of the $(3,2)$ type account for about 2/3 of the total, consistent with the fact that the 10.3% result in Eq. (2.41) is about 2/3 of the 14.8% result in Eq. (2.43).

2.4 Four classic problems

Let's now look at four classic probability problems. No book on probability would be complete without a discussion of the "Birthday Problem" and the "Game-Show Problem." Additionally, the "Prosecutor's Fallacy" and the "Boy/Girl Problem" are two other classics that are instructive to study in detail. All four of these problems have answers that might seem counterintuitive at first, but they eventually make sense if you think about them long enough!

After reading the statement of each problem, be sure to try solving it on your own before looking at the solution. If you can't solve it on your first try, set it aside and come back to it later. There's no hurry; the problem will still be there. There are only so many classic problems like these, so don't waste them. If you look at a solution too soon, the opportunity to solve it is gone, and it's never coming back. If you do eventually need to look at the solution, cover it up with a piece of paper and read one line at a time, to get a hint. That way, you can still (mostly) solve it on your own.

2.4.1 The Birthday Problem

We'll present the Birthday Problem first. Aside from being a very interesting problem, its unexpected result allows you to take advantage of unsuspecting people and win money on bets at parties (as long as they're large enough parties, as we'll see!).

Problem: How many people need to be in a room in order for there to be a greater than 1/2 probability that at least two of them have the same birthday? By "same birthday" we mean the same day of the year; the year may differ. Ignore leap years.

(At this point, as with all of the problems in this section, don't read any further until you've either solved the problem or thought hard about it for a long time.)

Solution: If there was ever a problem that called for the "art of not" strategy in Section 2.3.1, this is it. There are many different ways for there to be *at least* one common birthday (one pair, two pairs, one triple, etc.), and it is completely intractable to add up all of these individual probabilities. It is *much* easier (and even with the italics, this is a vast understatement) to calculate the probability that there *isn't* a common birthday, and then subtract this from 1 to obtain the probability that there *is* at least one common birthday.

The calculation of the probability that there *isn't* a common birthday proceeds as follows. Let there be n people in the room. We can imagine taking them one at a time and randomly plopping their names down on a calendar, with the (present) goal being that there are no common birthdays. The first name can go anywhere. But when we plop down the second name, there are only 364 "good" days left, because we don't want the day to coincide with the first name's day. The probability of success for the second name is therefore $364/365$. Then, when we plop down the third name, there are only 363 "good" days left (assuming that the first two people have different birthdays), because we don't want the day to coincide with either of the other two days. The probability of success for the third name is therefore $363/365$. Similarly, when we plop down the fourth name, there are only 362 "good" days left (assuming that the first three people have different birthdays). The probability of success for the fourth name is therefore $362/365$. And so on.

If there are n people in the room, the probability that all n birthdays are distinct (that is, there *isn't* a common birthday among any of the people; hence the superscript "no" below) therefore equals

$$P_n^{no} = 1 \cdot \frac{364}{365} \cdot \frac{363}{365} \cdot \frac{362}{365} \cdot \frac{361}{365} \cdot \ldots \cdot \frac{365 - (n-1)}{365}. \tag{2.45}$$

If you want, you can write the initial 1 here as $365/365$, to make things look nicer. Note that the last term involves $(n-1)$ and not n, because $(n-1)$ is the number of names that have already been plopped down. As a double check that this $(n-1)$ is correct, it works for small numbers like $n = 2$ and 3. You should always perform a simple check like this whenever you write down *any* expression involving a parameter such as n.

We now just have to multiply out the product in Eq. (2.45) to the point where it becomes smaller than $1/2$, so that the probability that there *is* a common birthday is larger than $1/2$. With a calculator, this is tedious, but not horribly painful. We find that $P_{22}^{no} = 0.524$ and $P_{23}^{no} = 0.493$. If P_n^{yes} is the probability that there *is* a common birthday among n people, then $P_n^{yes} = 1 - P_n^{no}$, so $P_{22}^{yes} = 0.476$ and $P_{23}^{yes} = 0.507$. Since our original goal was to have $P_n^{yes} > 1/2$ (or equivalently $P_n^{no} < 1/2$), we see that there must be at least 23 people in a room in order for there to be a greater than 50% chance that at least two of them have the same birthday. The probability in the $n = 23$ case is 50.7%.

The task of Problem 2.14 is to calculate the probability that among 23 people, *exactly* two of them have a common birthday. That is, there aren't two different pairs with common birthdays, or a triple with the same birthday, etc.

REMARK: The $n = 23$ answer to our problem is much smaller than most people would expect. As mentioned above, it therefore provides a nice betting opportunity. For $n = 30$, the probability of a common birthday increases to 70.6%, and most people would still find it hard to believe that among 30 people, there are probably two who have the same birthday. Table 2.4 lists various values of n and the probabilities, $P_n^{yes} = 1 - P_n^{no}$, that at least two people have a common birthday.

n	10	20	23	30	50	60	70	100
P_n^{yes}	11.7%	41.1%	50.7%	70.6%	97.0%	99.4%	99.92%	99.99997%

Table 2.4: Probability of a common birthday among n people.

Even for $n = 50$, most people would probably be happy to bet, at even odds, that no two people have the same birthday. But you'll win the bet 97% of the time.

One reason why many people can't believe the $n = 23$ result is that they're asking themselves a different question, namely, "How many people (in addition to me) need to be present in order for there to be at least a $1/2$ chance that someone else has *my* birthday?" The answer to this question is indeed much larger than 23. The probability that *no one* out of n people has a birthday on a *given day* is simply $(364/365)^n$, because each person has a 364/365 chance of not having that particular birthday. For $n = 252$, this is just over $1/2$. And for $n = 253$, it is just under $1/2$; it equals 0.4995. Therefore, you need to come across 253 other people in order for the probability to be greater than $1/2$ that at least one of them *does* have *your* birthday (or any other particular birthday). See Problem 2.16 for further discussion of this. ♣

2.4.2 The Game-Show Problem

We'll now discuss the Game-Show Problem. In addition to having a variety of common incorrect solutions, this problem also also a long history of people arguing vehemently in favor of those incorrect solutions.

Problem: A game-show host offers you the choice of three doors. Behind one of these doors is the grand prize, and behind the other two are goats. The host (who knows what is behind each of the doors) announces that after you select a door (without opening it), he will open one of the other two doors and purposefully reveal a goat. You select a door. The host then opens one of the other doors and reveals the promised goat. He then offers you the chance to switch your choice to the remaining door. To maximize the probability of winning the grand prize, should you switch or not? Or does it not matter?

Solution: We'll present three solutions, one right and two wrong. You should decide which one you think is correct before reading beyond the third solution. Cover up the page after the third solution with a piece of paper, so that you don't inadvertently see which one is correct.

- REASONING 1: Once the host reveals a goat, the prize must be behind one of the two remaining doors. Since the prize was randomly located to begin with, there must be equal chances that the prize is behind each of the two remaining doors. The probabilities are therefore both $1/2$, so it doesn't matter if you switch.

If you want, you can imagine a friend (who is aware of the whole procedure of the host announcing that he will open a door and reveal a goat) entering the room *after* the host opens the door. This person sees two identical unopened doors (he doesn't know which one you initially picked) and a goat. So for him there must be a 1/2 chance that the prize is behind each unopened door. The probabilities for you and your friend can't be any different, so you also say that each unopened door has a 1/2 chance of containing the prize. It therefore doesn't matter if you switch.

- REASONING 2: There is initially a 1/3 chance that the prize is behind any of the three doors. So if you don't switch, your probability of winning is 1/3. No actions taken by the host can change the fact that if you play a large number n of these games, then (roughly) $n/3$ of them will have the prize behind the door you initially pick.

  Likewise, if you switch to the other unopened door, there is a 1/3 chance that the prize is behind that door. (There is obviously a goat behind at least one of the other two doors, so the fact that the host reveals a goat doesn't tell you anything new.) Therefore, since the probability is 1/3 whether or not you switch, it doesn't matter if you switch.

- REASONING 3: As in the first paragraph of Reasoning 2, if you don't switch, your probability of winning is 1/3.

  However, if you switch, your probability of winning is greater than 1/3. It increases to 2/3. This can be seen as follows. Without loss of generality, assume that you pick the first door. (You can repeat the following reasoning for the other doors if you wish. It gives the same result.) There are three equally likely possibilities for what is behind the three doors: PGG, GPG, and GGP, where P denotes the prize and G denotes a goat. If you don't switch, then in only the first of these three cases do you win, so your odds of winning are 1/3 (consistent with the first paragraph of Reasoning 2). But if you do switch from the first door to the second or third, then in the first case PGG you lose, but in the other two cases you win, because the door not opened by the host has the prize. (The host has no choice but to reveal the G and leave the P unopened.) Therefore, since two out of the three equally likely cases yield success if you switch, your probability of winning if you switch is 2/3. So you do in fact want to switch.

Which of these three solutions is correct? Don't read any further until you've firmly decided which one you think is right.

The third solution is correct. The error in the first solution is the statement, "there must be equal chances that the prize is behind each of the two remaining doors." This is simply not true. The act of revealing a goat breaks the symmetry between the two remaining doors, as explained in the third solution. One door is the one you initially picked, while the other door is one of the two that you didn't pick. The fact that there are two possibilities doesn't mean that their probabilities have to be equal, of course!

The error in the supporting reasoning with your friend (who enters the room after the host opens the door) is the following. While it *is* true that both probabilities are 1/2 for your friend, they aren't both 1/2 for *you*. The statement, "the probabilities for you and your friend can't be any different," is false. You have information that your friend doesn't have; you know which of the two unopened doors is the one you initially picked and which is the door that the host chose to leave unopened. (And as seen in the third solution, this information yields probabilities of 1/3 and 2/3.) Your friend doesn't have this critical information. Both doors look the same to him. Probabilities can certainly be different for different people. If I flip a coin and peek and see a Heads, but I don't show you, then the probability of a Heads is 1/2 for you, but 1 for me.

The error in the second solution is that the act of revealing a goat *does* give you new information, as we just noted. This information tells you that the prize isn't behind that door, and it also distinguishes between the two remaining unopened doors. One is the door you initially picked, while the other is one of the two doors that you didn't initially pick. As seen in the third solution, this information has the effect of increasing the probability that the goat is behind the other door. Note that another reason why the second solution can't be correct is that the two probabilities of 1/3 don't add up to 1.

To sum up, it should be no surprise that the probabilities are different for the switching and non-switching strategies *after* the host opens a door (the probabilities are obviously the same, equal to 1/3, whether or not a switch is made *before* the host opens a door), because the host gave you some of the information he had about the locations of things.

REMARKS:

1. If you still doubt the validity of the third solution, imagine a situation with 1000 doors containing one prize and 999 goats. After you pick a door, the host opens 998 other doors and reveals 998 goats (and he said beforehand that he was going to do this). In this setup, if you don't switch, your chances of winning are 1/1000. But if you do switch, your chances of winning are 999/1000, which can be seen by listing out (or imagining listing out) the 1000 cases, as we did with the three PGG, GPG, and GGP cases in the third solution. It is clear that the switch should be made, because the *only* case where you lose after you switch is the case where you had initially picked the prize, and this happens only 1/1000 of the time.

   In short, a huge amount of information is gained by the revealing of 998 goats. There is initially a 999/1000 chance that the prize is somewhere behind the other 999 doors, and the host is kindly giving you the information of exactly which door it is (in the highly likely event that it is in fact one of the other 999).

2. The clause in the statement of the problem, "The host announces that after you select a door (without opening it), he will open one of the other two doors and purposefully reveal a goat," is crucial. If it is omitted, and it is simply stated that, "The host then opens one of the other doors and reveals a goat," then it is impossible to state a preferred strategy. If the host doesn't announce his actions beforehand, then for all you know, he *always* reveals a goat (in which case you should switch, as we saw above). Or he *randomly* opens a door and just happened to pick a goat (in which case it doesn't matter if you switch, as you can show in Problem 2.18). Or he opens a door and reveals a goat if and only if your initial door has the prize (in which case you definitely should

not switch). Or he could have one procedure on Tuesdays and another on Fridays, each of which depends on the color of the socks he's wearing. And so on.

3. As mentioned above, this problem is infamous for the intense arguments it lends itself to. There's nothing terrible about getting the wrong answer, nor is there anything terrible about not believing the correct answer for a while. But concerning arguments that drag on and on, it doesn't make any sense to argue about this problem for more than, say, 20 minutes, because at that point everyone should stop and just *play the game*! You can play a number of times with the switching strategy, and then a number of times with the non-switching strategy. Three coins with a dot on the bottom of one of them are all you need.[1] Not only will the actual game yield the correct answer (if you play enough times so that things average out), but the patterns that form will undoubtedly convince you of the correct reasoning (or reinforce it, if you're already comfortable with it). Arguing endlessly about an experiment, when you can actually *do* the experiment, is as silly as arguing endlessly about what's behind a door, when you can simply open the door.

4. For completeness, there is one subtlety we should mention here. In the second solution, we stated, "No actions taken by the host can change the fact that if you play a large number n of these games, then (roughly) $n/3$ of them will have the prize behind the door you initially pick." This part of the reasoning was correct; it was the "switching" part of the second solution that was incorrect. After doing Problem 2.18 (where the host randomly opens a door), you might disagree with the above statement, because it will turn out in that problem that the actions taken by the host *do* affect this $n/3$ result. However, the above statement is still correct for "*these* games" (the ones governed by the original statement of this problem). See the second remark in the solution to Problem 2.18 for further discussion. ♣

2.4.3 The Prosecutor's Fallacy

We now present one of the most classic problems/paradoxes in the subject of probability. This classic nature is due in no small part to the problem's critical relevance to the real world. After reading the statement of the problem below, you should think carefully and settle on an answer before looking at the solution. The discussion of conditional probability in Section 2.2.4 gives a hint at the answer.

Problem: Consider the following scenario. Detectives in a city, say, Boston (whose population we will assume to be one million), are working on a crime and have put together a description of the perpetrator, based on things such as height, a tattoo, a limp, an earring, etc. Let's assume that only one person in 10,000 fits the description. On a routine patrol the next day, police officers see a person fitting the description. This person is arrested and brought to trial based solely on the fact that he fits the description.

During the trial, the prosecutor tells the jury that since only one person in 10,000 fits the description (a true statement), it is highly unlikely (far beyond a reasonable doubt) that an innocent person fits the description (again a true statement); it is

---

[1]You actually don't need three objects. It's hard to find three exactly identical coins anyway. The "host" can simply roll a die, without showing the "contestant" the result. Rolling a 1 or 2 can mean that the prize is located behind the first door, a 3 or 4 the second, and a 5 or 6 the third. The game then basically involves calling out door numbers.

therefore highly unlikely that the defendant is innocent. If you were a member of the jury, would you cast a "guilty" vote? If yes, what is your level of confidence? If no, what is wrong with the prosecutor's reasoning?

Solution: We'll assume that we are concerned only with people living in Boston. There are one million such people, so if one person in 10,000 fits the description, this means that there are 100 people in Boston who fit it (one of whom is the perpetrator). When the police officers pick up someone fitting the description, this person could be any one of these 100 people. So the probability that the defendant in the courtroom is the actual perpetrator is only 1/100. In other words, there is a 99% chance that the person is innocent. A guilty verdict (based on the given evidence) would therefore be a horrible and tragic vote.

The above (correct) reasoning is fairly cut and dry, but it contradicts the prosecutor's reasoning. The prosecutor's reasoning must therefore be incorrect. But what exactly is wrong with it? It seems quite plausible at every stage. To isolate the flaw in the logic, let's list out the three separate statements the prosecutor made in his argument:

1. Only one person in 10,000 fits the description.

2. It is highly unlikely (far beyond a reasonable doubt) that an innocent person fits the description.

3. It is therefore highly unlikely that the defendant is innocent.

As we noted above when we posed the problem, the first two of these statements are true. Statement 1 is true by assumption, and Statement 2 is true basically because 1/10,000 is a small number. Let's be precise about this and work out the exact probability that an innocent person fits the description. Of the one million people in Boston, the number who fit the description is $(1/10,000)(10^6) = 100$. Of these 100 people, only one is guilty, so 99 are innocent. And the total number of innocent people is $10^6 - 1 = 999,999$. The probability that an innocent person fits the description is therefore

$$\frac{\text{innocent and fitting description}}{\text{innocent}} = \frac{99}{999,999} \approx 9.9 \cdot 10^{-5} \approx \frac{1}{10,000} \,. \qquad (2.46)$$

As expected, the probability is essentially equal to 1/10,000.

Now let's look at the third statement above. This is where the error is. This statement is false, because Statement 2 simply does not imply Statement 3. We know this because we have already calculated the probability that the defendant is innocent, namely 99%. This correct probability of 99% is vastly different from the incorrect probability of 1/10,000 that the prosecutor is trying to mislead you with. However, even though the correct result of 99% tells us that Statement 3 must be false, where exactly is the error? After all, at first glance Statement 3 *seems* to follow from Statement 2. The error is the confusion of conditional probabilities. In detail:

- Statement 2 deals with the probability of fitting the description, *given* innocence. The (true) statement is equivalent to, "*If* a person is innocent, *then* there is a very small probability that he fits the description." This probability is the conditional probability $P(D|I)$, with D for description and I for innocence.

- Statement 3 deals with the probability of innocence, *given* that the description is fit. The (false) statement is equivalent to, "*If* a person (such as the defendant) fits the description, *then* there is a very small probability that he is innocent." This probability is the conditional probability $P(I|D)$.

These two conditional probabilities are *not* the same. The error is the assumption (or implication, on the prosecutor's part) that they are. As we saw above, $P(D|I) = 99/999,999 \approx 0.0001$, whereas $P(I|D) = 0.99$. These two probabilities are markedly different.

Intuitively, $P(D|I)$ is very small because a very small fraction of the population (in particular, a very small fraction of the innocent people) fit the description. And $P(I|D)$ is very close to 1 because nearly everyone (in particular, nearly everyone who fits the description) is innocent. This state of affairs is indicated in Fig. 2.10. (This a just a rough figure; the areas aren't actually in the proper proportions.) The large oval represents the 999,999 innocent people, and the small oval represents the 100 people who fit the description.

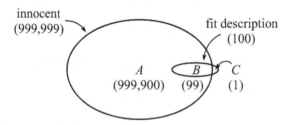

Figure 2.10: The different types of people in the prosecutor's fallacy.

There are three basic types of people in the figure: There are $A = 999,900$ innocent people who don't fit the description, $B = 99$ innocent people who do fit the description, and $C = 1$ guilty person who fits the description. (The fourth possibility – a guilty person who doesn't fit the description – doesn't exist.) The two conditional probabilities that are relevant in the above discussion are then

$$P(D|I) = \frac{B}{\text{innocent}} = \frac{B}{B+A} = \frac{99}{999,999},$$

$$P(I|D) = \frac{B}{\text{fit description}} = \frac{B}{B+C} = \frac{99}{100}. \qquad (2.47)$$

Both of these probabilities have B in numerator, because B represents the people who are innocent *and* fit the description. But the A in the first denominator is much larger than the C in second denominator. Or said in another way, B is a very small fraction of the innocent people (the large oval in Fig. 2.10), whereas it is a very large fraction of the people who fit the description (the small oval in Fig. 2.10).

The prosecutor's faulty reasoning has been used countless times in actual court cases, with tragic consequences. Innocent people have been convicted, and guilty people have walked free (the argument can work in that direction too). These consequences can't be blamed on the jury, of course. It is inevitable that many jurors will fail to spot the error in the reasoning. It would be silly to think that the entire population should be familiar with this issue in probability. Nor can the blame be put on the attorney making the argument. This person is either (1) overzealous and/or incompetent, or (2) entirely within his/her right to knowingly make an invalid argument (as distasteful as this may seem). In the end, the blame falls on either (1) the opposing attorney for failing to rebut the known logical fallacy, or (2) a legal system that in some cases doesn't allow a final rebuttal.

2.4.4 The Boy/Girl Problem

The well-known Boy/Girl Problem can be stated in many different ways, with answers that may or may not be the same. Three different formulations are presented below, and a fourth is given in Problem 2.19. Assume in all of them that any process involved in the scenario is completely random. That is, assume that any child is equally likely to be a boy or a girl (even though this isn't quite true in real life), and assume that there is nothing special about the person you're talking with, and assume that there are no correlations between children (as there are with identical twins), and so on.

Problem:

(a) You bump into a random person on the street who says, "I have two children. At least one of them is a boy." What is the probability that the other child is also a boy?

(b) You bump into a random person on the street who says, "I have two children. The older one is a boy." What is the probability that the other child is also a boy?

(c) You bump into a random person on the street who says, "I have two children, one of whom is this boy standing next to me." What is the probability that the other child is also a boy?

Solution:

(a) The key to all three of these formulations is to list out the various equally likely possibilities for the family's children, while taking into account only the "I have two children" information, and *not yet* the information about the boy. With B for boy and G for girl, the family in the present scenario in part (a) can be of four types (at least *before* the parent gives you information about the boy), each with probability 1/4:

$$\boxed{BB} \quad \boxed{BG} \quad \boxed{GB} \quad GG$$

Ignore the boxes for a moment. In each pair of letters, the first letter stands for the older child, and the second letter stands for the younger child.

Note that there are indeed four equally likely possibilities (BB, BG, GB, GG), as opposed to just three equally likely possibilities (BB, BG, GG), because the older child has a 50-50 chance of being a boy or a girl, as does the younger child. The BG and GB cases each get counted once, just as the HT and TH cases each get counted once when flipping two coins, where the four equally likely possibilities are HH, HT, TH, TT.

Under the assumption of general randomness stated in the problem, we are assuming that you are equally likely (at least *before* the parent gives you information about the boy) to bump into a parent of any one of the above four types of two-child families.

Let us *now* invoke the information that at least one child is a boy. This information tells us that you can't be talking with a GG parent. The parent must be a BB, BG, or GB parent, all equally likely. (They are equally likely, because they are all equivalent with regard to the "at least one of them is a boy" statement.) These are the boxed families in the above list. Of these three cases, only the BB case has the other child being a boy. The desired probability that the other child is a boy is therefore 1/3.

If you don't trust the reasoning in the preceding paragraph, just imagine performing many trials of the setup. This is always a good strategy when solving probability problems. Imagine that you encounter 1000 random parents of two children. You will encounter about 250 of each of the four types of parent. The 250 GG parents have nothing to do with the given setup, so we must discard them. Only the other 750 parents (BB, BG, GB) are able to provide the given information that at least one child is a boy. Of these 750 parents, 250 are of the BB type and thereby have a boy as the other child. The desired probability is therefore $250/750 = 1/3$.

(b) As in part (a), *before* the information about the boy is taken into account, there are four equally likely possibilities for the children (again ignore the boxes for a moment):

$$\boxed{\text{BB}} \quad \boxed{\text{BG}} \quad \text{GB} \quad \text{GG}$$

But once the parent tells you that the older child is a boy, the GB and GG cases are ruled out; remember that the first letter in each pair corresponds to the older child. So you must be talking with a BB or BG parent, both equally likely. Of these two cases, only the BB case has the other child being a boy. The desired probability that the other child is a boy is therefore 1/2.

(c) This version of the problem is a little trickier, because there are now *eight* equally likely possibilities (before the information about the boy is taken into account), instead of just four. This is true because for each of the four types of families in the above lists, the parent may choose to take either of the children for a walk (with equal probabilities, as we are assuming for everything). The eight equally likely possibilities are therefore shown in Fig. 2.5 (again ignore the boxes for a moment). The bold letter indicates the child you encounter.

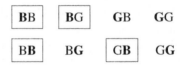

Table 2.5: The eight types of families, accounting for the child present.

Once the parent tells you that one of the children is the boy standing there, four of the eight possibilities are ruled out. Only the four boxed pairs in Fig. 2.5 (the ones with a bold **B**) satisfy the condition that the child standing there is a boy. Of these four (equally likely) possibilities, two of them have the other child being a boy. The desired probability that the other child is a boy is therefore 1/2.

REMARKS:

1. We used the given assumption of general randomness many times in the above solutions. One way to make things *non*random is to assume that the parent who is out for a walk is chosen randomly with equal 1/3 probabilities of being from BB families, or GG families, or one-boy-and-one-girl families. This is an artificial construction, because it means that a given BG or GB family (which together make up half of all two-child families) is less likely to be chosen than a given BB or GG family. This violates our assumption of general randomness. In this scenario, you can show that the answers to parts (a), (b), and (c) are 1/2, 2/3, and 2/3.

   Another way to make things nonrandom is to assume that in part (c) a girl is always chosen to go on the walk if the family has at least one girl. The answer to part (c) is then 1, because the only way a boy will be standing there is if both children are boys. On the other hand, if we assume that a boy is always chosen to go on the walk if the family has at least one boy, then the answer to part (c) is 1/3. This is true because for BB, the other child is a boy; and for both BG and GB (for which the boy is always chosen to go on the walk), the other child is a girl. Basically, the middle four pairs in Table 2.5 will all have a bold **B**, so they will all be boxed. There are countless ways to make things nonrandom, so unless we make an assumption of general randomness, there is no way to solve the problem.

2. Let's compare the scenarios in parts (a) and (b), to see exactly why the probabilities differ. In part (a), the parent's statement rules out the GG case. The BB, BG, and GB cases survive, with the BB families representing 1/3 of all of the possibilities. If the parent then changes the statement, "at least one of them is a boy" to "the older one is a boy," we are now in the realm of part (b). The GB case is now also ruled out (in addition to the GG case). So only the BB and BG cases survive, with the BB families representing 1/2 of all of the possibilities. This is why the probability jumps from 1/3 to 1/2 in going from part (a) to part (b). An additional group of families (GB) is ruled out.

   Let's now compare the scenarios in parts (a) and (c), to see exactly why the probabilities differ. As in the preceding paragraph, the parent's statement in part (a) rules out the GG case. If the parent then makes the additional statement "...and there he is over there next to that tree," we are now in the realm of part (c). Which additional families are ruled out? Well, in part (a), you could be talking with a parent in any of the families in Table 2.5 except the two GG entries. So there are six valid possibilities. But as soon as the parent adds the "and there he is" comment, the unboxed **GB** and **BG** entries are ruled out. So a larger fraction of the valid possibilities (now two out of four, instead of two out of six) have the other child being a boy.

3. Having gone through all of the above reasonings and the comparisons of the different cases, we should note that there is actually a much quicker way of obtaining the probabilities of $1/2$ in parts (b) and (c). If the parent says that the older child is a boy, or that one of the children is the boy standing next to her, then the parent is making a statement *solely about a particular child* (the older one, or the present one). The parent is saying nothing about the other child (the younger one, or the absent one). We therefore know nothing about that child. So by our assumption of general randomness, the other child is equally likely to be a boy or a girl. This should be contrasted with part (a). In that scenario, when the parent says that at least one child is a boy, the parent is *not* making a claim about a *specific child*, but rather about the *collective set* of the two children together. We are therefore not able to uniquely define the "other child" and simply say that the answer is $1/2$. The answer depends on both children together, and it turns out to be different from $1/2$ (namely $1/3$).

4. There is a subtlety in this problem that we should address: How does the parent decide what information to give you? A reasonable rule could be that in part (a) the parent says, "At least one child is a boy," if she is able to; otherwise she says, "At least one child is a girl." This is consistent with all of our above reasoning. But consider what happens if we tweak the rule so that now the parent says, "At least one child is a girl," if she is able to; otherwise she says, "At least one child is a boy." In this case, the answer to part (a) is 1, because the only parents making the "boy" statement are the BB parents. This minor tweak completely changes the problem.

   If you want to avoid this issue, you can rephrase part (a) as: You bump into a random person on the street and ask, "Do you have (exactly) two children? If so, is at least one of them a boy?" In the cases where the answers to both of these questions are "yes," what is the probability that the other child is also a boy? Alternatively, you can just remove the parent and pose the problem as: Consider all two-child families that have at least one boy. What is the probability that both children are boys? This phrasing isn't as catchy as the original, but it gets rid of the above issue.

5. In the various lists of types of families in the above solutions, only the boxed types were applicable. The unboxed ones didn't satisfy the conditions given in the statement of the problem, so we discarded them. This act of discarding the unboxed types is equivalent to using the conditional-probability statement in Eq. (2.5), which can be rearranged to say

$$P(B|A) = \frac{P(A \text{ and } B)}{P(A)}. \tag{2.48}$$

For example, in part (a) if we let $A = \{\text{at least 1 boy}\}$ and $B = \{2 \text{ boys}\}$, then we obtain

$$P\big((2 \text{ boys})|(\text{at least 1 boy})\big) = \frac{P\big((\text{at least 1 boy}) \text{ and } (2 \text{ boys})\big)}{P(\text{at least 1 boy})}. \tag{2.49}$$

The lefthand side of this equation is the probability we're trying to find. On the righthand side, we can rewrite $P((\text{at least 1 boy}) \text{ and } (2 \text{ boys}))$ as just $P(2 \text{ boys})$, because $\{2 \text{ boys}\}$ is a subset of $\{\text{at least 1 boy}\}$. So we have

$$P\big((2 \text{ boys})|(\text{at least 1 boy})\big) = \frac{P(2 \text{ boys})}{P(\text{at least 1 boy})} = \frac{1/4}{3/4} = \frac{1}{3}. \tag{2.50}$$

The preceding equations might look a bit intimidating, which is why we took a more intuitive route in the above solution to part (a), where we imagined doing 1000 trials and then discarding the 250 GG families. Discarding these families accomplishes the same thing as having the $P(\text{at least 1 boy})$ term in the denominator in Eq. (2.50); namely, they both signify that we are concerned only with families that have at least one boy. This remark leads us into the following section on Bayes' theorem.

6. If you thought that some of the answers to this problem were counterintuitive, then, well, you haven't seen anything yet! Tackle Problem 2.19 and you'll see why. ♣

2.5 Bayes' theorem

We now introduce Bayes' theorem, which gives a relation between certain conditional probabilities. The theorem is relevant to much of what we have been discussing in this chapter, particularly Section 2.4. We have technically already derived everything we need for the theorem (and we have actually already been using the theorem without realizing it), so the proof will be very quick. There are three common forms of the theorem. After we prove these, we'll do an example and then present a helpful way of thinking about the theorem in terms of pictures.

Theorem 2.1 (Bayes' theorem) *The "simple form" of Bayes' theorem is*

$$P(A|Z) = \frac{P(Z|A) \cdot P(A)}{P(Z)} \tag{2.51}$$

The "explicit form" is (with "$\sim A$" shorthand for "not A")

$$P(A|Z) = \frac{P(Z|A) \cdot P(A)}{P(Z|A) \cdot P(A) + P(Z|\sim A) \cdot P(\sim A)} \tag{2.52}$$

And the "general form" is

$$P(A_k|Z) = \frac{P(Z|A_k) \cdot P(A_k)}{\sum_i P(Z|A_i) \cdot P(A_i)} \tag{2.53}$$

where the A_i are a complete and mutually exclusive set of events. That is, every possible outcome belongs to one (hence the "complete") and only one (hence the "mutually exclusive") of the A_i.

Proof: The simple form of Bayes' theorem in Eq. (2.51) follows from what we noted back in Eq. (2.9). Since the order of A and Z doesn't matter in $P(A \text{ and } Z)$, we can write down two different expressions for this probability:

$$P(A \text{ and } Z) = P(A|Z) \cdot P(Z)$$
$$= P(Z|A) \cdot P(A). \tag{2.54}$$

If we equate the two righthand sides of these equations and divide through by $P(Z)$, we obtain Eq. (2.51).

The explicit form in Eq. (2.52) follows from the fact that the $P(Z)$ in the denominator of Eq. (2.51) can be written as

$$P(Z) = P(Z \text{ and } A) + P(Z \text{ and } \sim A)$$
$$= P(Z|A) \cdot P(A) + P(Z|\sim A) \cdot P(\sim A). \tag{2.55}$$

The first line here comes from the fact that every outcome is a member of either A or $\sim A$, and the second line comes from two applications of Eq. (2.54).

The general form in Eq. (2.53) is obtained by replacing the A in Eq. (2.51) with A_k and noting that

$$P(Z) = \sum_i P(Z \text{ and } A_i)$$
$$= \sum_i P(Z|A_i) \cdot P(A_i). \tag{2.56}$$

The first line here comes from the fact that every outcome is a member of exactly one of the A_i, and the second line comes from n applications (where n is the number of A_i) of Eq. (2.54). Note that Eq. (2.52) is a special case of Eq. (2.53), with $A_1 = A$ and $A_2 = \sim A$, and with $k = 1$ (so $A_k = A$). Note also that all of the numerators on the righthand sides of the three formulations of the theorem are equal to $P(A \text{ and } Z)$ or $P(A_k \text{ and } Z)$, from Eq. (2.54). ∎

As promised, these proofs were very quick. All we needed was Eq. (2.54) and the fact that $P(Z) = \sum_i P(Z \text{ and } A_i)$, which holds because the A_i are mutually exclusive and complete. However, even though the proofs were quick, and even though the theorem isn't anything we didn't already know (since we already knew the two ingredients in the preceding sentence), the theorem can still be a bit intimidating, especially the general form in Eq. (2.53). So we'll do an example to get some practice. But first some remarks.

REMARKS:

1. In Eq. (2.53) the $P(A_i)$ are known as the *prior* probabilities, the $P(Z|A_i)$ are known as the *conditional* probabilities, and $P(A_k|Z)$ is known as the *posterior* probability. The prior and conditional probabilities are the ones you are given (at least in this book; see the following remark), and the posterior probability is the one you are trying to find.

2. Since Bayes' theorem is simply a restatement of what we already know, you might be wondering what good it is and why it comes up so often when people talk about probability. Does it actually give us anything new? Well, yes and no. The theorem itself doesn't give us anything new, but the way in which it is *used* does.

   It would take many pages to do justice to this topic, but in a nutshell, there are two main types of probability reasoning. *Frequentist* reasoning (which is what we are using in this book) defines probability by imagining a large number of trials. In contrast, *Bayesian* reasoning doesn't require a large number of trials. The difference between these two reasonings shows up when one gets into statistical inference, that is, when one tries to estimate probabilities by gathering data (which we won't do in this book). In the end, the difference comes down to how one treats the prior probabilities $P(A_i)$

in Eq. (2.53). A frequentist considers them to be definite quantities (based on the frequencies obtained in large numbers of trials), whereas a Bayesian considers them to be unknowns whose values are given by specified distributions (determined in some manner). However, this difference is moot in this book, because we will always deal with situations where the prior probabilities take on definite values that are given. In this case, the frequentist and Bayesian reasonings are identical. They both boil down to Eq. (2.54). ♣

Let's now do an example. A common setup where Bayes' theorem is relevant involves false positives on a diagnostic test, so that's the setup we'll use here. After working through the example, we'll see how we can alternatively make use of a particularly helpful type of picture. There are many different probabilities that appear in Eq. (2.53), and it can be hard to remember what the theorem says or to get an intuitive feel for what's going on. In contrast, a quick glance at a figure such as Fig. 2.14 below makes it easy to remember the theorem and understand it intuitively.

Example (False positives): A hospital administers a test to see if a patient has a certain disease. Assume that we know the following three things:

- 2% of the overall population has the disease.

- If a person *does* have the disease, then the test has a 95% chance of correctly indicating that the person has it. (So 5% of the time, the test incorrectly indicates that the person doesn't have the disease.)

- If a person *does not* have the disease, then the test has a 10% chance of incorrectly indicating that the person has it; this is a "false positive" result. (So 90% of the time, the test correctly indicates that the person doesn't have the disease.)

The question we want to answer is: If a patient tests positive, what is the probability that they[2] actually have the disease?

We'll answer this question first by pretending that we haven't seen Bayes' theorem, and then by using the theorem. The reasoning will be exactly the same in both solutions, because in the first solution we'll actually be using Bayes' theorem without realizing it.

First solution: Imagine taking a large number of people (say, 1000) from the general population and testing them for the disease. A given person either has the disease or doesn't (two possibilities), and their test is either positive or negative (two possibilities). So there are $2 \cdot 2 = 4$ different types of people, with regard to the disease and the test. Let's make a probability tree to determine how many people of each type there are; see Fig. 2.11. The three given facts correspond to the three forks in the tree:

- The first fact tells us that of the given 1000 people, 2% (which is 20 people) have the disease (on average), while 98% (which is 980 people) don't have the disease.

- The second fact tells us that of the 20 people with the disease, 95% (which is 19 people) test positive, while 5% (which is 1 person) tests negative.

---

[2]I am using "they" as a gender-neutral singular pronoun, in protest of the present failing of the English language.

- The third fact tells us that of the 980 people without the disease, 10% (which is 98 people) test positive, while 90% (which is 882 people) test negative.

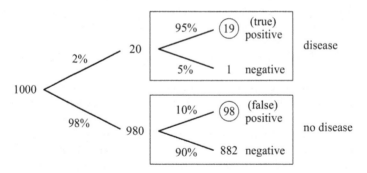

Figure 2.11: The probability tree for yes/no disease and positive/negative test.

The answer to the above question (namely, "If a patient tests positive, what is the probability that they actually have the disease?") can now simply be read off from the tree. The total number of people who test positive is the sum of the two circled numbers, which is $19 + 98 = 117$. And of these 117 people, only 19 have the disease. So our answer is

$$p = \frac{19}{19 + 98} = \frac{19}{117} = 16\%. \tag{2.57}$$

If we want to write this directly in terms of the given probabilities, then if we recall how we arrived at the numbers 19 and 98, we obtain

$$p = \frac{(0.95)(0.02)}{(0.95)(0.02) + (0.10)(0.98)} = 0.16. \tag{2.58}$$

Second solution: We'll use the "explicit form" of Bayes' theorem in Eq. (2.52), which is a special case of the "general form" in Eq. (2.53). In the notation of Eq. (2.52) we have

$$A = \text{have disease,}$$
$$\sim A = \text{don't have disease,}$$
$$Z = \text{test positive.} \tag{2.59}$$

Our goal is to calculate $P(A|Z)$, that is, the probability of having the disease, given a positive test. From the given facts in the three bullet points, we know that

$$P(A) = 0.02,$$
$$P(Z|A) = 0.95,$$
$$P(Z|\sim A) = 0.10. \tag{2.60}$$

Plugging these probabilities into Eq. (2.52) gives

$$P(A|Z) = \frac{P(Z|A) \cdot P(A)}{P(Z|A) \cdot P(A) + P(Z|\sim A) \cdot P(\sim A)}$$
$$= \frac{(0.95)(0.02)}{(0.95)(0.02) + (0.10)(0.98)} = 0.16, \tag{2.61}$$

in agreement with the first solution. This is the same expression as in Eq. (2.58), which is consistent with the fact that (as we mentioned above) our reasoning in the first solution was equivalent to using Bayes' theorem.

REMARK: We see that if a person tests positive, they have only a 16% chance of actually having the disease. This answer might seem surprisingly low. After all, the test seems fairly reliable; it gives the correct result 95% of the time if a person has the disease, and 90% of the time if a person doesn't have the disease. So how did we end up with an answer that is much smaller than either of these two percentages?

The explanation is that because the percentage of people with the disease is so tiny (2%), the small percentage (10%) of false positives among the non-disease people yields a *number* of false positives that is significantly larger than the *number* of true positives. Basically, 10% of 98% of 1000 (which is 98) is significantly larger than 95% of 2% of 1000 (which is 19). The 98 false positives dominate the 19 true positives. Although the 10% false-positive rate is small, it isn't small enough to prevent the smallness of the 2% disease rate from controlling the outcome. A takeaway from this discussion is that one must be very careful when testing for rare diseases. If the disease is very rare, then the test must be extremely accurate, otherwise a positive test isn't meaningful.

If we decrease the 10% percentage (that is, reduce the percentage of false positives) and/or increase the 2% percentage (that is, increase the percentage of people with the disease), then the answer to our original question will increase. That is, a larger fraction of the people who test positive will actually have the disease. For example, if we assume that 40% of the population have the disease (so 60% don't have it), and if we keep all the other percentages in the problem the same, then Eq. (2.58) becomes

$$p = \frac{(0.95)(0.40)}{(0.95)(0.40) + (0.10)(0.60)} = 0.86. \tag{2.62}$$

This probability is closer to 1 than in the original scenario, because if we have 1000 people, then the 60 (instead of the earlier 98) false positives are dominated by the 380 (instead of the earlier 19) true positives. You can verify these numbers.

In the limit where the 10% false-positive percentage in the original scenario goes to zero, or the 2% disease percentage goes to 100%, the number of false positives goes to zero. This is true because if 10% $\to$ 0% then the test never incorrectly says that a person has the disease when they don't; and if 2% $\to$ 100% then the entire population has the disease, so every positive test is a true one. In either of these limits, the answer to our question goes to 1 (or 100%); a positive test always correctly indicates the disease. ♣

In the first solution above, we calculated the various numbers and probabilities by using a probability tree. We can alternatively use a figure along the lines of Fig. 2.4. In the following discussion we'll pretend that we haven't seen Bayes' theorem, and then we'll circle back to the theorem and show in Fig. 2.14 how the different ingredients in the theorem correspond to the different parts of the figure.

Fig. 2.12 shows a pictorial representation of the probability tree in Fig. 2.11. The overall square represents the given 1000 people.[3] A vertical line divides the

---

[3]When drawing a figure like this, the area of a region can represent either the probability of being in that region, or the actual number of outcomes/people/etc. in that region. The usage should be clear from the context. We're using actual numbers here.

square into two rectangles – a very thin one on the left representing the 20 people with the disease, and a wide one on the right representing the 980 people without the disease. These two rectangles are further divided into the people who test positive (the shaded lower regions, with 19 and 98 people) or test negative (the unshaded upper regions, with 1 and 882 people). The desired probability of a person having the disease if they test positive equals the 19 true positives (the darkly shaded thin rectangle) divided by the total $19 + 98 = 117$ number of positives (both shaded regions).

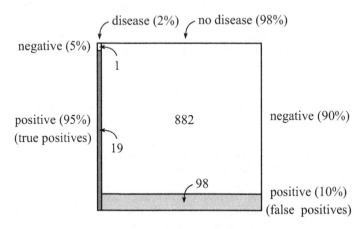

Figure 2.12: The probability square for yes/no disease and positive/negative test.

In Fig. 2.12 there are only two types of people in the population – those with the disease and those without it. As an example of a more general setup, let's consider how people commute to work. We'll assume that we are given the percentages of people who walk, bike, drive, take the bus, etc. And then for each of these types, we'll assume that we are also given the percentage who have a particular attribute – for example, the ability to play the guitar. We can then ask questions such as, "If we pick a random person (among those who commute to work) from the set of people who can play the guitar, what is the probability that this person walks to work?" If we compare this question to our earlier one involving the disease testing, we see that guitar playing is analogous to testing positive, and walking to work is analogous to having the disease. It's just that now we have many types of commuters instead of only two types of disease carriers (carriers or non carriers).

To answer the above question, we can draw a figure analogous to Fig. 2.12; see Fig. 2.13 with some made-up percentages for the various types of commuters. These percentages are undoubtedly completely unrealistic, but they're good enough for the sake of an example.

For simplicity, we'll assume that there are only four possible ways to commute to work. If the guitar players are represented by the shaded regions, then the answer to our question is obtained by dividing the area of the darkly shaded region (which represents the guitar players who are walkers) by the total area of all the shaded regions (which represents all of the guitar players). Mathematically, the preceding sentence is equivalent to dividing the first equality in Eq. (2.54) through by $P(Z)$

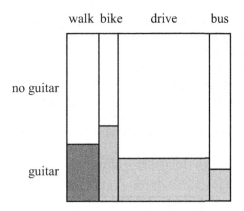

Figure 2.13: The probability square for a hypothetical commuting example.

and then letting A = "walk" and Z = "guitar":

$$P(\text{walk}|\text{guitar}) = \frac{P(\text{walk and guitar})}{P(\text{guitar})}$$
$$= \frac{\text{dark shaded area}}{\text{total shaded area}}. \tag{2.63}$$

Assuming that there are only four possible ways to commute to work, we need to be given eight pieces of information:

- We need to be given the four percentages of people who walk, bike, drive, or take the bus. (Actually, since these percentages must add up to 100%, there are only three independent bits of information here.) These percentages determine the relative widths of the vertical rectangles in Fig. 2.13. The analogous information in the "False positives" example was contained in the first bullet point on page 99 (the percentage of people who have the disease).

- For each of the four types of commuters, we need to be given the percentage who play the guitar. These four percentages determine the heights of the shaded areas within the vertical rectangles in Fig. 2.13. The analogous information in the "False positives" example was contained in the second and third bullet points on page 99.

Of course, if we are simply given the area of the darkly shaded region (which represents the number of guitar players who are walkers), and also the total area of all the shaded regions (which represents the total number of guitar players), then we can just divide the first of these two pieces of information by the second, and we're done. But in most situations, we're given the above eight (or whatever the relevant number is) pieces of information instead of these two, and the main task is to determine these two.

If you want to instead think in terms of a probability tree, as in Fig. 2.11, then in the present commuting example, the initial fork has four branches (for the walk/bike/drive/bus options), and then each of these four options splits into two possibilities (guitar or no guitar). We therefore end up with four circled numbers (the guitar players) instead of the two in Fig. 2.11, and we need to divide one of these (the one in the walking branch) by the sum of all four.

The interpretation of Bayes' theorem in terms of a figure like Fig. 2.13 is summarized in Fig. 2.14. In this figure, we are considering areas to represent probabilities instead of actual numbers (although either way is fine), because heights and widths then represent the relevant probabilities. It is invariably much more intuitive to think of the theorem in terms of a figure instead of algebraic manipulations, so when you think of Bayes' theorem, you'll probably want to think of Fig. 2.14.

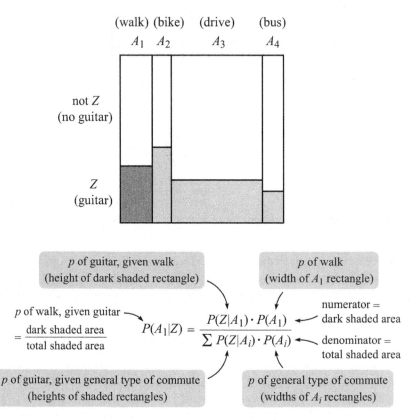

Figure 2.14: Pictorial representation of Bayes' theorem.

REMARKS:

1. It is often the case that you aren't given $P(Z)$ in the simple form of Bayes' theorem in Eq. (2.51), but instead need to calculate it via $\sum P(A_i) \cdot P(Z|A_i)$ or $P(Z|A) \cdot P(A) + P(Z| \sim A) \cdot P(\sim A)$, as we did in the "False positives" example. So the general form of Bayes' theorem in Eq. (2.53) or the explicit form in Eq. (2.52) is often the relevant one.

2. When using Bayes' theorem to calculate $P(A_1|Z)$, remember that in the notation of Fig. 2.14, the *first* letter A_1 in $P(A_1|Z)$ is one of the many A_i that divide up the *horizontal* span of the square, while the *second* letter Z is associated with the *vertical* span of the shaded areas.

3. In setups involving Bayes' theorem, there can be an arbitrary number n of the A_i columns in Fig. 2.14. (We've drawn the case with $n = 4$.) But each column is divided into only *two* regions, namely the Z region and the not-Z region. Of course, the not-Z region might very well be broken down into other regions, but that isn't relevant here.

   If you wish, you can think of there being only *two* columns, namely the A_1 column and the "not-A_1" column, which consists of all the other A_i. However, if you are given information for each of the A_i, then you will need to consider them separately. But after calculating all the relevant numbers, it is certainly fine to lump all the other A_i together into a single "not-A_1" column. Fig. 2.14 then becomes Fig. 2.15. The lightly shaded area here is the same as the total lightly shaded area in Fig. 2.14. Fig. 2.15 corresponds to the explicit form of Bayes' theorem in Eq. (2.52), while Fig. 2.14 corresponds to the general form in Eq. (2.53).

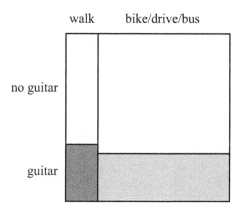

Figure 2.15: Grouping all of the nonwalkers together.

4. The essence of Bayes' theorem comes down to the fact that $P(A \text{ and } Z)$ can be written in the two different ways given in Eq. (2.54). In terms of Fig. 2.14, you can think of $P(A_1 \text{ and } Z)$, which is the area of the darkly shaded rectangle, in two different ways. It is a certain fraction (namely $P(A_1|Z)$) of the overall shaded area (namely $P(Z)$); this leads to the first equality in Eq. (2.54). And $P(A_1 \text{ and } Z)$ is also a certain fraction (namely $P(Z|A_1)$) of the leftmost (walking) rectangle area (namely $P(A_1)$); this leads to the second equality in Eq. (2.54).

   Said in another way, the *number* of guitar players who are walkers equals the *number* of walkers who are guitar players. This common number equals the area of the darkly shaded rectangle (which is the probability $P(A_1 \text{ and } Z)$) multiplied by the total number of people. Note that the first sentence above is *not* true (in general) if the word "number" is replaced by "fraction." That is, it is *not* true that the *fraction* of guitar players who are walkers equals the *fraction* of walkers who are guitar players. Equivalently, it is *not* true that $P(A_1|Z) = P(Z|A_1)$. Instead, these two conditional probabilities are related according to Eq. (2.51).

5. In Section 2.4 we solved the game-show problem, the prosecutor's fallacy, and the boy/girl problem without using Bayes' theorem. However, if we *had* used the theorem, the reasoning would have been basically the same, just as the reasoning that led to Eq. (2.58) in the "False positives" example was basically the same as the reasoning that led to Eq. (2.61). We chose to discuss the problems in Section 2.4 before discussing Bayes' theorem, so that it would be clear that the problems are still perfectly solvable

even if you've never heard of the theorem. If you want to solve the prosecutor's fallacy by explicitly using Bayes' theorem, see Problem 2.21. ♣

2.6 Stirling's formula

Stirling's formula gives an approximation to $n!$ that is valid for large n, in the sense that the larger n is, the better the approximation is. By "better," we mean that as n gets large, the approximation gets closer and closer to $n!$ in a *multiplicative* sense (as opposed to an additive sense). That is, the *ratio* of the approximation and $n!$ approaches 1. (The additive *difference* between the approximation and $n!$ gets larger and larger as n grows, but we don't care about that.) Stirling's formula is given by:

$$n! \approx n^n e^{-n} \sqrt{2\pi n} \qquad \text{(Stirling's formula)} \qquad (2.64)$$

Here e is the base of the natural logarithm, equal to $e \approx 2.71828$. See Appendix B for a discussion of e, often referred to as *Euler's number*. There are various proofs of Stirling's formula, but they generally involve calculus, so we'll just accept the formula here. It does indeed give an accurate approximation to $n!$ (an extremely accurate one, if n is large), as you can see from Table 2.6, where $S(n)$ stands for the $n^n e^{-n} \sqrt{2\pi n}$ Stirling approximation. Even if n is just 10, the approximation is off by only about 0.8%. And although there is never any need to use the formula for small numbers like 1 or 5, it works surprisingly well in those cases too.

n	$n!$	$S(n)$	$S(n)/n!$
1	1	0.922	0.922
5	120	118.0	0.983
10	$3.629 \cdot 10^6$	$3.599 \cdot 10^6$	0.992
100	$9.3326 \cdot 10^{157}$	$9.3249 \cdot 10^{157}$	0.9992
1000	$4.02387 \cdot 10^{2567}$	$4.02354 \cdot 10^{2567}$	0.99992

Table 2.6: Showing the accuracy of Stirling's formula.

You will note that for the powers of 10 in the table, the ratios of $S(n)$ to $n!$ all take the same form, namely decimals with an increasing number of 9's and then a 2. It's actually not a 2, because we rounded off, but it's essentially the same rounding off for all the numbers. This isn't a coincidence. It follows from a more accurate version of Stirling's formula, but we won't get into that here.

Stirling's formula will be critical in Chapter 5 when we talk about approximations to certain probability distributions. But for now, it is relevant when dealing with binomial coefficients of large numbers, because these binomial coefficients involve the factorials of large numbers. There are two main benefits to using Stirling's formula:

- Depending on the type of calculator you have, you might get an error message when you plug in the factorial of a number that is too big. Stirling's formula allows you to avoid this problem if you first simplify the expression that results from Stirling's formula (using the letter n to stand for the specific number you're dealing with), and *then* plug the simplified result into your calculator.

- If you use Stirling's formula and arrive at a simplified answer in terms of n (we'll call this a *symbolic* answer since it's written in terms of the symbol n instead of specific numbers), you can then plug in your specific value of n. Or you can plug in any other value, for that matter. The benefit of having a symbolic answer in terms of n is that you don't need to solve the problem from scratch every time you're given a new value of n. You simply need to plug the new value of n into your symbolic answer.

These two benefits are illustrated in the following example.

Example (50 out of 100): A coin is flipped 100 times. Calculate the probability of obtaining exactly 50 Heads.

Solution: In 100 flips, there are 2^{100} possible outcomes (all equally likely), of which $\binom{100}{50}$ have exactly 50 Heads. The probability of obtaining exactly 50 Heads is therefore

$$P(50) = \frac{1}{2^{100}} \binom{100}{50} = \frac{1}{2^{100}} \cdot \frac{100!}{50!\,50!}. \tag{2.65}$$

Now, although this is the correct answer, your calculator might not be able to handle the large factorials. But even if it can, let's use Stirling's formula so that we can produce a symbolic answer. To this end, we'll replace the number 50 with the letter n (and hence 100 with $2n$). In terms of n, we can write down the probability of obtaining exactly n Heads in $2n$ flips, and then we can use Stirling's formula (applied to both n and $2n$) to simplify the result. The first steps of this simplification will actually go in the wrong direction and create a big mess, but nearly everything will cancel out in the end. We obtain:

$$P(n) = \frac{1}{2^{2n}} \binom{2n}{n} = \frac{1}{2^{2n}} \cdot \frac{(2n)!}{n!\,n!} \approx \frac{1}{2^{2n}} \cdot \frac{(2n)^{2n} e^{-2n} \sqrt{2\pi(2n)}}{(n^n e^{-n} \sqrt{2\pi n})^2}$$

$$= \frac{1}{2^{2n}} \cdot \frac{2^{2n} n^{2n} e^{-2n} \cdot 2\sqrt{\pi n}}{n^{2n} e^{-2n} \cdot 2\pi n}$$

$$= \frac{1}{\sqrt{\pi n}}. \tag{2.66}$$

A simple answer indeed! And the "π" is a nice touch, too. In our specific case with $n = 50$, we have

$$P(50) \approx \frac{1}{\sqrt{\pi \cdot 50}} \approx 0.07979 \approx 8\%. \tag{2.67}$$

This is small, but not negligible. If we instead have $n = 500$, we obtain $P(500) \approx$ 2.5%. This is the probability of obtaining exactly 500 Heads in 1000 coin flips. As noted above, we can just plug in whatever number we want, and not have to redo the entire calculation!

The $1/\sqrt{\pi n}$ result in Eq. (2.66) is extremely clean. It is much simpler than the expression in Eq. (2.65), and *much* simpler than the expressions in the first two lines of Eq. (2.66). True, it's only an approximate result, but it's a good one. The exact result in Eq. (2.65) happens to be about 0.07959, so for $n = 50$ the ratio of the approximate result in Eq. (2.67) to the exact result is 1.0025. In other words, the approximation is off by only 0.25%. That's plenty good for most purposes.

When you derive a symbolic approximation like Eq. (2.66), you gain something and you lose something. You lose some truth, of course, because your answer technically isn't correct (although invariably its accuracy is quite sufficient). But you gain a great deal of information about how the answer depends on your input number, n. And along the same lines, you gain some aesthetics. The resulting symbolic answer is invariably nice and concise, so it allows you to easily see how the answer depends on n. For example, in our coin-flipping example, the expression in Eq. (2.66) is proportional to $1/\sqrt{n}$. This means that if we increase n by a factor of, say, 100, then $P(n)$ decreases by a factor of $\sqrt{100} = 10$. So without doing any work, we can quickly use the $P(50) \approx 8\%$ result to deduce that $P(5000) \approx 0.8\%$. In short, there is *far* more information contained in the symbolic result in Eq. (2.66) than in the numerical 8% result obtained directly from Eq. (2.65).

2.7 Summary

In this chapter we learned about probability. In particular, we learned:

- The probability of an event is defined to be the fraction of the time the event occurs in a very large number of identical trials. In many situations the possible outcomes are all equally likely, in which case the probability of a certain class of outcomes occurring is

$$p = \frac{\text{number of desired outcomes}}{\text{total number of possible outcomes}} \quad \text{(for equally likely outcomes)}$$
$$(2.68)$$

- The various "and" and "or" rules of probability are:

  1. For any two (possibly dependent) events,

  $$P(A \text{ and } B) = P(A) \cdot P(B|A). \qquad (2.69)$$

  2. In the special case of independent events, we have $P(B|A) = P(B)$, so Eq. (2.69) reduces to

  $$P(A \text{ and } B) = P(A) \cdot P(B). \qquad (2.70)$$

  3. For any two (possibly nonexclusive) events,

  $$P(A \text{ or } B) = P(A) + P(B) - P(A \text{ and } B). \qquad (2.71)$$

  4. In the special case of exclusive events, we have $P(A \text{ and } B) = 0$, so Eq. (2.71) reduces to

  $$P(A \text{ or } B) = P(A) + P(B). \qquad (2.72)$$

- *A* and *B* are independent events if any one of the following relations is true:

$$P(B|A) = P(B),$$
$$P(A|B) = P(A),$$
$$P(A \text{ and } B) = P(A) \cdot P(B). \qquad (2.73)$$

- The conditional probabilities $P(A|B)$ and $P(B|A)$ are not equal, in general.

- Two common ways to calculate probabilities are: (1) count up the number of desired outcomes, along with the total number of possible outcomes, and use Eq. (2.68) (assuming that the outcomes are equally likely), and (2) imagine things happening in succession (for example, picking seats or rolling dice), and then multiply the relevant probabilities. The results for some problems, in particular the Birthday Problem and the Game-Show Problem, might seem surprising at first, but you can avoid confusion by methodically using one (or both) of these strategies.

- Bayes' theorem takes a variety of forms; see Eqs. (2.51)–(2.53). The last of these is the "general form" of the theorem:

$$P(A_k|Z) = \frac{P(Z|A_k) \cdot P(A_k)}{\sum_i P(Z|A_i) \cdot P(A_i)}. \qquad (2.74)$$

The theorem tells us how the conditional probability $P(A_k|Z)$ is obtained from the set of conditional probabilities $P(Z|A_i)$.

- Stirling's formula, which gives an approximation to $n!$, takes the form,

$$n! \approx n^n e^{-n} \sqrt{2\pi n} \qquad \text{(Stirling's formula)} \qquad (2.75)$$

This approximation is very helpful for simplifying binomial coefficients. We will use it a great deal in Chapter 5.

2.8 Exercises

See **www.people.fas.harvard.edu/~djmorin/book.html** for a supply of problems without included solutions.

2.9 Problems

Section 2.1: Definition of probability

2.1. **Odds** *

If an event occurs with probability p, then the *odds* in favor of the event occurring are defined to be "p to $(1 - p)$." (And similarly, the odds *against* the event occurring are defined to be "$(1 - p)$ to p.") In other words, the odds are simply the ratio of the probabilities of the event occurring (namely p) and not occurring (namely $1-p$). It is customary to write "$p:(1-p)$" as shorthand for "p to $(1-p)$." (The odds are sometimes also written as the ratio $p/(1-p)$. But this fraction can look like a probability, which may cause confusion, so we'll avoid this notation.) In practice, the probabilities p and $1-p$ are usually multiplied through by the smallest number that turns them into integers. For example, odds of $1/3:2/3$ are generally written as $1:2$. Find the odds of the following events:

(a) Getting a Heads on a coin toss.

(b) Rolling a 5 on a die.

(c) Rolling a multiple of 2 or 3 on a die.

(d) Randomly picking a day of the week with more than six letters.

Section 2.2: The rules of probability

2.2. **Rules for three events** ∗∗

(a) Consider three events, A, B, and C. If they are all independent of each other, show that

$$P(A \text{ and } B \text{ and } C) = P(A) \cdot P(B) \cdot P(C). \qquad (2.76)$$

(b) If they are (possibly) dependent, show that

$$P(A \text{ and } B \text{ and } C) = P(A) \cdot P(B|A) \cdot P(C|A \text{ and } B). \qquad (2.77)$$

(c) If they are all mutually exclusive, show that

$$P(A \text{ or } B \text{ or } C) = P(A) + P(B) + P(C). \qquad (2.78)$$

(d) If they are (possibly) nonexclusive, show that

$$\begin{aligned} P(A \text{ or } B \text{ or } C) = {} & P(A) + P(B) + P(C) \\ & - P(A \text{ and } B) - P(A \text{ and } C) - P(B \text{ and } C) \\ & + P(A \text{ and } B \text{ and } C). \qquad (2.79) \end{aligned}$$

2.3. **"Or" rule for four events** ∗∗∗

Parts (a), (b), and (c) of Problem 2.2 generalize quickly to more than three events, but part (d) is trickier. Derive the "or" rule for four (possibly) nonexclusive events. That is, derive the rule analogous to Eq. (2.79).

2.4. **Red and blue balls** ∗

Show that the second expression in Eq. (2.9), with $A = \text{Red}_1$ and $B = \text{Blue}_2$, gives the correct result of $3/10$ for $P(\text{Red}_1 \text{ and } \text{Blue}_2)$ in the "balls in a box" example on page 64.

2.5. **Dependent events** ✶

Calculate the overall probability of B occurring in the scenario described by Fig. 2.16.

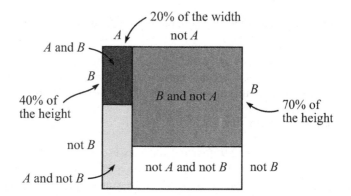

Figure 2.16: A hypothetical probability square.

2.6. **A single horizontal line** ✶

There is an asymmetry in Fig. 2.16. Because there is a single vertical line but two horizontal lines, it is easy to read off the $P(A)$ and $P(\text{not } A)$ probabilities, but not easy to read off the $P(B)$ and $P(\text{not } B)$ probabilities. Hence the calculation in Problem 2.5. Redraw Fig. 2.16 with a single horizontal line and two vertical lines (while keeping the areas (probabilities) of the four sub-rectangles the same, of course).

2.7. **Proofreading** ✶✶

Two people each proofread the same book. One person finds 100 errors, and the other finds 60. There are 20 errors common to both people. Assume that all errors are equally likely to be found (which is undoubtedly not true in practice), and also that the discovery of an error by one person is independent of the discovery of that error by the other person. Given these assumptions, roughly how many errors does the book have? *Hint:* Draw a picture similar to Fig. 2.1, and then find the probability of each person finding a given error.

Section 2.3: Examples

2.8. **Red balls, blue balls** ✶✶

Three boxes sit on a table. One box contains two red balls, another contains two blue balls, and the third contains one red ball and one blue ball. You choose one of the boxes at random, and then you draw a ball from that box. If it turns out to be a red ball, what is the probability that the other ball in the box is also red?

2.9. **Sock pairs** ∗∗

(a) Four red socks and four blue socks are in a drawer. You reach in and pull out two socks at random. What is the probability that you obtain a matching pair?

(b) Answer the same question, but now in the general case with n red socks and n blue socks.

(c) Presumably you answered the above questions by counting the relevant pairs of socks. Can you think of a quick probability argument, requiring no counting, that gives the answer to part (b) (and part (a))?

2.10. **Sock pairs, again** ∗∗

(a) As in Problem 2.9, four red socks and four blue socks are in a drawer. You reach in and pull out two socks at random. You then reach in and pull out two more socks (without looking at the socks in the first pair). What is the probability that the second pair you pull out is a matching pair? Answer this by calculating the probabilities, given that the first pair is (or is not) a matching pair.

(b) You should find that the answer to part (a) is the same as the answer to part (a) of Problem 2.9. Can you think of a quick probability argument, requiring no counting, that explains why this is the case? The reasoning will work in the general case with n red socks and n blue socks. And it will also work if you draw a third pair, or a fourth pair, etc. (without looking at any of the other pairs).

2.11. **At least one 6** ∗∗

Three dice are rolled. What is the probability of obtaining at least one 6? We solved this in Section 2.3.1, but your task here is to solve it the long way, by adding up the probabilities of obtaining exactly one, two, or three 6's.

2.12. **At least one 6, by the rules** ∗∗

Three dice are rolled. What is the probability of obtaining at least one 6? We solved this in Section 2.3.1, and again in Problem 2.11. But your task here is to solve it by using Eq. (2.79) from Problem 2.2, with each of the three letters in that formula standing for a 6 on each of the three dice.

2.13. **Rolling sixes** ∗∗

This problem was posed by Samuel Pepys to Isaac Newton in 1693 and is therefore known as the Newton-Pepys problem.

(a) 6 dice are rolled. What is the probability of obtaining at least one 6?

(b) 12 dice are rolled. What is the probability of obtaining at least two 6's?

(c) 18 dice are rolled. What is the probability of obtaining at least three 6's? Which of the above three probabilities is the largest?

Section 2.4: Four classic problems

2.14. Exactly one pair ∗∗

If there are 23 people in a room, what is the probability that *exactly* two of them have a common birthday? That is, we don't want two different pairs with common birthdays, or three people with a common birthday, etc.

2.15. My birthday ∗∗

(a) You are in a room with 100 other people. Let p be the probability that at least one of these 100 people has your birthday. Without doing any calculations, state whether p is larger, smaller, or equal to, 100/365.

(b) Now calculate the exact value of p.

2.16. My birthday, again ∗∗

We saw at the end of Section 2.4.1 that 253 is the answer to the question, "How many people (in addition to me) need to be present in order for there to be at least a 1/2 chance that someone else has *my* birthday?" We solved this by finding the smallest n for which $(364/365)^n$ is less than 1/2. Answer this question again, by making use of the approximation in Eq. (7.14) in Appendix C. What is the answer in the general case where there are N days in a year instead of 365? Assume that N is large.

2.17. My birthday, yet again ∗∗

With 253 other people in a room, what is the probability that *exactly* one of these people has your birthday? Exactly two? Exactly three?

2.18. A random game-show host ∗∗

Consider the following variation of the Game-Show Problem we discussed in Section 2.4.2. A game-show host offers you the choice of three doors. Behind one of these doors is the grand prize, and behind the other two are goats. The host announces that after you select a door (without opening it), he will *randomly* open one of the other two doors. You select a door. The host then randomly opens one of the other doors, and the result happens to be a goat. He then offers you the chance to switch your choice to the remaining door. Should you switch or not? Or does it not matter?

2.19. Boy/girl problem with general information ∗∗∗

This problem is an extension of the Boy/Girl Problem from Section 2.4.4. You should study that problem thoroughly before tackling this one. As in the original versions of the problem, assume that all processes are completely random. The new variation is the following:

You bump into a random person on the street who says, "I have two children. At least one of them is a boy whose birthday is in the summer." What is the probability that the other child is also a boy?

What if the clause is changed to, "whose birthday is on August 11th"? Or "who was born during a particular minute on August 11th"? Or more generally, "who has a particular characteristic that occurs with probability p"? *Hint*: Make a table of all of the various possibilities, analogous to the tables in Section 2.4.4.

Section 2.5: Bayes' theorem

2.20. **A second test** $**$

Consider the setup in the "False positives" example in Section 2.5. If we instead perform *two* successive tests on each person, what is the probability that a person who tests positive both times actually has the disease?

2.21. **Bayes' theorem for the prosecutor's fallacy** $**$

In Section 2.4.3 we discussed the prosecutor's fallacy. Explain the fallacy again here, but now by using Bayes' theorem. In particular, determine $P(I|D)$ (the probability of being innocent, given that the description is satisfied) by drawing a figure analogous to Fig. 2.14

2.22. **Black balls and white balls** $**$

One box contains two black balls, and another box contains one black ball and one white ball. You pick one of the boxes at random and draw a ball n times, with replacement after each draw. If a black ball is drawn all n times, what is the probability that you picked the box with two black balls?

2.10 Solutions

2.1. **Odds**

(a) The probability of getting a Heads is $1/2$, as is the probability of not getting a Heads. So the desired odds are $1/2:1/2$, or equivalently $1:1$. These are known as "even odds."

(b) The probability of rolling a 5 is $1/6$, and the probability of not rolling a 5 is $5/6$. So the desired odds are $1/6:5/6$, or equivalently $1:5$.

(c) There are four desired outcomes $(2, 3, 4, 6)$, so the "for" and "against" probabilities are $4/6$ and $2/6$, respectively. The desired odds are therefore $4/6:2/6$, or equivalently $2:1$.

(d) Tuesday, Wednesday, Thursday, and Saturday all have more than six letters, so the "for" and "against" probabilities are $4/7$ and $3/7$, respectively. The desired odds are therefore $4/7:3/7$, or equivalently $4:3$.

Note that to convert from odds to probability, the odds of $a:b$ in favor of an event occurring are equivalent to a probability of $a/(a+b)$ that the event occurs.

2.2. **Rules for three events**

(a) We can use the same type of reasoning that we used in Section 2.2.1. If we perform a large number of trials, then A occurs in a fraction $P(A)$ of them. (It is understood here that the words "on average" follow all statements of this form.)

And then B occurs in a fraction $P(B)$ *of these* trials, because the events are independent, which means that the occurrence of A doesn't affect the probability of B. So the fraction of the total number of trials where A and B both occur is $P(A) \cdot P(B)$. And then C occurs in a fraction $P(C)$ *of these* trials, because C is independent of A and B. So the fraction of the total number of trials where all three of A, B, and C occur is $P(A) \cdot P(B) \cdot P(C)$. The desired probability is therefore $P(A) \cdot P(B) \cdot P(C)$. If you want to visualize this geometrically, you'll need to use a cube instead of the square in Fig. 2.1.

This reasoning can easily be extended to an arbitrary number of independent events. The probability of all of the events occurring is simply the product of all of the individual probabilities.

(b) The reasoning in part (a) works again, with only slight modifications. If we perform a large number of trials, then A occurs in a fraction $P(A)$ of them. And then B occurs in a fraction $P(B|A)$ *of these* trials, by definition. So the fraction of the total number of trials where A and B both occur is $P(A) \cdot P(B|A)$. And then C occurs in a fraction $P(C|A \text{ and } B)$ *of these* trials, by definition. So the fraction of the total number of trials where all three of A, B, and C occur is $P(A) \cdot P(B|A) \cdot P(C|A \text{ and } B)$. The desired probability is therefore $P(A) \cdot P(B|A) \cdot P(C|A \text{ and } B)$.

Again, this reasoning can easily be extended to an arbitrary number of (possibly) dependent events. For four events, we just need to tack on the factor $P(D|A \text{ and } B \text{ and } C)$, and so on.

(c) Since the events are all mutually exclusive, we don't have to worry about any double counting. The total number of trials where A or B or C occurs is simply the sum of the number of trials where A occurs, plus the number where B occurs, plus the number where C occurs. The same statement must be true if we substitute the word "fraction" for "number," because the fractions are related to the numbers via division by the total number of trials. And since the fractions are the probabilities, we end up with the desired result, $P(A \text{ or } B \text{ or } C) = P(A) + P(B) + P(C)$. If there are more events, we simply have more terms in the sum.

(d) This rule is more involved than the preceding three. Let's think of the probabilities in terms of areas, as we did in Section 2.2.2. The generic situation for three events is shown in Fig. 2.17. For simplicity, we've chosen the three regions to be circles with the same size, but this of course isn't necessary. The various overlap regions are shown, with the juxtaposition of two letters standing for their intersection. So AB means "A and B." The labels might appear to suggest otherwise, but remember that A includes the whole circle, and not just the white part. Similarly, AB includes the dark ABC region too, and not just the lighter region where the AB label is.

Our goal is to determine the total area contained in the three circles, because this represents the probability of "A or B or C." We can add up the areas of the A, B, and C circles, but then we need to subtract off the areas that we double counted. These areas are the pairwise overlaps of the circles, that is, AB, AC, and BC (remember that each of these regions includes the dark ABC region in the middle). At this point, we've correctly counted all of the white and light gray regions exactly once. But what about the ABC region in the middle? We counted it three times in the A, B, and C regions, but then we subtracted it off three times in the AB, AC, and BC regions. So at the moment, we haven't counted it at all. We therefore need to add it on once. Then every part of the

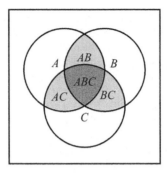

Figure 2.17: Venn diagram for three nonexclusive events.

union of the circles will be counted exactly once. The total area is therefore

$$\text{Total area} = A + B + C - AB - AC - BC + ABC, \qquad (2.80)$$

where we are using the regions' labels to stand for their areas. Translating this from a statement about areas to a statement about probabilities yields the desired result,

$$P(A \text{ or } B \text{ or } C) = P(A) + P(B) + P(C)$$
$$- P(A \text{ and } B) - P(A \text{ and } C) - P(B \text{ and } C)$$
$$+ P(A \text{ and } B \text{ and } C). \qquad (2.81)$$

2.3. "Or" rule for four events

As in Problem 2.2(d), we'll discuss things in terms of areas. If we add up the areas of four regions, A, B, C, and D, then we have double counted the pairwise overlaps, so we need to subtract these off. There are six of these regions: AB, AC, AD, BC, BD, and CD. But then what about the triple overlaps, such as ABC? We counted ABC three times in the A, B, and C regions, but then we subtracted it off three times in the AB, AC, and BC regions. So at the moment, we haven't counted it at all. We therefore need to add it on once. (This is the same reasoning as in Problem 2.2(d).) Likewise for ABD, ACD, and BCD. Finally, what about the quadruple overlap region, $ABCD$? We counted this four times in the single regions (like A), then we subtracted it off six times in the double regions (like AB), and then we added it on four times in the triple regions (like ABC). So at the moment, we have counted it $4 - 6 + 4 = 2$ times. Since we want to count it only one time, we need to subtract it off once. The total area is therefore

$$\text{Total area} = A + B + C + D$$
$$- AB - AC - AD - BC - BD - CD$$
$$+ ABC + ABD + ACD + BCD$$
$$- ABCD. \qquad (2.82)$$

Writing this in terms of probabilities gives the result,

$$
\begin{aligned}
P(A \text{ or } B \text{ or } C \text{ or } D) = {} & P(A) + P(B) + P(C) + P(D) \\
& - P(A \text{ and } B) - P(A \text{ and } C) - P(A \text{ and } D) \\
& \quad - P(B \text{ and } C) - P(B \text{ and } D) - P(C \text{ and } D) \\
& + P(A \text{ and } B \text{ and } C) + P(A \text{ and } B \text{ and } D) \\
& \quad + P(A \text{ and } C \text{ and } D) + P(B \text{ and } C \text{ and } D) \\
& - P(A \text{ and } B \text{ and } C \text{ and } D).
\end{aligned} \tag{2.83}
$$

REMARK: You might think that it's a bit of a coincidence that at every stage, we either overcounted or undercounted each region *once*. Equivalently, the coefficient of every term in Eqs. (2.82) and (2.83) is ±1. The same thing is true in the case of three events in Eqs. (2.80) and (2.81). Likewise in the case of two events in Eq. (2.18), and trivially in the case of one event. Is it also true for larger numbers of events? Indeed it is, and the binomial expansion is the key to understanding why.

We won't go through every step, but if you want to think about it, the main points to realize are: First, the numbers 4, 6, and 4 in the above counting in the four-event case are actually the binomial coefficients $\binom{4}{1}$, $\binom{4}{2}$, $\binom{4}{3}$. This makes sense because, for example, the number of regions of double overlap (like AB) that contain the region $ABCD$ is simply the number of ways to pick two letters from four letters, which is $\binom{4}{2}$. Second, the "alternating sum" $\binom{4}{1} - \binom{4}{2} + \binom{4}{3}$ equals 2 (which means that we have overcounted the $ABCD$ region by one time), because this is what you obtain when you expand the righthand side of $0 = (1-1)^4$ with the binomial expansion. (This is a nice little trick.) And third, you can show how this generalizes to a larger number n of events. For even n, the alternating sum of the relevant binomial coefficients is 2, as we just saw for $n = 4$. For odd n, the alternating sum is zero, which means that we have undercounted by one time. (The relevant binomial coefficients are all but the first and last in the expansion of $(1-1)^n$, and these two coefficients are either 1 and 1 for even n, or 1 and -1 for odd n.) For example, $\binom{5}{1} - \binom{5}{2} + \binom{5}{3} - \binom{5}{4} = 0$. This "alternating sum" rule for counting is known as the *inclusion–exclusion principle*. ♣

2.4. Red and blue balls

By counting the various kinds of pairs in Table 2.1, we find $P(\text{Blue}_2) = 12/20 = 3/5$ (by looking at all 20 pairs), and $P(\text{Red}_1|\text{Blue}_2) = 6/12 = 1/2$ (by looking at only the 12 pairs below the horizontal line). So we have

$$
\begin{aligned}
P(\text{Red}_1 \text{ and } \text{Blue}_2) &= P(\text{Blue}_2) \cdot P(\text{Red}_1|\text{Blue}_2) \\
&= \frac{3}{5} \cdot \frac{1}{2} = \frac{3}{10},
\end{aligned} \tag{2.84}
$$

in agreement with Eq. (2.10). As mentioned in the third remark on page 66, it still makes sense to talk about $P(\text{Red}_1|\text{Blue}_2)$, even though the second pick happens after the first pick.

2.5. Dependent events

FIRST SOLUTION: This problem is equivalent to finding the fraction of the total area that lies above the horizontal line segments in Fig. 2.16. The upper left region is $40\% = 2/5$ of the area that lies to the left of the vertical line, which itself is $20\% = 1/5$ of the total area. And the upper right region is $70\% = 7/10$ of the area that lies to the right of the vertical line, which itself is $80\% = 4/5$ of the total area. The fraction of the total area that lies above the horizontal line segments is therefore

$$
\frac{1}{5} \cdot \frac{2}{5} + \frac{4}{5} \cdot \frac{7}{10} = \frac{2}{25} + \frac{14}{25} = \frac{16}{25} = 64\%. \tag{2.85}
$$

SECOND SOLUTION: We'll use the rule in Eq. (2.5) twice. First, note that

$$P(B) = P(A \text{ and } B) + P((\text{not } A) \text{ and } B). \tag{2.86}$$

This is true because either A happens or it doesn't. We can apply Eq. (2.5) to each of the two terms in Eq. (2.86) to obtain

$$P(B) = P(A) \cdot P(B|A) + P(\text{not } A) \cdot P(B| \text{not } A)$$
$$= \frac{1}{5} \cdot \frac{2}{5} + \frac{4}{5} \cdot \frac{7}{10} = \frac{2}{25} + \frac{14}{25} = \frac{16}{25} = 64\%, \tag{2.87}$$

which is exactly the same equation as in the first solution. This is no surprise, of course, because the two solutions are actually same. They are simply presented in a different language. Comparing the solutions makes it clear how conditional probabilities like $P(B|A)$ are related to fractional areas.

2.6. A single horizontal line

As usual, let the total area of the square in Fig. 2.16 be 1. Then from the given lengths along the sides of the square, we find that the upper two areas (probabilities) are 0.08 and 0.56, for a total of 0.64; this is $P(B)$. And the lower two areas are 0.12 and 0.24, for a total of 0.36; this is $P(\text{not } B)$. The single horizontal line in Fig. 2.18 must therefore be 64% of the way down from the top of the square. And the two vertical lines must be $0.08/0.64 = 12.5\%$ and $0.12/0.36 = 33.3\%$ of the way from the left side. The four areas are the same (by construction) as in Fig. 2.16. It's just that in Fig. 2.18, the $P(B) = 0.64$ probability is clear by simply looking at the figure. If we wanted to calculate $P(A)$ from Fig. 2.18, we would have to do a calculation analogous to the one we did in Problem 2.5.

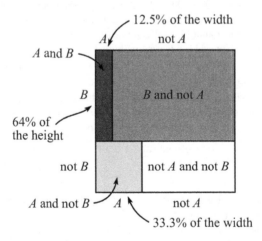

Figure 2.18: Redrawing Fig. 2.16 with a single horizontal line.

2.7. Proofreading

The breakdown of the errors is shown in Fig. 2.19. If the two people are labeled A and B, then 20 errors are found by both A and B, 80 are found by A but not B, and 40 are found by B but not A.

If we consider only the 100 errors found by A, we see that 20 of them are found by B, which is a 1/5 fraction. Since we are assuming that B finding a given error is

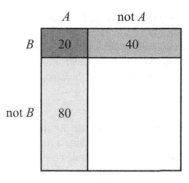

Figure 2.19: Breakdown of errors found by A and B.

independent of A finding it, we see that if B finds $1/5$ of the errors found by A, then he must find $1/5$ of the complete set of errors (on average). So $1/5$ is the probability that B finds any given error. Therefore, since we know that B found a total of 60 errors, the total number N of errors in the book must be given by $60/N = 1/5 \implies N = 300$. The unshaded region in Fig. 2.19 therefore represents $300 - 80 - 20 - 40 = 160$ errors. This is the number that both people missed.

We can also do things the other way around. If we consider only the 60 errors found by B, we see that 20 of them are found by A, which is a $1/3$ fraction. By the same reasoning as above, this $1/3$ is the probability that A finds any given error. And since we know that A found a total of 100 errors, the total number N must be given by $100/N = 1/3 \implies N = 300$, as above.

Another method (although in the end it's the same as the above methods) is the following. Let the area of the unshaded region in Fig. 2.19 be x. Then if we look at how the areas of the two vertical rectangles are divided by the horizontal line, we see that the ratio of x to 40 must equal the ratio of 80 to 20. So $x = 160$, as we found above. Alternatively, if we look at how the areas of the two horizontal rectangles are divided by the vertical line, we see that the ratio of x to 80 must equal the ratio of 40 to 20. So again, $x = 160$.

It is quite fascinating that you can get a sense of the total number of errors just by comparing the results of two readers' independent proofreadings. There is no need to actually find all the errors and count them up, if you only want to make a rough estimate. The larger the numbers involved, the better the estimate, in a multiplicative sense.

2.8. Red balls, blue balls

Let's ignore for a moment the fact that you happen to draw a red ball. Without this condition, there are six equally likely results of the process; you are equally likely to draw any of the six balls in the boxes. This fact can be argued by symmetry (there is nothing special about any of the balls). Or you can break down the probabilities: you have a $1/3$ chance of drawing a given box, and then a $1/2$ chance of drawing a given ball in that box. So all of the probabilities are equal to $(1/3)(1/2) = 1/6$.

Let's use the numbers 1 through 6 to label the balls: 1 and 2 are the two red balls in the first box, 3 and 4 are the two blue balls in the second box, and 5 and 6 are, respectively, the red and blue balls in the third box. If you play n games, where n is large, you will obtain approximately $n/6$ of each of the numbers 1 through 6.

Let's now invoke the fact that you draw a red ball. This means that the $n/2$ games where you draw a blue ball (3, 4, and 6) aren't relevant. Only the $n/2$ games where you draw a red ball (1, 2, and 5) are relevant. And *of these games*, 2/3 have the ball coming from the first box (in which case the other ball is red), and 1/3 have the ball coming from the third box (in which case the other ball is blue). The desired probability that the other ball is red is therefore 2/3.

REMARKS:

1. The statement of the problem asks for the probability that the other ball in the box is also red, given that you draw a red ball. Since the word "probability" is used, it is understood that we must consider a large number of trials and look at what happens, on average, in these trials. Although the setup in the problem mentions only one trial, we must consider many. The given question, namely "If it turns out to be a red ball, what is the probability that the other ball in the box is also red?," is really just shorthand for the question, "If you run a large number of trials and look only at the ones where the drawn ball is red, in what fraction of these trials is the other ball in the box also red?"

2. In the statement of the problem, the clause, "You choose one of the boxes at random," is critical. Consider the alternative question: "Someone gives you a box containing either two red balls, two blue balls, or one of each. You draw a ball from this box. If it turns out to be a red ball, what is the probability that the other ball in the box is also red?" This question is unanswerable, because for all you know, the person always gives you a box with two red balls. Or perhaps she always gives you a box with one ball of each color, and you just happened to pick the red ball. Maybe it's 90% the former and 10% the latter, or maybe it depends on the day of the week. There is no way to tell what happens in a large number of trials. Even if you *do* perform a large number of trials and throw away the ones where you pick a blue ball, there is still no way to determine the probability associated with a future trial, because at any point the person might change her rules for the type of box she gives you.

3. What if, instead of three equally likely boxes sitting on the table, we have a single box and we color each of the two balls red or blue, based on coin tosses? There are then four equally likely possibilities for the contents of the box: RR, RB, BR, and BB. We therefore effectively have four equally likely boxes instead of three. You can show, with a quick modification of our original reasoning, that the answer is now 1/2 instead of 2/3.

   This result of 1/2 makes intuitive sense, due to the following alternative reasoning. Imagine picking a ball, without looking at it. The other ball has a 1/2 chance of being red, because its color is determined by a coin flip. Now look at the ball you picked. The other ball still has a 1/2 chance of being red, because your act of looking at the ball you picked can't change the color of the other ball. Therefore, if the ball you picked is red, then the other ball has a 1/2 chance of being red. Of course, the same thing is true if the ball you picked is blue, but those trials don't have anything to do with the given setup where you pick a red ball. ♣

2.9. Sock pairs

(a) The total number of possible pairs that you can draw from the eight socks in the drawer is $\binom{8}{2} = 28$. The number of ways that you can draw a red pair from the four red socks is $\binom{4}{2} = 6$. Likewise for the four blue socks. So there are

12 ways in all that you can draw a matching pair. The desired probability is therefore $12/28 = 3/7$.

(b) If there are now n red and n blue socks in the drawer, the total number of possible pairs that you can draw is $\binom{2n}{2} = 2n(2n - 1)/2$. The number of ways that you can draw a red pair from the n red socks is $\binom{n}{2} = n(n - 1)/2$. Likewise for the n blue socks. So there are $n(n - 1)$ ways in all that you can draw a matching pair. The desired probability is therefore

$$\frac{n(n - 1)}{2n(2n - 1)/2} = \frac{n - 1}{2n - 1}. \tag{2.88}$$

If $n = 4$, this yields the probability of $3/7$ that we obtained in part (a).

(c) For the quick probability argument, imagine drawing the two socks in succession. The first sock is either red or blue. Whichever color it is, there are now $n - 1$ socks remaining of that color. And there are $2n - 1$ socks remaining in all. So the probability that the second sock has the same color as the first is $(n - 1)/(2n - 1)$.

For large n, this result approaches $1/2$. This makes sense because if n is large, the removal of the first sock from the drawer only negligibly changes the distribution of socks from 50-50. So you're basically flipping a coin with the second sock.

2.10. **Sock pairs, again**

(a) We know from Problem 2.9 that there is a $3/7$ probability of obtaining a matching first pair, and hence a $4/7$ probability of obtaining a non-matching first pair. So there is a $3/7$ probability that we are left with two socks of one color and four of the other, and there is a $4/7$ probability that we are left with three socks of each color.

In the first of these two cases, there are $\binom{6}{2} = 15$ possible pairs we can draw for our second pair, of which $\binom{2}{2} + \binom{4}{2} = 1 + 6 = 7$ are matching pairs. The probability that the second pair is matching, given that the first pair is matching (which happens with probability $3/7$), is therefore $7/15$.

Similarly, in the second of the two cases, there are again $\binom{6}{2} = 15$ possible pairs we can draw for our second pair, of which $\binom{3}{2} + \binom{3}{2} = 3 + 3 = 6$ are matching pairs. The probability that the second pair is matching, given that the first pair *isn't* matching (which happens with probability $4/7$), is therefore $6/15$.

The desired probability (that the second pair is matching) is therefore

$$\frac{3}{7} \cdot \frac{7}{15} + \frac{4}{7} \cdot \frac{6}{15} = \frac{21 + 24}{105} = \frac{3}{7}. \tag{2.89}$$

You can apply the same reasoning to the general case with n red and n blue socks, but it gets a bit messy. In any event, there is no need to work through the algebra, because there is a much quicker line of reasoning in part (b) below.

(b) We'll be general from the start here. That is, we'll assume that we have n socks of each color, and that we successively draw n pairs until there are no socks left in the drawer. We claim that all n pairs have the same $(n-1)/(2n-1)$ probability of matching, assuming that we haven't looked at any of the other pairs yet. This assumption is important; we must not have any knowledge of the other pairs. If we *do* have knowledge, then this affects the probabilities for future pairs. For example, in part (a) above, we saw that if the first pair is matching, the second pair has a 7/15 chance of matching. But if the first pair isn't matching, the second pair has a 6/15 chance of matching.

Imagine drawing the $2n$ socks in succession and lining them up on a table. We can label them as s_1, s_2, s_3, ..., s_{2n}. We can then divide them into n pairs, (s_1, s_2), (s_3, s_4), ..., (s_{2n-1}, s_{2n}). If we ask for the probability that, say, the third pair (socks s_5 and s_6) is matching (assuming we haven't looked at any of the other pairs), we can *now* imagine looking at this particular pair. And if we look at s_5 first and then at s_6, we can use the reasoning in part (c) of Problem 2.9 to say that the probability of a matching pair is $(n-1)/(2n-1)$. This reasoning works for any of the n pairs; there is nothing special about a specific pair (assuming we haven't looked at any of the other pairs). All pairs therefore have equal $(n-1)/(2n-1)$ probabilities of being matching pairs.

The point here is that if you don't look at the pairs you've already picked, then for all practical purposes the present pair you're picking is the first pair. The order in which you draw the pairs therefore doesn't matter, so the desired probabilities are all equal.

2.11. **At least one 6**

The probability of obtaining exactly one 6 equals $\binom{3}{1} \cdot (1/6)(5/6)^2$, because there are $\binom{3}{1} = 3$ ways to pick which die is the 6. And then given this choice, there is a 1/6 chance that the die is in fact a 6, and a $(5/6)^2$ chance that both of the other dice are not 6's.

The probability of obtaining exactly two 6's equals $\binom{3}{2} \cdot (1/6)^2(5/6)$, because there are $\binom{3}{2} = 3$ ways to pick which two dice are the 6's. And then given this choice, there is a $(1/6)^2$ chance that they are in fact both 6's, and a 5/6 chance that the other die is not a 6.

The probability of obtaining exactly three 6's equals $\binom{3}{3} \cdot (1/6)^3$, because there is just $\binom{3}{3} = 1$ way for all three dice to be 6's. And then there is a $(1/6)^3$ chance that they are in fact all 6's.

The total probability of obtaining at least one six is therefore

$$\binom{3}{1} \cdot \left(\frac{1}{6}\right)\left(\frac{5}{6}\right)^2 + \binom{3}{2} \cdot \left(\frac{1}{6}\right)^2\left(\frac{5}{6}\right) + \binom{3}{3} \cdot \left(\frac{1}{6}\right)^3 = \frac{75}{216} + \frac{15}{216} + \frac{1}{216}$$

$$= \frac{91}{216}, \tag{2.90}$$

in agreement with the result in Section 2.3.1.

REMARK: If we add this result to the probability of obtaining zero 6's, which is $(5/6)^3$, the sum is 1, because we have now taken into account every possible outcome. This fact was what we used to solve the problem the quick way in Section 2.3.1, after all. But let's pretend that we don't know the sum is 1, and let's verify this explicitly. If we write $(5/6)^3$ suggestively as $\binom{3}{0} \cdot (5/6)^3$, then our goal is to show that

$$\binom{3}{0} \cdot \left(\frac{5}{6}\right)^3 + \binom{3}{1} \cdot \left(\frac{1}{6}\right)\left(\frac{5}{6}\right)^2 + \binom{3}{2} \cdot \left(\frac{1}{6}\right)^2\left(\frac{5}{6}\right) + \binom{3}{3} \cdot \left(\frac{1}{6}\right)^3 = 1. \tag{2.91}$$

This is indeed a true statement, because the lefthand side is simply the binomial expansion of $(5/6 + 1/6)^3 = 1$. This makes it clear why the sum of the probabilities of the various outcomes will still be 1, even if we have, say, an eight-sided die (again, forgetting that we know intuitively that the sum must be 1). The only difference is that we now have the expression $(7/8 + 1/8)^3 = 1$, which is still true. And any other exponent (that is, any other number of rolls) will also yield a sum of 1, as we know it must. ♣

2.12. At least one 6, by the rules

We'll copy Eq. (2.79) here:

$$\begin{aligned} P(A \text{ or } B \text{ or } C) = &\ P(A) + P(B) + P(C) \\ &- P(A \text{ and } B) - P(A \text{ and } C) - P(B \text{ and } C) \\ &+ P(A \text{ and } B \text{ and } C). \end{aligned} \tag{2.92}$$

The lefthand side of this equation is the probability of obtaining at least one 6. (Remember that the "or" is the "inclusive or.") So our task is to evaluate the righthand side, which involves three different types of terms.

The probability of obtaining a 6 on any given die (without caring what happens with the other two dice) is $1/6$, so

$$P(A) = P(B) = P(C) = \frac{1}{6}. \tag{2.93}$$

The probability of obtaining 6's on two given dice (without caring what happens with the third die) is $(1/6)^2$, so

$$P(A \text{ and } B) = P(A \text{ and } C) = P(B \text{ and } C) = \frac{1}{36}. \tag{2.94}$$

The probability of obtaining 6's on all three dice is $(1/6)^3$, so

$$P(A \text{ and } B \text{ and } C) = \frac{1}{216}. \tag{2.95}$$

Eq. (2.92) therefore gives the probability of obtaining at least one 6 as

$$3 \cdot \frac{1}{6} - 3 \cdot \frac{1}{36} + \frac{1}{216} = \frac{108 - 18 + 1}{216} = \frac{91}{216}, \tag{2.96}$$

in agreement with the result in Section 2.3.1 and Problem 2.11.

2.13. Rolling sixes

(a) In all three parts of this problem, there are far fewer ways to fail to obtain the specified number of 6's than to succeed. So we'll calculate the probability of failure and then subtract that from 1 to obtain the probability of success.

If 6 dice are rolled, the probability of obtaining zero 6's is $(5/6)^6$. The probability of obtaining at least one 6 is therefore

$$1 - \left(\frac{5}{6}\right)^6 = 0.665. \tag{2.97}$$

(b) If 12 dice are rolled, the probability of obtaining zero 6's is $(5/6)^{12}$, and the probability of obtaining exactly one 6 is $\binom{12}{1}(1/6)^1(5/6)^{11}$, because there are $\binom{12}{1}$ possibilities for the one die that shows a 6. The probability of obtaining at least two 6's is therefore

$$1 - \left(\frac{5}{6}\right)^{12} - \binom{12}{1}\left(\frac{1}{6}\right)^1\left(\frac{5}{6}\right)^{11} = 0.619. \tag{2.98}$$

(c) Similarly, if 18 dice are rolled, the probability of obtaining zero 6's is $(5/6)^{18}$, the probability of obtaining exactly one 6 is $\binom{18}{1}(1/6)^1(5/6)^{17}$, and the probability of obtaining exactly two 6's is $\binom{18}{2}(1/6)^2(5/6)^{16}$. The probability of obtaining at least three 6's is therefore

$$1 - \left(\frac{5}{6}\right)^{18} - \binom{18}{1}\left(\frac{1}{6}\right)^1\left(\frac{5}{6}\right)^{17} - \binom{18}{2}\left(\frac{1}{6}\right)^2\left(\frac{5}{6}\right)^{16} = 0.597. \qquad (2.99)$$

We see that the probability in part (a) is the largest.

REMARK: We can also pose the problem with larger numbers of rolls. For example, if 600 dice are rolled, what is the probability of obtaining at least 100 6's? Or more generally, if $6n$ dice are rolled, what is the probability of obtaining at least n 6's? From the same type of reasoning as above, the answer in the general case is

$$1 - \sum_{k=0}^{n-1} \binom{6n}{k}\left(\frac{1}{6}\right)^k\left(\frac{5}{6}\right)^{6n-k}. \qquad (2.100)$$

For large n, it is intractable to evaluate this sum by hand. But it's easy to use a computer to evaluate it for any n. For $n = 10$, 100, and 1000 we obtain probabilities of, respectively, 0.554, 0.517, and 0.505. These probabilities decrease with n, and they appear to approach the nice simple answer of $1/2$ in the $n \to \infty$ limit. See Problem 5.2 for an explanation of where this $1/2$ comes from. ♣

2.14. **Exactly one pair**

There are $\binom{23}{2}$ possible pairs that can have the common birthday. Let's look at one particular pair and calculate the probability that these two people have a common birthday, while everyone else has a *unique* birthday. We'll then multiply this result by $\binom{23}{2}$ to account for all the possible pairs.

The probability that a given pair has a common birthday is $1/365$, because the first person's birthday can be chosen to be any day, and then the second person has a $1/365$ chance of matching that day. We then need the 21 other people to have 21 different birthdays, none of which is the same as the pair's birthday. The first of these people can end up in any of the remaining 364 days; this happens with probability $364/365$. The second of these people can end up in any of the remaining 363 days; this happens with probability $363/365$. And so on, until the 21st of these people can end up in any of the remaining 344 days; this happens with probability $344/365$.

The total probability that exactly one pair has a common birthday is therefore

$$\binom{23}{2} \cdot \frac{1}{365} \cdot \frac{364}{365} \cdot \frac{363}{365} \cdot \frac{362}{365} \cdot \cdots \cdot \frac{344}{365}. \qquad (2.101)$$

Multiplying this out gives $0.363 = 36.3\%$. This is smaller than the "at least one common birthday" result of 50.7% that we found in Section 2.4.1 for 23 people, as it must be. The remaining $50.7\% - 36.3\% = 14.4\%$ probability corresponds to occurrences of two different pairs with common birthdays, or three people with a common birthday, etc.

2.15. **My birthday**

(a) p is smaller than $100/365$. If the events "Person A having your birthday" and "Person B having your birthday," etc., were all mutually exclusive, then p would be equal to 100/365. But these events are not mutually exclusive, because it is certainly possible for two (or more) of the people to have your birthday. These multiple-event probabilities are counted twice (or more) in the naive 100/365 result. So they must be subtracted off in order to obtain the correct probability. The correct probability is therefore smaller than 100/365.

Note that if we replace the number 100 here by 365 (or anything larger), then the "smaller" answer is obvious, because the probability p is certainly smaller than 365/365 = 1. This suggests (although it doesn't prove) that the answer for the number 100 (or any other number) is "smaller." The one exception is where 100 is replaced by 1, that is, where there is only one other person in the room. In this case we don't have to worry about double counting any probabilities, so the answer is exactly 1/365.

(b) The probability that *no one* out of the 100 people has your birthday equals $(364/365)^{100}$. The probability that at least one of them *does* have your birthday is therefore

$$p = 1 - \left(\frac{364}{365}\right)^{100} = 0.24. \tag{2.102}$$

This is indeed smaller than 100/365 = 0.27. It is only slightly smaller, though, because the multiple-event probabilities are small.

2.16. My birthday, again

We may as well be general right from the start and assume that there are N days in a year. We can eventually set $N = 365$. If there are N days in a year, then the probability that *no one* out of n people has your birthday equals $(1 - 1/N)^n$. This is an exact expression, but we can simplify it by making use of the approximation in Eq. (7.14), namely $(1 + a)^n \approx e^{na}$. With $a \equiv -1/N$ here, $(1 - 1/N)^n$ becomes

$$\left(1 - \frac{1}{N}\right)^n \approx e^{-n/N}. \tag{2.103}$$

Our goal is to have this probability be smaller than 1/2, so that the probability that someone *does* have your birthday is larger than 1/2. Taking the log of both sides of $e^{-n/N} < 1/2$ gives

$$-\frac{n}{N} < \ln\left(\frac{1}{2}\right) \implies -\frac{n}{N} < -\ln 2 \implies \frac{n}{N} > \ln 2 \tag{2.104}$$

$$\implies n > N \ln 2 \approx (0.693)N.$$

Therefore, if $n > N \ln 2$, it is more likely than not that at least one of the n people has your birthday. For $N = 365$, we find that $N \ln 2$ is slightly less than 253, so this agrees with the (exact) result we obtained by simply taking the nth power of 364/365. Since $\ln 2$ is very close to 0.7, a quick approximation to the answer to this problem is $(0.7)N$.

2.17. My birthday, yet again

ONE PERSON: The probability that a specific person has your birthday is 1/365. Since we want *exactly* one person to have your birthday, we want none of the other 252 people to have it; this occurs with probability $(364/365)^{252}$. There are 253 ways to pick the specific person who has your birthday, so the total probability that exactly one of the 253 people has your birthday is

$$253 \cdot \frac{1}{365} \cdot \left(\frac{364}{365}\right)^{252} = 0.347. \tag{2.105}$$

Two PEOPLE: The probability that two specific people have your birthday is $(1/365)^2$. The probability that none of the other 251 people have your birthday is $(364/365)^{251}$. There are $\binom{253}{2}$ ways to pick the two specific people who have your birthday, so the total probability that exactly two of the 253 people have your birthday is

$$\binom{253}{2}\left(\frac{1}{365}\right)^2\left(\frac{364}{365}\right)^{251} = 0.120. \tag{2.106}$$

THREE PEOPLE: By similar reasoning, the probability that exactly three of the 253 people have your birthday is

$$\binom{253}{3}\left(\frac{1}{365}\right)^3\left(\frac{364}{365}\right)^{250} = 0.0276. \tag{2.107}$$

The pattern is clear. The probability that exactly k people have your birthday is

$$P(k) = \binom{253}{k}\left(\frac{1}{365}\right)^k\left(\frac{364}{365}\right)^{253-k}. \tag{2.108}$$

For $k = 0$, this gives the $(364/365)^{253} \approx 1/2$ probability (obtained at the end of Section 2.4.1 and in Problem 2.16) that no one has your birthday. Note that the $P(k)$ probabilities are simply the terms in the binomial expansion:

$$\left(\frac{1}{365} + \frac{364}{365}\right)^{253} = \sum_{k=0}^{253}\binom{253}{k}\left(\frac{1}{365}\right)^k\left(\frac{364}{365}\right)^{253-k}. \tag{2.109}$$

Since the lefthand side of this equation equals 1, we see that the sum of the $P(k)$ also equals 1. This must be the case, of course, because the number of other people who have your birthday has to be *something*.

2.18. A random game-show host

We'll solve this problem by listing out the various possibilities. Without loss of generality, assume that you pick the first door. (You can repeat the following reasoning for the other doors if you wish. It gives the same result.) There are three equally likely possibilities for what is behind the three doors: PGG, GPG, and GGP, where P denotes the prize and G denotes a goat. For each of these three possibilities, since you picked the first door, the host opens either the second or third door (with equal probabilities). So there are six equally likely results of his actions. These are shown in Fig. 2.7, with the bold letters signifying the object revealed.

	PGG	GPG	GGP
open 2nd door	P**G**G	G**P**G	G**G**P
open 3rd door	PG**G**	GP**G**	GG**P**

Table 2.7: There are six equally likely scenarios with a randomly opened door, assuming that you pick the first door.

We now note that the two results where the prize is revealed (the crossed-out GPG and GGP results) are not relevant to this problem, because we are told that the host happens to reveal a goat. Only the four other results are relevant:

PGG   PGG   GPG   GGP

They are all still equally likely, so their probabilities must each be 1/4. We see that if you *don't* switch from the first door, you win on the first two of these results and lose on the second two. And if you *do* switch, you lose on the first two and win on the second two. So either way, your probability of winning is 1/2. It therefore doesn't matter if you switch.

REMARKS:

1. In the original version of the problem in Section 2.4.2, the probability of winning was 2/3 if you switched. How can it possibly decrease to 1/2 in the present random version, when in both versions the exact same thing happened, namely the host revealed a goat?

   The difference is due to the two cases where the host reveals the prize in the random version (the **GPG** and **GGP** cases). You don't benefit from these cases in the random version, because we are told in the statement of the problem that they don't exist. But in the original version, they represent guaranteed success if you switch, because the host is forced to open the other door, which is a goat.

   But still you may say, "If there are two setups, and if I pick, say, the first door in each, and if the host reveals a goat in each (by prediction in one case, and by random pick in the other), then *exactly the same thing happens in both setups.* How can the resulting probabilities (for winning on a switch) be different?" The answer is that although the two outcomes are the same, probabilities have nothing to do with *two* setups. Probabilities are defined only for a *large number* of setups. And if you play a large number of these pairs of games (prediction in one, random pick in the other), then in 1/3 of the pairs the host will reveal different things (a goat in the prediction version and the prize in the random version). These cases yield success in the original prediction version, but they are irrelevant in the random version. They are effectively thrown away there.

2. We will now address the issue mentioned in the fourth remark in Section 2.4.2. We correctly stated in Section 2.4.2 that in the original version of the problem, "No actions taken by the host can change the fact that if you play a large number n of these games, then (roughly) $n/3$ of them will have the prize behind the door you initially pick." However, in the present random version of the problem, something *does* affect the probability that the prize is behind the door you initially pick. It is now 1/2 instead of 1/3. So can something affect this probability or not?

   Well, yes and no. If *all* of the n games are considered (as in the original version), then $n/3$ of them have the prize behind the initial door, and that's that. However, the random version of the problem involves throwing away 1/3 of the games (the ones where the host reveals the prize), because it is assumed in the statement of the problem that the host happens to reveal a goat. So for the *remaining games* (which are 2/3 of the initial total, hence $2n/3$), 1/2 of them now have the prize behind your initial door.

   If you play a large number n of games of each version (including the $n/3$ games that are thrown away in the random version), then the actual *number* of games that have the prize behind your initial door is the same, namely $n/3$. It's just that in the original version this number can be thought of as 1/3 of n, whereas in the random version it can be thought of as 1/2 of $2n/3$. So in the end, the thing that influences the probability (that the initial door you pick has the prize) and changes it from 1/3 to 1/2 isn't the opening of a door, but rather the throwing away of 1/3 of the games. Since no games are thrown away in the original version, the above statement in quotes is correct (with the key phrase being "*these* games").

3. As with the original version of the problem, if you find yourself arguing about the answer for an excessive amount of time, you should just *play the game* a bunch of times (at least a few dozen, to get good enough statistics). The randomness can be determined by a coin toss. As mentioned above, you will end up throwing away 1/3 of the games (the ones where the host reveals the prize). ♣

2.19. **Boy/girl problem with general information**

Let's be general right from the start and consider the case where the boy has a particular characteristic that occurs with probability p. (So $p = 1/4$ if the characteristic is a summer birthday.) As in all of the versions of this problem in Section 2.4.4, we'll list out the various possibilities in a table, *before* the parent's additional information (beyond "I have two children") is taken into account. It is still the case that the BB, BG, GB, and GG types of two-child families are all equally likely, with a 1/4 probability for each. We are again ordering the children in a given pair by age; the first letter is associated with the older child. But we could just as well order them by, say, height or shoe size.

In the present version of the problem, there are now various different subtypes within each type of family, depending on whether or not the children have the given characteristic (which occurs with probability p). For example, if we look at the BB types, there are four possibilities for the occurrence(s) of the characteristic. With "y" standing for "yes, the child has the characteristic," and "n" standing for "no, the child doesn't have the characteristic," the four possibilities are B_yB_y, B_yB_n, B_nB_y, and B_nB_n. (In the second possibility here, for example, the older boy has the characteristic, and the younger boy doesn't.) Since y occurs with probability p, we know that n occurs with probability $1 - p$. The probabilities associated with each of the four possibilities are therefore equal to the 1/4 probability that BB occurs, multiplied by, respectively, p^2, $p(1 - p)$, $(1 - p)p$, and $(1 - p)^2$.

The same reasoning holds with the BG, GB, and GG types, so we obtain a total of $4 \cdot 4 = 16$ distinct possibilities. These are listed in Table 2.8 (ignore the boxes for a moment). The four subtypes in any given row all have the same occurrence(s) of the characteristic, so they all have the same probability; this probability is listed on the right. The subtypes in the middle two rows all have equal probabilities. As mentioned above, in the case where the given characteristic is "having a birthday in the summer," p equals 1/4. So the probabilities associated with the four rows in that case are equal to 1/4 multiplied by, respectively, 1/16, 3/16, 3/16, and 9/16.

Before the parent gives you the additional information, all 16 of the subtypes in the table are possible. But after the statement is made that there is at least one boy with the given characteristic (that is, there is at least one B_y in the pair of children), only seven subtypes remain. These are indicted with boxes. The other nine subtypes are ruled out.

We now simply observe that the three boxes in the left-most column in the table have the other child being a boy, while the four other boxes in the second and third columns have the other child being a girl. The desired probability that the other child is a boy is therefore equal to the sum of the probabilities of the left three boxes, divided by the sum of the probabilities of all seven boxes. This gives (ignoring the common factor of 1/4 in all of the probabilities)

$$P_{BB} = \frac{p^2 + 2 \cdot p(1 - p)}{3 \cdot p^2 + 4 \cdot p(1 - p)} = \frac{2p - p^2}{4p - p^2} = \frac{2 - p}{4 - p}. \tag{2.110}$$

	BB	BG	GB	GG	Probability
yy	B_yB_y	B_yG_y	G_yB_y	G_yG_y	$(1/4) \cdot p^2$
yn	B_yB_n	B_yG_n	G_yB_n	G_yG_n	$(1/4) \cdot p(1-p)$
ny	B_nB_y	B_nG_y	G_nB_y	G_nG_y	$(1/4) \cdot p(1-p)$
nn	B_nB_n	B_nG_n	G_nB_n	G_nG_n	$(1/4) \cdot (1-p)^2$

Table 2.8: The 16 types of families.

In the case where the given characteristic is "having a birthday in the summer," p equals $1/4$. Plugging this into Eq. (2.110) gives the probability that the other child is also a boy as $P_{BB} = 7/15 = 0.467$.

If the given characteristic is "having a birthday on August 11th," then $p = 1/365$, which yields $P_{BB} = 729/1459 = 0.4997 \approx 1/2$.

If the given characteristic is "being born during a particular minute on August 11th," then p is essentially equal to zero, so Eq. (2.110) tells us that P_{BB} is essentially equal to $1/2$. This makes sense, because if $p = 0$, then the $p(1-p)$ probability for the middle two rows in Table 2.8 is much larger than the p^2 probability for the top row. Of course, *all* of these probabilities are very small in the small-p limit, but p^2 is much smaller than $p(1-p) \approx p$ when p is small. So we can ignore the top row. We are then left with four boxes, two of which are BB and two of which are BG/GB. The desired probability therefore equals $1/2$.

Another somewhat special case is $p = 1/2$. (You can imagine that every child flips a coin, and we're concerned with the children who get Heads.) In this case we have $p = 1-p$, so all of the probabilities in the righthand column in Table 2.8 are equal. All 16 entries in the table therefore have equal probabilities (namely $1/16$). Determining probabilities is then just a matter of counting boxes, so the answer to the problem is $3/7$, because three of the seven boxes are of the BB type.

REMARKS:

1. The above $P_{BB} \approx 1/2$ result in the $p \approx 0$ case leads to the following puzzle. Let's say that you bump into a random person on the street who says, "I have two children. At least one of them is a boy." At this stage, you know that the probability that the other child is also a boy is $1/3$, from part (a) of the original problem in Section 2.4.4. But if the parent then adds, "...who was born during a particular minute on August 11th," then we just found that the probability that the other child is also a boy jumps to (essentially) $1/2$. Why exactly did this jump take place?

In the original scenario in Section 2.4.4, there were three equally likely possibilities after the parent gave the additional information, namely BB, BG, and GB. Only $1/3$ of these cases (namely BB) had the other child being a boy. In the new scenario (with $p \approx 0$), there are four equally likely possibilities after the parent gives the additional information, namely B_yB_n, B_nB_y, B_yG_n, and G_nB_y. (As mentioned above, we're ignoring the top row in Table 2.8 since $p \approx 0$.)

So in the new scenario, 1/2 of these cases (the two BB cases) have the other child being a boy. The critical point here is that BB now counts *twice*, whereas it counted only once in the original scenario. This is due to the fact that a BB parent is *twice as likely* (compared with a BG or GB parent) to be able to say that a boy was born during a particular minute on August 11th, because with two boys there are two chances to achieve this highly improbable characteristic. In contrast, a BB parent is *no more likely* (compared with a BG or GB parent) to be able to say simply that at least one child is a boy.

2. In the other extreme where the given characteristic is "being born on *any* day," we have $p = 1$. (This clearly isn't much of a characteristic, since it is satisfied by everyone.) So Eq. (2.110) gives $P_{BB} = 1/3$. In this $p = 1$ case, only the entries in the top row in Table 2.8 have nonzero probabilities. We are therefore in the realm of the first scenario in Section 2.4.4, where we started off with the four types of families (BB, BG, GB, GG) and then ruled out the GG type, yielding a probability of 1/3. It makes sense that the 1/3 answer in the $p = 1$ case is the same as the 1/3 answer in the first scenario in Section 2.4.4, because the "being born on *any* day" statement provides no additional information. So the setup is equivalent to the first scenario in Section 2.4.4, where the parent provided no additional information (beyond the fact that one child was a boy). ♣

2.20. **A second test**

The relevant probability tree is obtained by simply tacking on one more iteration of branches to Fig. 2.11. The result is shown in Fig. 2.20. (We've again arbitrarily started with 1000 people.) We are concerned only with the two numbers 18.05 and 9.8, because these are the only numbers associated with positive results for both tests (labeled as "++"). The desired probability is therefore

$$p = \frac{18.05}{18.05 + 9.8} = 64.8\%. \tag{2.111}$$

This is significantly larger than the result of 16% in the original example in Section 2.5.

Note that since we are concerned only with two of the final eight numbers, there was actually no need to draw the entire probability tree. The two relevant numbers are obtained from the products,

$$(1000)(0.02)(0.95)(0.95) = 18.05,$$
$$(1000)(0.98)(0.1)(0.1) = 9.8. \tag{2.112}$$

These products make it clear how to proceed in the general case of n tests. If we perform n successive tests on each person, then the probability that a person who tests positive all n times actually has the disease is

$$p = \frac{(0.02)(0.95)^n}{(0.02)(0.95)^n + (0.98)(0.1)^n}. \tag{2.113}$$

If $n = 1$ then $p = 0.16$, as we found in the original example. If, say, $n = 4$, then $p = 99.4$. Here the smallness of the $(0.1)^n$ factor in Eq. (2.113) wins out over the smallness of the 0.02 factor. In this case, although not many people have the disease, the number of people who falsely test positive all four times is even smaller. If n is large, then p is essentially equal to 1.

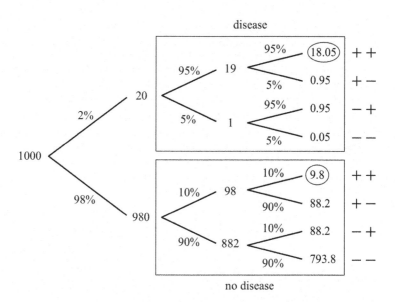

Figure 2.20: The probability tree for two tests.

2.21. **Bayes' theorem for the prosecutor's fallacy**

A given person is either innocent or guilty, and either fits the description or doesn't. Our goal is to find $P(I|D)$. From the second remark at the end of Section 2.5, we want the horizontal span of our square to be associated with the innocent and guilty possibilities, and the vertical span to be associated with the description or not-description possibilities. The result is shown in Fig. 2.21. This figure contains the same information as Fig. 2.10, but in rectangular instead of oval form.

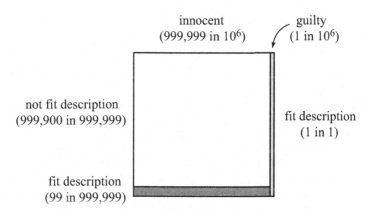

Figure 2.21: The probability square for the prosecutor's fallacy.

We haven't draw things to scale, because if we did, both of the shaded rectangles would be too thin to see. The thin vertical rectangle represents the single guilty person, and the rest of the square represents the 999,999 innocent people. The guilty person fits the description, so the entire thin vertical rectangle is shaded. (As in Fig. 2.10, the fourth possible group of people – guilty and not fitting the description – has zero people in it.) Only about 0.01% of the innocent people fit the description, so the darkly shaded rectangle is very squat. The desired probability $P(I|D)$ equals the number of people in the darkly shaded region (namely 99) divided by the total number of people in both shaded regions (namely 99 + 1). So the desired probability $P(I|D)$ of being innocent, given that the description is satisfied, equals $99/100 = 0.99$.

As we mentioned in Section 2.4.3, $P(I|D)$ (which is close to 1) is *not* equal to $P(D|I)$ (which is close to 0). The former is the ratio of the darkly shaded area to the total shaded area, while the latter is the ratio of the darkly shaded area to the area of the entire left vertical rectangle (the whole square minus the one guilty person).

If you want to use the simple form of Bayes' theorem in Eq. (2.51), instead of using the probability square in Fig. 2.21, you can write

$$P(I|D) = \frac{P(D|I) \cdot P(I)}{P(D)} = \frac{\dfrac{99}{999,999} \cdot \dfrac{999,999}{10^6}}{\dfrac{1}{10^4}} = \frac{99}{100}, \tag{2.114}$$

as desired. You can verify that the various probabilities we used here are correct.

2.22. Black balls and white balls

Our goal is to calculate the probability that the box you pick is the one with two black balls, given that all n draws are black. We'll denote this probability by $P(B_2|nB)$. The two possibilities for the box you pick are the B_2 box with two black balls, and the B_1 box with one black ball. So with this notation, the general form (which is the same as the explicit form in this case) of Bayes' theorem in Eq. (2.53) gives

$$P(B_2|nB) = \frac{P(nB|B_2) \cdot P(B_2)}{P(nB|B_2) \cdot P(B_2) + P(nB|B_1) \cdot P(B_1)}. \tag{2.115}$$

We are given that $P(B_1) = P(B_2) = 1/2$, and we also know that $P(nB|B_2) = 1$ and $P(nB|B_1) = (1/2)^n$. So Bayes' theorem gives

$$\begin{aligned} P(B_2|nB) &= \frac{1 \cdot (1/2)}{1 \cdot (1/2) + (1/2)^n \cdot (1/2)} \\ &= \frac{1}{1 + 1/2^n} = \frac{2^n}{2^n + 1}. \end{aligned} \tag{2.116}$$

If $n = 1$, then $P(B_2|nB) = 2/3$. And if $n = 10$, then $P(B_2|nB) = 1024/1025 \approx 99.9\%$.

If you want to solve the problem without explicitly using Bayes' theorem, the math turns out to be essentially the same. Imagine doing a large number N of trials of the given process. On average, you will pick each of the two boxes $N/2$ times. All n draws will be black in all of the $N/2$ cases where you pick B_2. But all n draws will be black in only $1/2^n$ of the $N/2$ cases where you pick B_1. The other $1 - 1/2^n$ fraction of the cases (where you draw at least one white ball) aren't relevant here. You are therefore dealing with the B_2 box in $N/2$ of the $N/2 + (N/2)/2^n$ times that you draw n black balls. The desired probability is then

$$\begin{aligned} P(B_2|nB) &= \frac{N/2}{N/2 + (N/2)/2^n} \\ &= \frac{1}{1 + 1/2^n}, \end{aligned} \tag{2.117}$$

as above.

Chapter 3

Expectation values

We begin this chapter by introducing in Section 3.1 the important concept of an *expectation value*. Roughly speaking, the expectation value is a fancy name for the average. In Section 3.2 we discuss the *variance*, which is a particular type of expectation value related to the square of the result of a random process. Section 3.3 covers the *standard deviation*, which is defined to be the square root of the variance. The standard deviation gives a rough measure of the spread of the outcomes of a random process. A special kind of standard deviation is the *standard deviation of the mean*, discussed in Section 3.4. This is the standard deviation of the average of a particular number of trials of a random process. We will see that the standard deviation of the mean is smaller than the standard deviation of just one trial of a random process. This fact leads to the *law of large numbers*, which we will discuss in detail in Chapter 5. Section 3.5 covers the *sample variance*, which gives a proper estimate (based on a sample set of numbers) of the true variance of a probability distribution. This section is rather mathematical and can be skipped on a first reading.

3.1 Expectation value

Consider a variable that can take on certain numerical values with certain probabilities. Such a variable is appropriately called a *random variable*. For example, the number of Heads that can arise in two coin tosses is a random variable, and it can take on the values of 0, 1, and 2. A random variable is usually denoted with an uppercase letter, such as X, while the actual values that the variable can take on are denoted with lowercase letters, such as x. So we say, "The number of Heads that can arise in two coin tosses is a random variable X, and the values that X can take on are $x_1 = 0$, $x_2 = 1$, and $x_3 = 2$." Note that the subscript here is an index starting with 1, and not the number of Heads.

The possible outcomes of a random process must be *numerical* if we are to use the term "random variable." So, for example, we don't use the term "random variable" to describe the possible outcomes of a coin toss if these outcomes are Heads and Tails. But we *do* use this term if we assign, say, the number 1 to Heads and the number 0 to Tails. Many of the examples in Chapter 2 involved random variables (for example, rolling dice or counting the number of Heads in a given number of coin tosses), even though we waited until now to define what a random variable is.

The probabilities of the three possible outcomes for X in the above example of two coin tosses are $P(x_1) = 1/4$, $P(x_2) = 1/2$, and $P(x_3) = 1/4$, because the four possible outcomes (HH, HT, TH, TT) are all equally likely. (The collection of these probabilities is called the *probability distribution* for X.) We'll talk at length about probability distributions in Chapter 4, but for now all you need to know is that a probability distribution is simply the collective information about how the total probability (which is always 1) is distributed among the various possible outcomes.

The *expectation value* (or *expected value*) of a random variable X is the expected average obtained in a large number of trials of the process. So in some sense, the expectation value is just the average. However, these two terms have different usages. The *average* is generally associated with trials that have already taken place, for example: the average number of points per game a player scored in last year's basketball season was 14. In contrast, the *expectation value* refers to the average that you would expect to obtain in trials yet to be carried out, for example: the expectation value of the number of Heads you will obtain in 10 coin tosses is 5. A third word meaning roughly the same thing is the *mean*. This can be used in either of the above two contexts (past or future trials).

The expectation value (or the mean) of a random variable X is denoted by either $E(X)$ or μ_X. However, if there is only one random variable at hand (and hence no possibility of confusion), we often don't bother writing the subscript X in μ_X. So the various notations we'll use are:

$$\text{Expectation value:} \quad E(X) \equiv \mu_X \equiv \mu. \qquad (3.1)$$

As an example of an expectation value, consider the roll of a die. Since the numbers 1 through 6 are all equally probable, the expectation value is just their average, which is $(1 + 2 + 3 + 4 + 5 + 6)/6 = 3.5$. Of course, if you roll one die, there is no chance that you will actually obtain a 3.5, because you can roll only the integers 1 through 6. But this is irrelevant as far as the expectation value goes, because we're concerned only with the expected *average* value of a large number of trials. An expectation value of 3.5 is simply a way of saying that if you roll a die 1000 times and add up all the results, you should get a total of about 3500. Again, it is extremely unlikely (but not impossible in this case) that you will get a total of *exactly* 3500, but this doesn't matter when dealing with the expectation value.

The colloquial use of the word "expected" can cause some confusion, because you might think that the expected value is the value that is most likely to occur. This is *not* the case. If we have a process with four equally likely outcomes, $1, 2, 2, 7$, then even though 2 is the most likely value, the "expected value" is the average of the numbers, which is 3, which never occurs.

In order for an expectation value to exist, we need each possible outcome to be associated with a number, as is always the case for a random variable, by definition. If there are no actual numbers involved, then it is impossible to form the average (or actually the *weighted average*; see Eq. (3.4) below). For example, let's say we draw a card from a deck, and let's assume that we're concerned only with its suit. It makes no sense to talk about the expected value of the suit, because it makes no sense to take an average of a heart, diamond, spade, and club. If, however, we assign "suit values" of 1 through 4, respectively, to these suits (so that we're now dealing with an actual random variable – the suit value), then it *does* make sense to talk about the expected value of the suit value, and it happens to be 2.5 (the average of 1 though 4).

The above example with the rolled die consisted of six equally likely outcomes, so we found the expectation value by simply taking the average of the six outcomes. But what if the outcomes have different probabilities? For example, what if we have three balls in a box, two labeled with a "1" and one labeled with a "4"? If we pick a ball, what is the expectation value of the resulting number? (We'll denote this number by the random variable X.)

To answer this, imagine performing a large number of trials of the process. Let's be general and denote this large number by n. Since the probability of picking a 1 is 2/3, we expect about $(2/3)n$ of the numbers to be a 1. Likewise, about $(1/3)n$ of the numbers should be a 4. The total sum of all the numbers should therefore be about $(2/3)n \cdot 1 + (1/3)n \cdot 4$. To obtain the expected average, we just divide this result by n, which gives

$$E(X) = \frac{(2/3)n \cdot 1 + (1/3)n \cdot 4}{n} = \frac{2}{3} \cdot 1 + \frac{1}{3} \cdot 4 = 2. \tag{3.2}$$

Note that the n's canceled out, so the result is independent of n. This is how it should be, because the expected average value shouldn't depend on the exact hypothetical number of trials you do.

In general, if a random variable X has two possible outcomes x_1 and x_2 instead of 1 and 4, and if the associated probabilities are p_1 and p_2 instead of 2/3 and 1/3, then the same reasoning as above gives the expectation value as

$$E(X) = \frac{(p_1 n) \cdot x_1 + (p_2 n) \cdot x_2}{n}$$
$$= p_1 x_1 + p_2 x_2. \tag{3.3}$$

What if we have more than two possible outcomes? The same reasoning works again, but now with more terms in the sum. You can quickly verify (by again imagining a large number of trials, n) that if the outcomes are $x_1, x_2, \ldots, x_m$, and if the associated probabilities are $p_1, p_2, \ldots, p_m$, then the expectation value is

$$\boxed{E(X) = p_1 x_1 + p_2 x_2 + \cdots + p_m x_m} \tag{3.4}$$

This is called the *weighted average* of the outcomes, because each outcome is weighted (that is, multiplied) by its probability. This weighting has the effect of making outcomes with larger probabilities contribute more to the expectation value. This makes sense, because these outcomes occur more often, so they should influence the average more than outcomes that occur less often.

Eq. (3.4) involves a discrete sum, because we're assuming here that our random variable takes on a discrete set of values. If we have a continuous random variable (we'll discuss these in Chapter 4), then the sum in Eq. (3.4) is replaced by an integral.

Example 1 (Expected number of Heads): If you flip a coin four times, what is the expected value of the number of Heads you obtain?

Solution: Without doing any work, we know that the expected number of Heads is 2, because half of the coins will be Heads and half will be Tails, on average.

We can also solve the problem by using Eq. (3.4). By looking at the 16 equally likely outcomes in Table 1.6 in Section 1.3, the probabilities of obtaining 0, 1, 2, 3, or 4 Heads are, respectively, 1/16, 4/16, 6/16, 4/16, and 1/16. So Eq. (3.4) gives the expectation value of the number of Heads as

$$\frac{1}{16} \cdot 0 + \frac{4}{16} \cdot 1 + \frac{6}{16} \cdot 2 + \frac{4}{16} \cdot 3 + \frac{1}{16} \cdot 4 = \frac{32}{16} = 2. \tag{3.5}$$

Example 2 (Flip until Heads): If you flip a coin until you get a Heads, what is the expected total number of coins you flip?

Solution: There is a 1/2 chance that you immediately get a Heads, in which case you flip only one coin. There is a $1/2 \cdot 1/2 = 1/4$ chance that you get a Tails and then a Heads, in which case you flip two coins. There is a $1/2 \cdot 1/2 \cdot 1/2 = 1/8$ chance that you get a Tails, then another Tails, then a Heads, in which case you flip three coins. And so on. The expectation value of the total number of coins is therefore

$$\frac{1}{2} \cdot 1 + \frac{1}{4} \cdot 2 + \frac{1}{8} \cdot 3 + \frac{1}{16} \cdot 4 + \frac{1}{32} \cdot 5 + \cdots . \tag{3.6}$$

This sum has an infinite number of terms, although they eventually become negligibly small. The sum is a little tricky to calculate; see Problem 3.1. However, if you use a calculator to add up the first dozen or so terms, it becomes clear that the sum approaches 2. You are encouraged to convince yourself of this result experimentally, by doing a reasonably large number of trials, say, 50.

Let's now prove a handy theorem involving the sum of two random variables, although you might think the theorem is so obvious that there's no need to prove it.

Theorem 3.1 *The expectation value of the sum of two random variables equals the sum of the expectation values of the two variables. That is,*

$$\boxed{E(X + Y) = E(X) + E(Y)} \tag{3.7}$$

Proof: Imagine performing a large number n of trials to experimentally determine $E(X+Y)$. Each trial involves picking values of X and Y and then forming their sum $X + Y$. That is, you pick values x_1 and y_1 and form the sum $x_1 + y_1$. Then you pick values x_2 and y_2 and form the sum $x_2 + y_2$. You keep doing this a total of n times, where n is large. In the $n \to \infty$ limit, the average value that you obtain for $X + Y$

equals the expectation value of $X + Y$. So (with the $n \to \infty$ limit understood)

$$E(X + Y) \equiv \frac{1}{n} \sum (x_i + y_i)$$

$$= \frac{1}{n} \sum x_i + \frac{1}{n} \sum y_i$$

$$\equiv E(X) + E(Y), \tag{3.8}$$

as desired. ∎

This theorem is intuitive. X simply contributes $E(X)$ to the average, and Y contributes $E(Y)$. Note that we made no assumption about the independence of X and Y in the proof. They can be independent or dependent, and the theorem still holds.

Having just used the word "independent," we should define what we mean by this. Two variables are *independent random variables* if the value of one variable doesn't affect the probability distribution of the other. For example, if X and Y are the results of the rolls of two dice, then X and Y are independent. If you know that the left die shows a 5, then the probability distribution for the right die still consists of six equal probabilities of $1/6$. Mathematically, the random variables X and Y are independent if

$$P(x|y) = P(x) \qquad \text{(independent random variables),} \tag{3.9}$$

for any values of x and y. Likewise with X and Y switched. (More formally, Eq. (3.9) can be written as $P(X = x|Y = y) = P(X = x)$.) This definition of independent *random variables* is similar to the definition of independent *events* given near the start of Section 2.2.1 and in Eq. (2.12). But the definition for random variables is more general. Two variables are independent if *any* event (that is, any outcome or set of outcomes) associated with one variable is independent of *any* event associated with the other variable. Alternatively, we can say that two random variables X and Y are independent if

$$P(x, y) = P(x)P(y) \qquad \text{(independent random variables),} \tag{3.10}$$

for any values of x and y. (More formally, Eq. (3.10) can be written as $P(X = x \text{ and } Y = y) = P(X = x) \cdot P(Y = y)$.) The equivalence of Eqs. (3.9) and (3.10) is exactly analogous to the equivalence of Eqs. (2.12) and (2.13).

Example: Let X take on the values 1 and 2 with equal probabilities of $1/2$, and let Y take on the values 1, 2, and 3 with equal probabilities of $1/3$. Assume that X and Y are independent. Find $E(X + Y)$ by explicitly using Eq. (3.4), and then verify that Eq. (3.7) holds.

Solution: We first quickly note that $E(X) = 1.5$ and $E(Y) = 2$. To use Eq. (3.4) to calculate $E(X + Y)$, we must first determine the various p_i probabilities. If $X = 1$, the three possible values of $X + Y$ are 2, 3, and 4. And if $X = 2$, the three possible values of $X + Y$ are 3, 4, and 5. Because X and Y are independent, all six of these combinations have probabilities of $(1/2)(1/3) = 1/6$, from Eq. (3.10). So we have

$$P(2) = \frac{1}{6}, \quad P(3) = \frac{2}{6}, \quad P(4) = \frac{2}{6}, \quad P(5) = \frac{1}{6}. \qquad (3.11)$$

Eq. (3.4) then gives the expectation value of $X + Y$ as

$$E(X + Y) = \frac{1}{6} \cdot 2 + \frac{2}{6} \cdot 3 + \frac{2}{6} \cdot 4 + \frac{1}{6} \cdot 5 = \frac{21}{6} = 3.5. \qquad (3.12)$$

This is indeed equal to $E(X) + E(Y) = 1.5 + 2 = 3.5$, as Eq. (3.7) claims.

If X and Y are instead *dependent*, then we can't apply Eq. (3.4) without being told what the dependence is, because there is no way to determine the p_i's in Eq. (3.4) without knowing the specific dependence. But $E(X + Y) = E(X) + E(Y)$ will *still hold in any case*. See Problem 3.3 for an example.

Using the same kind of reasoning as in the proof of Theorem 3.1, you can quickly show that

$$E(aX + bY + c) = aE(X) + bE(Y) + c, \qquad (3.13)$$

where a, b, and c are numerical constants. The result in Theorem 3.1 is the special case where $a = 1$, $b = 1$, and $c = 0$. Likewise, similar reasoning in the case of many random variables gives

$$E(a_1 X_1 + a_2 X_2 + \cdots + a_n X_n) = a_1 E(X_1) + a_2 E(X_2) + \cdots + a_n E(X_n), \quad (3.14)$$

as you would expect. You can add on a constant c here, too.

A special case of Eq. (3.14) arises when we perform n trials of the same process. In this case, the n random variables X_i are all associated with the same probability distribution. That is, the X_i are *identically distributed* random variables. For example, the X_i might all refer to the rolling of a die. With each a_i chosen to be 1, Eq. (3.14) then implies

$$E(X_1 + X_2 + \cdots + X_n) = nE(X). \qquad (3.15)$$

We could just as well pick any particular i and write $E(X_i)$ on the righthand side here. The expectation values $E(X_i)$ are all equal, because the X_i are all associated with the same probability distribution. But for simplicity we are using the generic letter X to stand for the random variable associated with the given probability distribution.

REMARK: A word on notation: Since the X_i all come from the same distribution (that is, they are identically distributed), it is tempting to replace all of them with the same letter (say, X) and write the lefthand side of Eq. (3.15) as $E(nX)$. This is incorrect. The random variable nX is *not* the same as the random variable $X_1 + X_2 + \cdots + X_n$. The former involves picking *one* value from the given distribution and multiplying it by n, whereas the latter involves picking n different (or at least generally different) values and adding them up. The results of these processes are not the same. They do happen to have the same expectation value, so Eq. (3.15) would still be true with $E(nX)$ on the lefthand side. But the two processes have different spreads of the values around the common expectation value $nE(X)$, as we'll see in Section 3.2. Also, if you roll ten dice, for example, then nX must be a multiple of 10 (from 10 to 60), whereas the sum of ten X_i values can be any integer from 10 to 60. ♣

You often apply the result in Eq. (3.15) without even knowing it. For example, let's say we flip a coin and define our random variable X to be 1 if we get Heads and 0 if we get Tails. These occur with equal probabilities, so the expectation value of X is $E(X) = (1/2) \cdot 1 + (1/2) \cdot 0 = 1/2$. If we then flip 100 coins, Eq. (3.15) tells us that the expectation value of $X_1 + X_2 + \cdots + X_{100}$ (that is, the expected number of Heads in the 100 flips) is $100 \cdot E(X) = 50$, which is probably what you would have thought anyway, without using Eq. (3.15).

However, you shouldn't get carried away with this type of reasoning, because Eq. (3.14) holds only for *linear* combinations of the random variables. It is *not* true, for example, that $E(1/X) = 1/E(X)$ or that $E(X^2) = (E(X))^2$. You can verify these non-equalities in the case where X is the result of a die roll. You can show that $E(1/X) \approx 0.41$, whereas $1/E(X) = 1/(3.5) \approx 0.29$. Similarly, you can show that $E(X^2) \approx 15.2$, whereas $(E(X))^2 = 3.5^2 = 12.25$.

Theorem 3.1 and its corollaries deal with *sums* of random variables. Let's now prove a theorem involving the *product* of random variables.

Theorem 3.2 *The expectation value of the product of two __independent__ random variables equals the product of the expectation values of the two variables. That is,*

$$\boxed{E(XY) = E(X) \cdot E(Y)} \qquad \text{(independent variables)} \qquad (3.16)$$

Note that this theorem (concerning the product XY) requires that X and Y be independent, unlike Theorem 3.1 (concerning the sum $X + Y$).

Proof: The product XY is itself a random variable, and it takes on the values $x_i y_j$, where i runs through the n_X possible values of X, and j runs through the n_Y possible values of Y. There are therefore $n_X n_Y$ possible values of the product $x_i y_i$. Starting with Eq. (3.4) and then applying Eq. (3.10), the expectation value of XY is

$$
\begin{aligned}
E(XY) &= \sum_{i=1}^{n_X} \sum_{j=1}^{n_Y} P(x_i, y_j) \cdot x_i y_j \\
&= \sum_{i=1}^{n_X} \sum_{j=1}^{n_Y} P(x_i) P(y_j) \cdot x_i y_j \\
&= \left(\sum_{i=1}^{n_X} P(x_i) \cdot x_i \right) \left(\sum_{j=1}^{n_Y} P(y_j) \cdot y_j \right) \\
&= E(X) \cdot E(Y). \qquad (3.17)
\end{aligned}
$$

The use of Eq. (3.10), which is valid only for independent random variables, is what allowed us to break up the sum in the first line here into the product of two separate sums. ∎

Example: Let X be the result of a coin flip where we assign the value 2 to Heads and 1 to Tails. And let Y be the result of another (independent) coin flip where we assign the value 4 to Heads and 3 to Tails. Then $E(X) = 3/2$ and $E(Y) = 7/2$.

Let's explicitly calculate $E(XY)$, to show that it equals $E(X)E(Y)$. There are four equally likely outcomes for the random variable XY:

$$2 \cdot 4 = 8, \qquad 2 \cdot 3 = 6, \qquad 1 \cdot 4 = 4, \qquad 1 \cdot 3 = 3. \tag{3.18}$$

$E(XY)$ is the average of these numbers, so $E(XY) = 21/4$. And this is indeed equal to the product $E(X)E(Y)$, as Eq. (3.16) claims.

As an example of a setup involving *dependent* random variables, where Eq. (3.16) does *not* hold, consider again the above two coins. But let's now stipulate that the second coin always shows the same side as the first coin. So the values of 2 and 4 are always paired together, as are the values 1 and 3. There are now only two (equally likely) outcomes for XY, namely $2 \cdot 4 = 8$ and $1 \cdot 3 = 3$. The expectation value of XY is then $11/2$, which is *not* equal to $E(X)E(Y) = 21/4$.

The expectation value plays an important role in betting and decision making, because it is the amount of money you should be willing to pay up front in order to have a "fair game." By this we mean the following. Consider a game in which you can win various amounts of money, based on the various possible outcomes. For example, let's say that you roll a die and that your winnings equal the resulting number (in dollars). How much money should you be willing to pay to play this game? Also, how much money should the "house" (the people running the game) be willing to charge you for the opportunity to play the game? You certainly shouldn't pay, say, $6 each time you play it, because at best you will break even, and most of the time you will lose money. On average, you will win the average of the numbers 1 through 6, which is $3.50. So this is the most that you should be willing to pay for each trial of the game. If you pay more than this, then you will lose money on average. Conversely, the "house" should charge you *at least* $3.50 to play the game each time, because otherwise it will lose money on average.

Putting these two results together, we see that $3.50 is the amount the game should cost *if the goal is to have a fair game*, that is, a game where neither side wins any money on average. Of course, in games run by casinos and such, things are arranged so that you pay more than the expectation value. So on average the house wins, which is consistent with the fact that casinos stay in business.

Note the italics in the previous paragraph. These are important, because when real-life considerations are taken into account, there might very well be goals that supersede the goal of having a fair game. The above discussion should therefore *not* be taken to imply that you should *always* play a game if the fee is smaller than the expectation value, or that you should *never* play a game if the fee is larger than the expectation value. It depends on the circumstances. See Problem 3.4 for a discussion of this.

3.2 Variance

In the preceding section, we defined the expectation value $E(X)$ as the expected average value obtained in many trials of a random variable X. In addition to $E(X)$, there are other expectation values that are associated with a random variable X. For example, we can calculate $E(X^2)$, which is the expectation value of the *square* of the value of X. If we're rolling a die, the square of the outcome can take on the values of 1^2, 2^2, ..., 6^2 (all equally likely). $E(X^2)$ is the average of these six values, which is $91/6 = 15.17$. We can also calculate other expectation values, such as $E(X^7)$ or $E(2X^3 - 8X^5)$, although arbitrary ones like these aren't of much use.

A slight modification of $E(X^2)$ that turns out to be extremely useful in probability and statistics is the *variance*. It is denoted by $\text{Var}(X)$ and defined to be

$$\boxed{\text{Var}(X) \equiv E[(X - \mu)^2]} \qquad (\text{where } \mu \equiv E[X]) \qquad (3.19)$$

In words: the variance of a random variable X is the expectation value of the square of the difference between X and the mean μ (which itself is the expectation value of X). We're using μ here (without bothering with the subscript X) instead of $E(X)$, to make the above equation and future ones less cluttered.

When calculating the variance $E[(X - \mu)^2]$, Eq. (3.4) still applies. It's just that the X values are replaced with the $(X - \mu)^2$ values. $E[(X - \mu)^2]$ is the same type of quantity as $E(X^2)$, except that we're measuring the values of X relative to the expectation value μ. That's what we're doing when we take the difference $X - \mu$. The examples below should make things clear.

In addition to "$\text{Var}(X)$," the variance is also denoted by σ_X^2 (or just σ^2), due to the definition of the standard deviation, σ, below in Section 3.3. When talking about the variance, sometimes people say "the variance of a random variable," and sometimes they say "the variance of a probability distribution." These mean the same thing.

Example 1 (Die roll): The expectation value of the six equally likely outcomes of a die roll is $\mu = 3.5$. The variance is therefore

$$\begin{aligned}
\text{Var}(X) &= E[(X - 3.5)^2] \\
&= \frac{1}{6}\Big[(1 - 3.5)^2 + (2 - 3.5)^2 + (3 - 3.5)^2 \\
&\qquad + (4 - 3.5)^2 + (5 - 3.5)^2 + (6 - 3.5)^2\Big] \\
&= \frac{1}{6}[6.25 + 2.25 + 0.25 + 0.25 + 2.25 + 6.25] \\
&= 2.92.
\end{aligned} \qquad (3.20)$$

Example 2 (Coin flip): Consider a coin flip where we assign the value 1 to Heads and 0 to Tails. The expectation value of these two equally likely outcomes is $\mu = 1/2$, so the variance is

$$\mathrm{Var}(X) = E[(X - 1/2)^2]$$

$$= \frac{1}{2}\left[(1 - 1/2)^2 + (0 - 1/2)^2\right] = \frac{1}{4}. \tag{3.21}$$

Example 3 (Biased coin): Consider a biased coin, where the probability of getting Heads is p and the probability of getting Tails is $1 - p \equiv q$. If we again assign the value 1 to Heads and 0 to Tails, then the expectation value is $\mu = p \cdot 1 + (1 - p) \cdot 0 = p$. The variance is therefore

$$\mathrm{Var}(X) = E[(X - p)^2]$$

$$= p \cdot (1 - p)^2 + (1 - p) \cdot (0 - p)^2$$

$$= p(1 - p)[(1 - p) + p]$$

$$= p(1 - p) \equiv pq. \tag{3.22}$$

As you can see in the above examples, the steps in finding the variance are:

1. Find the mean.

2. Find all the differences from the mean.

3. Square each of these differences.

4. Find the expectation value of these squares.

The variance of a random variable is related to how much the outcomes are spread out away from the mean. Note well that the variance in Eq. (3.19) involves *first* squaring the differences from the mean, and *then* finding the expectation value of these squares. If instead you first find the expectation value of the differences from the mean, and then square the result, you will obtain zero. This is true because

$$(E(X - \mu))^2 = (E(X) - \mu)^2 = (\mu - \mu)^2 = 0. \tag{3.23}$$

We would obtain zero here even without the squaring operation, of course.

The variance depends only on the spread of the outcomes relative to the mean, and not on the mean itself. For example, if we relabel the faces on a die by adding 100, so that they are now 101 through 106, then the mean changes significantly to 103.5. But the variance remains at the 2.92 value we found in the first example above, because all of the differences from the mean are the same as for a normal die.

If a is a numerical constant, then the variance of aX equals $a^2\mathrm{Var}(X)$. This follows from the definition of the variance in Eq. (3.19), along with the result in

Eq. (3.13). The latter tells us that $E(aX) = aE(X) \equiv a\mu$, so the former (along with another application of the latter) gives

$$\text{Var}(aX) = E\left[\left((aX) - (a\mu)\right)^2\right] = E\left[a^2(X - \mu)^2\right]$$
$$= a^2 E\left[(X - \mu)^2\right] = a^2\text{Var}(X), \tag{3.24}$$

as desired.

The variance of the sum of two *independent* variables turns out to be the sum of the variances of the two variables, as we show in the following theorem. Due to the nonlinearity of X in $E[(X - \mu)^2]$, it isn't so obvious that the variances should simply add linearly. But they indeed do.

Theorem 3.3 *Let X and Y be two <u>independent</u> random variables. Then*

$$\boxed{\text{Var}(X + Y) = \text{Var}(X) + \text{Var}(Y)} \qquad \text{(independent variables)} \qquad (3.25)$$

Proof: We know from Eq. (3.7) that the mean of $X + Y$ is $\mu_X + \mu_Y$. So

$$\text{Var}(X + Y) = E\left[\left((X + Y) - (\mu_X + \mu_Y)\right)^2\right] \tag{3.26}$$
$$= E\left[\left((X - \mu_X) + (Y - \mu_Y)\right)^2\right]$$
$$= E\left[(X - \mu_X)^2\right] + 2E\left[(X - \mu_X)(Y - \mu_Y)\right] + E\left[(Y - \mu_Y)^2\right]$$
$$= \text{Var}(X) + 0 + \text{Var}(Y).$$

The zero here arises from the fact that X and Y (and hence $X - \mu_X$ and $Y - \mu_Y$) are independent variables, which from Eq. (3.16) implies that the expectation value of the product equals the product of the expectation values. That is,

$$E[(X - \mu_X)(Y - \mu_Y)] = E(X - \mu_X) \cdot E(Y - \mu_Y)$$
$$= (E(X) - \mu_X) \cdot (E(Y) - \mu_Y)$$
$$= (\mu_X - \mu_X) \cdot (\mu_Y - \mu_Y)$$
$$= 0. \quad \blacksquare \tag{3.27}$$

Example (Two coins): Let's verify that Eq. (3.25) holds if we define X and Y to each be the result of independent coin flips where we assign the value 1 to Heads and 0 to Tails. The random variable $X + Y$ takes on the values of 0, 1, and 2 with probabilities 1/4, 1/2, and 1/4, respectively. The expectation value of $X + Y$ is 1, so the variance is

$$\text{Var}(X + Y) = \frac{1}{4}\left[(0 - 1)^2\right] + \frac{1}{2}\left[(1 - 1)^2\right] + \frac{1}{4}\left[(2 - 1)^2\right] = \frac{1}{2}. \tag{3.28}$$

And we know from Eq. (3.21) that the variance of each single coin flip is $\text{Var}(X) = \text{Var}(Y) = 1/4$. So it is indeed true that $\text{Var}(X + Y) = \text{Var}(X) + \text{Var}(Y)$.

As an example of a setup involving *dependent* random variables, where Eq. (3.25) does *not* hold, consider again the above two coins. But let's now stipulate that the second coin always shows the same side as the first coin. So the 1's are always paired together, as are the 0's. There are then only two (equally likely) outcomes for $X + Y$, namely $0 + 0 = 0$ and $1 + 1 = 2$. The expectation value of $X + Y$ is 1, so the variance is

$$\text{Var}(X + Y) = \frac{1}{2}\left[(0 - 1)^2\right] + \frac{1}{2}\left[(2 - 1)^2\right] = 1, \tag{3.29}$$

which is *not* equal to $\text{Var}(X) + \text{Var}(Y) = 1/2$.

Repeated application of Eq. (3.25) gives the variance of the sum of an arbitrary number of independent variables as

$$\text{Var}(X_1 + X_2 + \cdots + X_n) = \text{Var}(X_1) + \text{Var}(X_2) + \cdots + \text{Var}(X_n). \tag{3.30}$$

By "repeated application" we mean the following. Let the Y in Eq. (3.25) be equal to X_n, and let the X be the sum of X_1 through X_{n-1}. This gives

$$\text{Var}(X_1 + X_2 + \cdots + X_n) = \text{Var}(X_1 + X_2 + \cdots + X_{n-1}) + \text{Var}(X_n). \tag{3.31}$$

Then repeat the process with $Y \equiv X_{n-1}$ and with X equal to the sum of X_1 through X_{n-2}. And so on. This eventually yields Eq. (3.30).

If all of the X_i are *independent and identically distributed* random variables (i.i.d. variables, for short), then Eq. (3.30) gives

$$\boxed{\text{Var}(X_1 + X_2 + \cdots + X_n) = n\text{Var}(X)} \qquad \text{(i.i.d. variables)} \tag{3.32}$$

where X represents any one of the X_i. For example, we can flip a coin n times and write down the total number of Heads obtained. (In doing this, we're effectively assigning the value 1 to Heads and 0 to Tails.) This sum of n independent and identically distributed coin flips is the binomial process we discussed in Section 1.8. Since we know the "1/4" result in Eq. (3.21) for the variance of a single flip, Eq. (3.32) gives the variance of the binomial process as $n\text{Var}(X) = n/4$. More generally, if we have a biased coin with $P(\text{Heads}) = p$ and $P(\text{Tails}) = 1 - p \equiv q$, then the combination of Eqs. (3.22) and (3.32) tells us that the variance of the number of Heads in n flips is

$$\boxed{\text{Var}(\text{Heads in } n \text{ flips}) = npq} \qquad \text{(biased coin)} \tag{3.33}$$

REMARK: As mentioned in the remark following Eq. (3.15), the sum $X_1 + X_2 + \cdots + X_n$ in Eq. (3.32) is not the same as nX. Although the random variables X_i are all *identically distributed*, that certainly doesn't mean that their *values* are identical. The values of the X_i will generally be different. So when forming the sum, we can't just take one of the values and multiply it by n. Although the expectation-value statement in Eq. (3.15) happens to remain true if we replace the sum $X_1 + X_2 + \cdots + X_n$ with nX, the variance statement in Eq. (3.32) does *not* remain true. From Eq. (3.24), the variance of nX equals $n^2\text{Var}(X)$, which isn't the same as the $n\text{Var}(X)$ result in Eq. (3.32). ♣

When dealing with the product of two random variables, it turns out that the equation analogous to Eq. (3.16) for expectation values does *not* hold for variances, even if X and Y are independent. That is, it is *not* true that $\text{Var}(XY) =$

$\text{Var}(X)\text{Var}(Y)$. See Problem 3.6 for an example showing that this equality doesn't hold.

It is often useful to write the variance, which we defined in Eq. (3.19), in the following alternative form:

$$\text{Var}(X) = E(X^2) - \mu^2 \qquad (3.34)$$

That is, the variance equals the expectation value of the square, minus the square of the expectation value. This can be demonstrated as follows. Starting with Eq. (3.19), we have

$$\begin{aligned}
\text{Var}(X) &= E[(X - \mu)^2] \\
&= E[X^2 - 2\mu \cdot X + \mu^2] \\
&= E(X^2) - 2\mu \cdot E(X) + \mu^2 \\
&= E(X^2) - 2\mu^2 + \mu^2 \\
&= E(X^2) - \mu^2,
\end{aligned} \qquad (3.35)$$

as desired. We have used the fact that $E(X)$ means the same thing as μ. And we have used Eq. (3.13) (which says that the expectation value of the sum equals the sum of the expectation values) to go from the second line to the third line. You can quickly verify that this expression for $\text{Var}(X)$ gives the same variances that we found in the three examples near the beginning of this section; see Problem 3.7.

Variance of a set of numbers

In the above discussion of the variance, the definition in Eq. (3.19) was based on a random variable X with a given probability distribution. We can, however, also define the variance for an arbitrary set of numbers, even if they don't have anything to do with a probability distribution. Given an arbitrary set S of n numbers, $x_1, \ldots, x_n$, let their average (or mean) be denoted by $\bar{x}$. We'll also occasionally use $\langle x \rangle$ to denote the average:

$$\text{Average}: \quad \bar{x} \equiv \langle x \rangle \equiv \frac{1}{n} \sum_1^n x_i. \qquad (3.36)$$

Then the variance of the set S is defined to be

$$\text{Var}(S) \equiv \frac{1}{n} \sum_1^n (x_i - \bar{x})^2 \qquad \text{(for a set } S \text{ of numbers)} \qquad (3.37)$$

In words: the variance of the set S is the average value of the square of the difference from the mean. Note the slight difference between the preceding sentence and the sentence following Eq. (3.19). That sentence involved the "*expectation* value of the square...," whereas the present sentence involves the "*average* value of the square...." This distinction is due to the fact that (as we noted near the beginning of

Section 3.1) the term "expectation value" is relevant to a probability distribution for a random variable X. If you are instead simply given a set S of numbers, then you can take their average, but it doesn't make sense to talk about an expectation value, because there are no future trials for which you can expect anything. (Technically, if you are imagining that the set S of numbers came from a probability distribution, then you can talk about the best guess for the expectation value of the distribution. But we won't get into that here.)

As an example, if we have the set S of four numbers, 2.3, 5.6, 3.8, and 4.7, then the average is 4.1, so the variance is

$$\text{Var}(S) = \frac{1}{4}\left[(2.3 - 4.1)^2 + (5.6 - 4.1)^2 + (3.8 - 4.1)^2 + (4.7 - 4.1)^2\right]$$
$$= 1.485. \qquad\qquad (3.38)$$

Note that all of the numbers are weighted equally here. This isn't the case (in general) when calculating the variance in Eq. (3.19).

Later on in Section 3.5 we'll encounter a slightly modified version of Eq. (3.37) called the "sample variance," which has an $n - 1$ instead of an n in the denominator.

3.3   Standard deviation

The *standard deviation* of a random variable (or equivalently, of a probability distribution) is defined to be the square root of the variance:

$$\boxed{\sigma_X \equiv \sqrt{\text{Var}(X)}} \qquad\qquad (3.39)$$

As with the mean μ, the subscript X is usually dropped if there is no ambiguity about which random variable we are referring to. With the definition in Eq. (3.39), we can write the variance as σ_X^2. You will often see this notation for the variance, since it is quicker to write than $\text{Var}(X)$, and even quicker if you drop the subscript X. Like the variance, the standard deviation gives a rough measure of how much the outcomes are spread out away from the mean. We'll draw some pictures below that demonstrate this.

From Eqs. (3.19) and (3.34), we can write the standard deviation in two equivalent ways:

$$\boxed{\sigma = \sqrt{E[(X - \mu)^2]} = \sqrt{E(X^2) - \mu^2}} \qquad\qquad (3.40)$$

Using the first of these forms, the steps in finding the standard deviation are the same as in finding the variance, with a square root tacked on the end:

1. Find the mean.

2. Find all the differences from the mean.

3. Square each of these differences.

4. Find the expectation value of these squares.

5. Take the square root of this expectation value.

As with the variance, the standard deviation depends only on the spread relative to the mean, and not on the mean itself. If we relabel the faces on a die by adding 100, so that they are now 101 through 106, then the mean changes significantly to 103.5, but the standard deviation remains at $\sqrt{2.92} = 1.71$ (using the 2.92 value for the variance in Eq. (3.20)).

Since the standard deviation is simply the square root of the variance, we can quickly translate all of the statements we made about the variance in Section 3.2 into statements about the standard deviation. Let's list them out.

- From Eq. (3.24) the standard deviation of aX is just a times the standard deviation of X:

$$\sigma_{aX} = a\sigma_X. \tag{3.41}$$

- If X and Y are two *independent* random variables, then Eq. (3.25) becomes

$$\boxed{\sigma_{X+Y}^2 = \sigma_X^2 + \sigma_Y^2} \qquad \text{(independent variables)} \tag{3.42}$$

This is the statement that standard deviations "add in quadrature" for independent variables.

- The more general statement in Eq. (3.30) can similarly be rewritten as (again only for independent variables)

$$\sigma_{X_1+X_2+\cdots+X_n}^2 = \sigma_{X_1}^2 + \sigma_{X_2}^2 + \cdots + \sigma_{X_n}^2. \tag{3.43}$$

Taking the square root of Eq. (3.43) gives (again only for independent variables):

$$\sigma_{X_1+X_2+\cdots+X_n} = \sqrt{\sigma_{X_1}^2 + \sigma_{X_2}^2 + \cdots + \sigma_{X_n}^2}. \tag{3.44}$$

- If all of the X_i are *independent and identically distributed* random variables, then Eq. (3.44) becomes

$$\boxed{\sigma_{X_1+X_2+\cdots+X_n} = \sqrt{n}\,\sigma_X} \qquad \text{(i.i.d. variables)} \tag{3.45}$$

- From Eq. (3.22) the standard deviation of a single flip of a biased coin (with Heads equalling 1 and Tails equalling 0) is

$$\sigma = \sqrt{pq}. \tag{3.46}$$

- If we flip the biased coin n times, then from either Eq. (3.33) or Eq. (3.45), the standard deviation of the number of Heads is

$$\boxed{\sigma = \sqrt{npq}} \qquad \text{(n biased coins)} \tag{3.47}$$

For a fair coin ($p = q = 1/2$), this equals

$$\boxed{\sigma = \sqrt{n/4}} \qquad (n \text{ fair coins}) \qquad (3.48)$$

For example, the standard deviation of the number of Heads in $n = 100$ fair coin flips is $\sigma = \sqrt{100/4} = 5$. This is a handy fact to remember. The standard deviations for other numbers of flips can then quickly be determined by using the fact that σ is proportional to $\sqrt{n}$. For example, 1000 is 10 times 100, and $\sqrt{10} \approx 3$, so the σ for $n = 1000$ flips is about $3 \cdot 5 = 15$. (It's actually more like 16.) Similarly, 10,000 is 100 times 100, and $\sqrt{100} = 10$, so the σ for $n = 10,000$ flips is $10 \cdot 5 = 50$.

- In terms of σ, Eq. (3.34) becomes

$$\sigma^2 = E(X^2) - \mu^2. \qquad (3.49)$$

If we solve for $E(X^2)$ here, we see that the expectation value of the square of a random variable X is

$$\boxed{E(X^2) = \sigma^2 + \mu^2} \qquad (3.50)$$

This result checks in two limits. First, if $\mu = 0$ then Eq. (3.50) says that σ^2 (which is the variance) equals $E(X^2)$. This agrees with what Eq. (3.19) says when μ equals zero. Second, if $\sigma = 0$ then Eq. (3.50) says that $E(X^2)$ equals μ^2. This makes sense, because if $\sigma = 0$ then there is no spread in the possible outcomes. That is, there is only one possible outcome, which must then be μ, by definition; the expectation value of one number is simply that number. So $E(X^2) = \mu^2$.

As mentioned above, the standard deviation (like the variance) gives a rough measure of how much the outcomes are spread out away from the mean. This measure is actually a much more appropriate one than the variance's measure, because whereas the units of the variance are the same as X^2, the units of the standard deviation are the same as X. It therefore makes sense to draw the standard deviation in the same figure as the plot of the probability distribution for the various outcomes (with the X values lying on the horizontal axis). We'll talk much more about plots of probability distributions in Chapter 4, but for now we're concerned only with what the standard deviation looks like when superimposed on the plot.

Example: Fig. 3.1 shows four examples of the standard deviation superimposed on the probability distribution. The commentary on each plot is as follows.

- FIRST PLOT: For a die roll, the probability of each of the six numbers is 1/6. And since the variance in Eq. (3.20) is 2.92, the standard deviation is $\sigma = \sqrt{2.92} = 1.71$. This is the rough spread of the outcomes, relative to the mean (which is 3.5). Some outcomes lie inside the range of $\pm\sigma$ around the mean, and some lie outside.

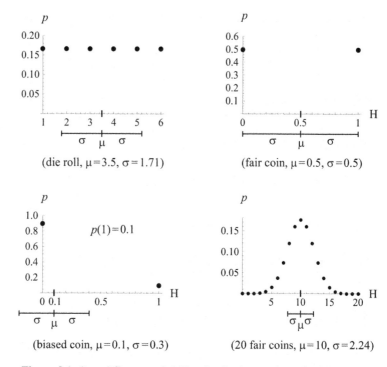

Figure 3.1: Four different probability distributions and standard deviations.

- SECOND PLOT: For a fair coin flip (with Heads = 1 and Tails = 0), Eq. (3.48) gives the standard deviation as $\sigma = 1/2$. Both outcomes therefore lie right at the $\pm\sigma$ locations relative to the mean (which is $1/2$). This makes sense; all of the outcomes (there are only two of them) are a distance of $1/2$ from the mean.

- THIRD PLOT: For a biased coin flip (again with Heads = 1 and Tails = 0), we have assumed that the probabilities are $p = 1/10$ for Heads and $1 - p = 9/10 \equiv q$ for Tails. So Eq. (3.46) gives the standard deviation as $\sigma = \sqrt{(1/10)(9/10)} = 3/10$. As noted prior to Eq. (3.22), the mean of the roll is p, which is $1/10$ here. The outcome of 0 lies inside the range of $\pm\sigma$ around the mean, while the outcome of 1 lies (far) outside.

- FOURTH PLOT: For n flips of a fair coin, Eq. (3.48) gives the standard deviation of the number of Heads as

$$\sigma = \frac{\sqrt{n}}{2} . \tag{3.51}$$

If we pick n to be 20, then we have $\sigma = \sqrt{20}/2 = 2.24$. Five outcomes lie inside the range of $\pm\sigma$ around the mean (which is 10), while the other 16 lie outside. Although there are more outcomes outside, and additionally some of them are far away from the mean, their probabilities are small, so they don't have an overwhelming influence on $\sigma \equiv \sqrt{E[(X - \mu)^2]}$.

In all of the above cases, σ gives a rough measure of the spread of the outcomes. More precisely, Eq. (3.40) tells us that σ is the square root of the expectation value of the square of the distance from the mean. This is a mouthful, so you might be wondering – if we want to get a rough idea of the spread of the various outcomes, why don't we use something simpler? For example, we could just calculate the expected *distance* from the mean, that is, $E(|X - \mu|)$. The absolute value bars here produce the various distances (which are nonnegative quantities, by definition). Although this is a perfectly reasonable definition, it is also a messy one. Quantities involving absolute values are somewhat artificial, because if $|x - \mu|$ is negative, then we have to throw in a minus sign by hand and say that $|x - \mu| = -(x - \mu)$. In contrast, the square of a quantity (which always yields a nonnegative number) is a very natural thing. Additionally, the standard deviation defined by Eq. (3.39) has some nice properties, one of which is Eq. (3.42). An analogous statement (with or without the squares) wouldn't hold in general if we defined σ as $E(|X - \mu|)$. A quick counterexample is provided by two independent coin flips (with Heads = 1 and Tails = 0), as you can verify. The "σ" for each flip would be $1/2$, and the "σ" for the sum of the flips would also be $1/2$.

3.4   Standard deviation of the mean

Consider the fourth plot in Fig. 3.1, which shows the probability distribution and the standard deviation for the number of Heads in 20 fair coin tosses. What if we are instead concerned not with the *total* number of Heads in 20 coin tosses, but rather with the *average* number of Heads per toss (averaged over the 20 tosses)? For example, if we happen to get 12 Heads in the 20 tosses (which, by looking at the plot, has a probability of about 12%), then the average number of Heads per toss is obtained by dividing 12 by 20, yielding $12/20 = 0.6$.

In the same manner, to obtain the entire probability distribution for the *average* number of Heads per toss, we simply need to keep the same dots in the plot in Fig. 3.1, but divide the numbers on the x axis by 20. This gives the probability distribution shown in Fig. 3.2(a). An average of 0.5 Heads per toss is of course the most likely average, and it occurs with a probability of about 18% (the same as the probability of getting a total of 10 Heads in 20 tosses).

The standard deviation of the *total* number of Heads that appear in n tosses is the $\sigma_{\text{tot}} = \sqrt{n}/2$ result in Eq. (3.51). The standard deviation of the *average* number of Heads per toss is therefore

$$\sigma_{\text{avg}} = \frac{\sigma_{\text{tot}}}{n} = \frac{\sqrt{n}/2}{n} = \frac{1}{2\sqrt{n}}. \tag{3.52}$$

This is true because if X_{tot} represents the total number of Heads in n flips, and if X_{avg} represents the average number of Heads per flip, then $X_{\text{avg}} = X_{\text{tot}}/n$. Eq. (3.41) then gives $\sigma_{\text{avg}} = \sigma_{\text{tot}}/n$. Equivalently, a given span of the x axis in Fig. 3.2(a) has only $1/n$ (with $n = 20$) the length of the corresponding span in the fourth plot in Fig. 3.1. So the spread in Fig. 3.2(a) is $1/n$ times the spread in Fig. 3.1. With $n = 20$, Eq. (3.52) gives the standard deviation of the average (which is usually called the

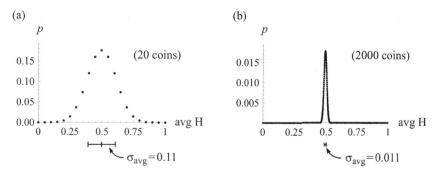

Figure 3.2: The probability distribution for the *average* number of Heads per toss, for 20 or 2000 coin tosses.

standard deviation of the mean) as $\sigma_{\text{avg}} = 1/(2\sqrt{20}) = 0.11$. This is indicated in Fig. 3.2(a); the bar shown has a total length of $2\sigma_{\text{avg}}$.

Let's repeat the above analysis, but now with 2000 tosses. The probability distribution for the total number of Heads is peaked around 1000, of course. From Eq. (3.51) the standard deviation of the *total* number of Heads that appear in 2000 tosses is $\sigma_{\text{tot}} = \sqrt{2000}/2 = 22.4$. To obtain the probability distribution for the *average* number of Heads per toss (which is peaked around 0.5), we just need to divide all the numbers on the x axis (associated with the total number of Heads) by 2000, analogous to what we did with 20 tosses, in going from the fourth plot in Fig. 3.1 to Fig. 3.2(a). The resulting probability distribution for the average number of Heads per toss is shown in Fig. 3.2(b). From Eq. (3.52) the standard deviation of the average number of Heads per toss is $\sigma_{\text{avg}} = 1/(2\sqrt{2000}) = 0.011$. This is indicated in the figure; the (very short) bar shown has a total length of $2\sigma_{\text{avg}}$.

Since 2000 and 20 differ by a factor of 100, the σ_{tot} for 2000 tosses is 10 times larger than the σ_{tot} for 20 tosses, because the result in Eq. (3.51) is proportional to $\sqrt{n}$. But σ_{avg} is 10 times smaller, because the σ_{avg} in Eq. (3.52) is proportional to $1/\sqrt{n}$. The latter of these two facts is why the bump of points in Fig. 3.2(b) is much thinner than the bump in Fig. 3.2(a). The σ_{avg} that we have drawn in Fig. 3.2(b) is barely long enough to be noticeable. But even if you don't calculate and compare the standard deviations of the two plots in Fig. 3.2, it is obvious that the bump is much thinner in Fig. 3.2(b).

Let's recap what we've learned. Although σ_{tot} is *larger* (by a factor of 10) in the $n = 2000$ case, σ_{avg} is *smaller* (by a factor of 10) in the $n = 2000$ case. The first of these results deals with the *absolute* (or additive) deviation σ_{tot} from the expected value of $n/2$, while the second deals with the *fractional* (or multiplicative) deviation σ_{avg} from the expected value of $1/2$. The point here is that although the absolute deviation σ_{tot} grows with n (which is intuitive), it does so in a manner that is only proportional to $\sqrt{n}$. So when this deviation is divided by n when calculating the *average* number of Heads, the fractional deviation σ_{avg} ends up being proportional to $1/\sqrt{n}$, which *decreases* with n (which might not be so intuitive).

Note that although the *expectation value* of the average number of Heads per toss is independent of the number of tosses (it is always 0.5 for a fair coin, as it is in the two plots in Fig. 3.2(b)), the *distribution* of the average number of Heads per toss *does* depend on the number of tosses. That is, the shapes of the two curves in Fig. 3.2 are different (on the same scale from 0 to 1 on the x axis). For example, in the case of 20 tosses, you have a reasonable chance of obtaining an average that is 0.6 or more. But in the case of 2000 tosses, you are *extremely* unlikely to obtain such an average.

Let us formalize the above results with the following theorem.

Theorem 3.4 *Consider a random variable X with standard deviation σ. We make no assumptions about the shape of the probability distribution. Let $\overline{X}$ be the random variable formed by taking the average of n independent trials of the random variable X. Then the standard deviation of $\overline{X}$ is given by $\sigma_{\overline{X}} = \sigma_X / \sqrt{n}$, which is often written in the slightly more succinct form,*

$$\boxed{\sigma_{\overline{X}} = \frac{\sigma}{\sqrt{n}}}$$ (standard deviation of the mean)   (3.53)

This is the standard notation, although technically the letter n should appear as a label somewhere on the lefthand side of the equation, because the standard deviation of $\overline{X}$ depends on the number n of trials that you are averaging over.

Proof: Let the n independent trials of the variable X be labeled $X_1, X_2, \ldots, X_n$. (So the X_i are independent and identically distributed random variables.) Then $\overline{X}$ is given by

$$\overline{X} \equiv \frac{X_1 + X_2 + \cdots + X_n}{n}.$$ (3.54)

From Eq. (3.41) the standard deviation of $\overline{X}$ equals $1/n$ times the standard deviation of $X_1 + X_2 + \cdots + X_n$. But from Eq. (3.45) the latter is $\sqrt{n}\sigma$. The standard deviation of $\overline{X}$ is therefore $\sqrt{n}\sigma/n = \sigma/\sqrt{n}$, as desired. ∎

In short (as we've mentioned a number of times), the above proof comes down to the fact that Eq. (3.45) says that the standard deviation of the *sum* of n independent and identical trials grows with n, but only like $\sqrt{n}$. When we take the average and divide by n, we obtain a standard deviation of the mean that is smaller than the original σ by a factor of $\sqrt{n}$.

More generally, if we are concerned with the average of n *different* random variables with different standard deviations, we can use Eqs. (3.41) and (3.44) to say

$$\sigma_{\overline{X}} \equiv \sigma_{\frac{X_1+X_2+\cdots+X_n}{n}}$$

$$= \frac{1}{n}\sigma_{X_1+X_2+\cdots+X_n}$$

$$= \frac{\sqrt{\sigma_{X_1}^2 + \sigma_{X_2}^2 + \cdots + \sigma_{X_n}^2}}{n}.$$ (3.55)

This reduces to Eq. (3.53) when all of the σ_{X_i} are equal (even if the distributions aren't the same).

The thinness of the curve in Fig. 3.2(b), which is a consequence of the $\sqrt{n}$ in the denominator in Eq. (3.53), is consistent with the "law of large numbers." This law says that if you perform a very large number of trials, the observed average will likely be very close to the theoretically predicted average. In a little more detail: many probability distributions, such as the ones in Fig. 3.2, are essentially Gaussian (or "normal" or "bell-curve") in shape. And it can be shown numerically that for a Gaussian distribution, the probability of lying within one standard deviation from the mean (that is, in the range $\mu \pm \sigma$) is 68%, the probability of lying within two standard deviations from the mean is 95%, and the probability of lying within three standard deviations from the mean is 99.7%. For wider ranges, the probability is effectively 1, for most practical purposes. This is why we mentioned above that for 2000 coin tosses, the average number of Heads per toss is *extremely* unlikely to be 0.6 or larger. Since 0.6 exceeds the mean 0.5 by 0.1, and since $\sigma_{\text{avg}} = 0.011$ in this case, we see that 0.6 is about nine standard deviations above the mean. The probability of being more than nine standard deviations above the mean is utterly negligible (it's about 10^{-19}).

We threw around a number of terms and results in the preceding paragraph. We'll eventually get to these. Section 4.8 covers the Gaussian distribution, and Chapter 5 covers the law of large numbers and the central limit theorem. This theorem explains why many probability distributions are approximately Gaussian.

Example (Rolling 10,000 dice):

(a) 10,000 dice are rolled. What is the expectation value of the total number of 6's that appear? What is the standard deviation of this number?

(b) What is the expectation value of the *average* number of 6's that appear per roll? What is the standard deviation of this average?

(c) Do you think you have a reasonable chance of getting a 6 on at least 20% of the rolls?

Solution:

(a) The probability of getting a 6 on a given roll is $p = 1/6$, so the expected total number of 6's that appear in the 10,000 rolls is $(1/6) \cdot (10,000) = 1667$. To find the standard deviation of the total number of 6's, we can assign the value 1 to a roll of 6, and a value of 0 to the five other rolls. Since $p = 1/6$, we're effectively flipping a biased coin that has a $p = 1/6$ chance of success. From Eq. (3.47) the standard deviation of the total number of 6's that come up in 10,000 rolls is

$$\sigma_{\text{tot}} = \sqrt{npq} = \sqrt{(10,000)(1/6)(5/6)} = 37. \tag{3.56}$$

(b) The expectation value of the average number of 6's that appear per roll equals $1667/10,000 = 1/6$, of course. The standard deviation of the average is obtained from the standard deviation of the total number of 6's (given in Eq. (3.56)) by dividing by 10,000 (just as we divided by n in the discussion of Fig. 3.2). So we obtain $\sigma_{\text{avg}} = 37/10,000 = 0.0037$.

Alternatively, the standard deviation of the average (mean) is obtained from the standard deviation of a single roll by using Eq. (3.53). Eq. (3.46) gives the standard deviation of a single roll as $\sigma_{single} = \sqrt{pq} = \sqrt{(1/6)(5/6)} = 0.37$. So Eq. (3.53) gives

$$\sigma_{avg} = \frac{\sigma_{single}}{\sqrt{n}} = \frac{0.37}{\sqrt{10,000}} = 0.0037. \tag{3.57}$$

Note that three different σ's have appeared in this problem:

$$\sigma_{single} = 0.37, \qquad \sigma_{tot} = 37, \qquad \sigma_{avg} = 0.0037. \tag{3.58}$$

σ_{tot} is obtained from σ_{single} by multiplying by $\sqrt{n}$ (see Eq. (3.45) or Eq. (3.47)), while σ_{avg} is obtained from σ_{single} by dividing by $\sqrt{n}$ (see Eq. (3.53)). Consistent with these relations, σ_{avg} is obtained from σ_{tot} by dividing by n (see Eq. (3.52)), because averaging involves dividing by n.

(c) You do *not* have a reasonable chance of getting a 6 on at least 20% of the rolls. This is true because 20% of the rolls corresponds to 2000 6's, which is 333 more than the expected number 1667. And 333 is 9 times the standard deviation $\sigma_{tot} = 37$. The probability of a random process ending up at least nine standard deviations above the mean is utterly negligible, as we noted in the discussion preceding this example. Fig. 3.3 shows the probability distribution for the range of $\pm 4\sigma_{tot}$ around the mean. It is clear from the figure that even if we had posed the question with 18% (which corresponds to 1800 rolls, which is about $(3.6)\sigma_{tot}$ above the mean) in place of 20%, the answer would still be that you do not have a reasonable chance of getting a 6 on at least 18% of the rolls. The probability is about 0.016%.

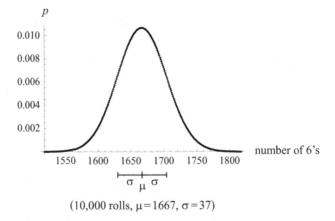

(10,000 rolls, $\mu=1667$, $\sigma=37$)

Figure 3.3: The probability distribution for the number of 6's in 10,000 dice rolls.

Alternatively, we can answer the question by working in terms of percentages. The standard deviation of $\sigma_{avg} = 0.0037$ is equivalent to 0.37%. The difference between 20% and 16.7% is 3.3%, which is 9 times the standard deviation of 0.37%. The probability is therefore negligibly small.

REMARK: Interestingly, the probability curve in Fig. 3.3 looks quite symmetric around the maximum. You might find this surprising, given that the probabilities of rolling a 6 or not a 6 (namely 1/6 and 5/6) aren't equal. In the special case where the two probabilities in the binomial process are both equal to $1/2$ (as they are in Fig. 3.2), it is clear that the probability curve should be symmetric. But when they aren't equal, there isn't a simple argument that explains the symmetry. And indeed, for small numbers of dice rolls, the curve definitely isn't symmetric. And also when the whole range (from 0 to 10,000) is included, the curve definitely isn't symmetric (the bump is off to the left at the 1667 mark). But for large numbers of rolls, the curve is approximately symmetric in the region near the maximum (where the probability is nonnegligible). We'll show in Section 5.1 why this is the case. ♣

3.5 Sample variance

[This section is rather mathematical and can be skipped on a first reading.]

Our goal in this section is to produce an estimate for the standard deviation of a probability distribution, given only a collection of randomly chosen values from the distribution. Up to this point in the book, we have been answering questions involving known distributions. In this section we'll switch things up by starting with some data and then trying to determine the probability distribution (or at least one aspect of it, namely the standard deviation). We are foraging into the subject of statistics here, so this section technically belongs more in a statistics book than a probability book.[1] However, we are including it here partly because it provides a nice excuse to get some practice with expectation values, and partly because students often find the factor of $n - 1$ in the "sample variance" below in Eq. (3.73) mysterious. We hope to remedy this.

Recall that for a given probability distribution (or equivalently, for a given random variable X), the variance is defined in Eq. (3.19) as

$$\text{Var}(X) \equiv E[(X - \mu)^2], \quad \text{where} \quad \mu \equiv E[X]. \tag{3.59}$$

We usually write $\text{Var}(X)$ as σ^2, because the standard deviation σ is defined to be the square root of the variance. As we noted in Eq. (3.35), the variance also takes the (often more convenient) form of $\sigma^2 = E(X^2) - \mu^2$. Since μ is a constant, we were able to take it outside the E operation in the middle term when going from the second to third line in Eq. (3.35). In all of our past calculations, the probability distribution was assumed to be given, so both μ and σ were known quantities.

Consider now a setup where we are working with a probability distribution $P(x)$ that we don't know anything about. It may be discrete, or it may be continuous. We

---

[1]Although the words "probability" and "statistics" are often used interchangeably in a colloquial sense, there is a difference between the two. In a nutshell, the difference comes down to what you are given and what you are trying to find. *Probability* involves using (perhaps after deriving theoretically) a probability distribution to predict the likelihood of future outcomes, whereas *statistics* involves using observed outcomes to deduce the properties of the underlying probability distribution (invariably with the eventual goal of using probability to predict the likelihood of future outcomes). Said in a slightly different way, probability takes you from theory to experiment, and statistics takes you from experiment to theory.

don't know the functional form of $P(x)$, the mean μ, the standard deviation σ, or anything else. These quantities do exist, of course; there is a definite distribution with definite properties. It's just that we don't know what they are. Let's say that we try to calculate the variance σ^2 by picking a random sample of n numbers (call them $x_1, x_2, \ldots, x_n$) and finding their average $\overline{x}$, and then finding the average value of $(x_i - \overline{x})^2$. That is, we calculate

$$\boxed{\tilde{s}^2 \equiv \frac{1}{n} \sum_1^n (x_i - \overline{x})^2} \qquad \text{where} \qquad \overline{x} \equiv \frac{1}{n} \sum_1^n x_i. \qquad (3.60)$$

Note that we cannot use the mean μ of the distribution in place of the average $\overline{x}$ of our n numbers, because we don't know what μ is. Although $\overline{x}$ is likely to be close to μ (if n is large), it is unlikely to be exactly equal to μ, because there will be random effects due to the finite size of n.

A word on notation: The $\tilde{s}^2$ in Eq. (3.60) means exactly the same thing as Var(S) in Eq. (3.37). We have switched notation to $\tilde{s}^2$ simply because it is quicker to write. We are using $\tilde{s}$ instead of σ, so that we don't confuse the sum $(1/n) \sum_1^n (x_i - \overline{x})^2$ with the actual variance σ^2 of the distribution. σ^2 involves a theoretical expectation value over the entire distribution, not just a particular set of n numbers. As with $\overline{x}$ and μ, although $\tilde{s}^2$ is likely to be close to σ^2 (if n is large), it is unlikely to be exactly equal to σ^2. We are using a tilde in $\tilde{s}$ to distinguish it from the plain letter s, which is reserved for the "sample variance" in Eq. (3.73) below. (Some people make this distinction by using an uppercase S for $\tilde{s}$.) We should technically be putting a subscript n on both $\tilde{s}$ and s (and $\overline{x}$), because these quantities depend on the specific set of n numbers. But we have omitted it to keep the calculations below from getting too cluttered.

If we want to reduce the effects of the finite size of n, in order to make $\tilde{s}^2$ be as close as possible to the actual σ^2 of the distribution, there are two reasonable things we can do. First, we can take the $n \to \infty$ limit. This will in fact give the actual σ^2 of the distribution, as we will show below. But let's leave this option aside for now. A second strategy is to imagine picking a huge number N of sets of n numbers from the distribution, and then taking the average of the N values of $\tilde{s}^2$, each of which is itself an average of the n numbers $(x_i - \overline{x})^2$. Will this average in the $N \to \infty$ limit get rid of any effects of the finite size of n and yield the actual σ^2 for the distribution? It turns out, somewhat surprisingly, that it will *not*. Instead it will yield, as we will show below in Theorem 3.5, an average value of $\tilde{s}^2$ equal to

$$\tilde{s}^2_{\text{avg}} = \frac{(n-1)\sigma^2}{n}. \qquad (3.61)$$

For any finite value of n, this expression for $\tilde{s}^2_{\text{avg}}$ is smaller than the actual variance σ^2 of the distribution. But $\tilde{s}^2_{\text{avg}}$ does approach σ^2 in the $n \to \infty$ limit.

Note that when we talk about taking the average of $\tilde{s}^2$ over a huge number $N \to \infty$ of trials, we are equivalently talking about the *expectation value* of $\tilde{s}^2$. This is how an expectation value is defined. This expectation value is (using the fact that the expectation value of the sum equals the sum of the expectation values; see

Eq. (3.7))

$$E[\tilde{s}^2] = E\left[\frac{1}{n}\sum_1^n (X_i - \overline{X})^2\right] = \frac{1}{n}\sum_1^n E\left[(X_i - \overline{X})^2\right], \quad (3.62)$$

where $\overline{X} \equiv (1/n)\sum_1^n X_i$.

In going from Eq. (3.60) to Eq. (3.62) and taking the expectation value of $\tilde{s}^2$, we have made an important change in notation from the lowercase x_i to the uppercase X_i. A lowercase x_i refers to a *specific value* of the random variable X_i (where the X_i are independent and identically distributed random variables, all associated with the given probability distribution). There is nothing random about x_i; it has a definite value. It would therefore be of no use to take the expectation value of $(1/n)\sum_1^n (x_i - \overline{x})^2$. More precisely, we could take the expectation value if we wanted to, but it would simply yield the same definite value of $(1/n)\sum_1^n (x_i - \overline{x})^2$, just as the expectation value of the specific number 173.92 is 173.92. In contrast, the random variable X_i can take on many values, and when taking an expectation value of an expression involving X_i, it is understood that we are averaging over a large number N of trials involving (generally) many different x_i values of X_i.

Before we present the general proof that $E[\tilde{s}^2] = (n-1)\sigma^2/n$, let's demonstrate that this result holds in the special case of $n = 2$, just to make it more believable.

Example: In the $n = 2$ case, show that $E[\tilde{s}^2] = \sigma^2/2$.

Solution: If $n = 2$, then we have two independent and identically distributed random variables, X_1 and X_2. The sum in Eq. (3.62) therefore has only two terms in it, so we obtain

$$E[\tilde{s}^2] = \frac{1}{2}\sum_1^2 E\left[(X_i - \overline{X})^2\right]$$

$$= \frac{1}{2}\left(E\left[\left(X_1 - \frac{X_1 + X_2}{2}\right)^2\right] + E\left[\left(X_2 - \frac{X_1 + X_2}{2}\right)^2\right]\right). \quad (3.63)$$

The terms in parentheses are equal to $\pm(X_1 - X_2)/2$. The overall sign doesn't matter, because these quantities are squared. The two expectation values are therefore the same, so we end up with (using the fact that the expectation value of the sum equals the sum of the expectation values)

$$E[\tilde{s}^2] = \frac{1}{4}E\left[(X_1 - X_2)^2\right] = \frac{1}{4}\left(E[X_1^2] + E[X_2^2] - 2E[X_1 X_2]\right). \quad (3.64)$$

Let's look at the two types of terms here. Since X_1 and X_2 are identically distributed, the $E[X_1^2]$ and $E[X_2^2]$ terms are equal. And from Eq. (3.50) this common value is $E[X^2] = \sigma^2 + \mu^2$. (Remember that σ and μ exist and have definite values, even though we don't know what they are.) For the $E[X_1 X_2]$ term, since X_1 and X_2 are independent variables, Eq. (3.16) tells us that $E[X_1 X_2] = E[X_1]E[X_2] = \mu \cdot \mu = \mu^2$. Plugging these results into Eq. (3.64) gives

$$E[\tilde{s}^2] = \frac{1}{4}\left(2 \cdot (\sigma^2 + \mu^2) - 2\mu^2\right) = \frac{\sigma^2}{2}. \quad (3.65)$$

This is consistent with the $E[\tilde{s}^2] = (n-1)\sigma^2/n$ result (with $n = 2$) that we will show below. We therefore see that if you want to use $E[\tilde{s}^2]$ (when $n = 2$) as an approximation for the actual variance σ^2 of the given distribution, you will be off by a factor of $1/2$. This isn't a very good approximation! We will discuss below why $E[\tilde{s}^2]$ is always an *underestimate* of σ^2, for any finite value of n.

Note that the μ terms canceled out in Eq. (3.65). In retrospect, we know that this must be the case (for any value of n, not just $n = 2$), because the original sum in Eq. (3.62) is independent of μ. This is true because if we shift all of the X_i values by the same amount, then the average $\overline{X}$ also shifts by this amount, so the differences $X_i - \overline{X}$ are unchanged.

Let's now prove the general result. The proof is a bit mathematical, but the final result will be well worth it. As mentioned at the beginning of this section, we'll get some good practice with expectation values here.

Theorem 3.5 *The expectation value of $\tilde{s}^2$ (where $\tilde{s}^2$ is given in Eq. (3.62)) equals $(n-1)/n$ times the actual variance σ^2 of the distribution. That is,*

$$E[\tilde{s}^2] \equiv E\left[\frac{1}{n}\sum_1^n (X_i - \overline{X})^2\right] = \frac{(n-1)\sigma^2}{n} \tag{3.66}$$

Proof: If we expand the square in Eq. (3.66), we obtain

$$E[\tilde{s}^2] = \frac{1}{n}E\left[\sum_1^n X_i^2 - 2\left(\sum_1^n X_i\right)\overline{X} + n\overline{X}^2\right]. \tag{3.67}$$

But $\sum_1^n X_i$ equals $n\overline{X}$, by the definition of $\overline{X}$. We therefore have (using the fact that the expectation value of the sum equals the sum of the expectation values)

$$E[\tilde{s}^2] = \frac{1}{n}\left[\sum_1^n E[X_i^2] - 2E[(n\overline{X})\overline{X}] + nE[\overline{X}^2]\right]. \tag{3.68}$$

As in the above example with $n = 2$, the $E[X_i^2]$ terms are all equal, because the X_i are identically distributed variables. We'll label the common value as $E[X^2]$. We have n such terms, so

$$E[\tilde{s}^2] = \frac{1}{n}\left(nE[X^2] - 2nE[\overline{X}^2] + nE[\overline{X}^2]\right)$$
$$= E[X^2] - E[\overline{X}^2]. \tag{3.69}$$

This result is similar to the result in Eq. (3.35). There is, however, a critical difference. $\overline{X}$ is now a random variable (being the average of the n random variables X_i), whereas the μ in Eq. (3.35) was a constant.

Eq. (3.69) contains two terms that we need to evaluate. The $E[X^2]$ term is simple. From Eq. (3.50) we have

$$E[X^2] = \sigma^2 + \mu^2. \tag{3.70}$$

The $E[\overline{X}^2]$ term is a bit more involved. $\overline{X}$ equals $(1/n)\sum_1^n X_i$, so $\overline{X}^2$ equals $1/n^2$ times the square of the sum of the X_i. When the sum $(X_1 + X_2 + \cdots + X_n)$ is squared, there will be n terms like X_1^2, X_2^2, etc., which are all identically distributed with a common expectation value of $E[X^2]$. And there will be $\binom{n}{2} = n(n-1)/2$ cross terms like $2X_1X_2$, $2X_1X_3$, $2X_2X_3$, etc., which are again all identically distributed, with a common expectation value of, say, $E[X_1X_2]$. We therefore have

$$\begin{aligned} E[\overline{X}^2] &= \frac{1}{n^2}\left(nE[X^2] + \frac{n(n-1)}{2}E[2X_1X_2]\right) \\ &= \frac{1}{n^2}\left(nE[X^2] + n(n-1)E[X_1]E[X_2]\right) \\ &= \frac{1}{n^2}\left(n(\sigma^2 + \mu^2) + n(n-1)\mu^2\right) \\ &= \frac{\sigma^2}{n} + \mu^2. \end{aligned} \tag{3.71}$$

As in the above example, we have used the fact that the X_i's are independent random variables, which allows us to write $E[X_1X_2] = E[X_1]E[X_2] = \mu^2$. But X_i isn't independent of itself, of course. That is why $E[X^2]$ isn't equal to $E[X]E[X] = \mu^2$. Instead, it is equal to $\sigma^2 + \mu^2$.

Substituting Eqs. (3.70) and (3.71) into Eq. (3.69) gives

$$E[\tilde{s}^2] = (\sigma^2 + \mu^2) - \left(\frac{\sigma^2}{n} + \mu^2\right) = \left(\frac{n-1}{n}\right)\sigma^2, \tag{3.72}$$

as desired. As noted in the $n = 2$ example above, the μ dependence drops out. ∎

In Eq. (3.71) we chose to derive the value of $E[\overline{X}^2]$ from scratch by working through some math, because this type of calculation will be helpful if you want to tackle Problem 3.12. However, there is a much quicker way to find $E[\overline{X}^2]$. From Eq. (3.50) we know that the expectation value of the square of a random variable equals the square of the mean plus the square of the standard deviation. With $\overline{X} \equiv (X_1 + X_2 + \cdots + X_n)/n$ as our random variable, the mean is μ, of course. And from Eq. (3.53) the standard deviation is $\sigma/\sqrt{n}$. The $\sigma^2/n + \mu^2$ result in Eq. (3.71) then immediately follows.

Let's recap what the above theorem implies. If you want to determine the true variance σ^2 of an unknown distribution by picking numbers, you have two main options:

- You can pick a huge set of numbers, because in the $n \to \infty$ limit, $\tilde{s}^2$ approaches σ^2. This is due to two effects. First, the $(n-1)/n$ factor in Eq. (3.72) approaches 1 in the $n \to \infty$ limit, so $E[\tilde{s}^2]$ equals σ^2. And second, the result from Problem 3.12 tells us that the spread of the values of $\tilde{s}^2$ around its expected value (which is σ^2 in the $n \to \infty$ limit) goes to zero in the $n \to \infty$ limit. So $\tilde{s}^2$ is essentially guaranteed to be equal to σ^2.

- You can pick a set with a "normal" size n (say, 20, although a small number like 2 will work fine too) and calculate the variance $\tilde{s}^2$ of the set of n numbers.

You can then repeat this process a huge number $N \to \infty$ of times and take the average of the N variances you have calculated. From Eq. (3.53), the standard deviation of this average will be proportional to $1/\sqrt{N}$ and will therefore be very small. The average will therefore be very close to the expected value of $\tilde{s}^2$, which from Eq. (3.72) is $(n-1)\sigma^2/n$. This is always an underestimate of the actual σ^2 of the distribution. But if you multiply by $n/(n-1)$, then you will obtain σ^2.

In the above proof, we proved mathematically that $E[\tilde{s}^2] = (n-1)\sigma^2/n$. But is there an intuitive way of at least seeing why $E[\tilde{s}^2]$ is *smaller* than σ^2, leaving aside the exact $(n-1)/n$ factor? Indeed there is, and in the end it comes down to the fact that $\overline{X}$ is a random variable instead of a constant. In Eq. (3.66) the consequence of this is that if we look at specific x_i values of the X_i random variables, then the $(x_1 - \overline{x})^2$ term, for example, is smaller (on average) than $(x_1 - \mu)^2$. This is true because $\overline{x}$ involves x_1, which implies that if x_1 is, say, large, then the mean will be shifted upward slightly toward x_1. (The average of the *other* $n-1$ numbers equals μ, on average. So the average of *all* of the n numbers including x_1 must lie a little closer to x_1, on average.) This effect is most pronounced for small n, such as $n = 2$. Another line of reasoning involves looking at Eqs. (3.71) and (3.72). $E[\overline{X}^2]$ is larger than μ^2 (by an amount σ^2/n), due to the fact that the value of $\overline{X}$ generally differs slightly from μ. The square of this difference contributes to $E[\overline{X}^2]$; see the paragraph immediately following the proof. So a number larger than μ^2 is subtracted off in Eq. (3.72).

As mentioned above, a quick corollary of Theorem 3.5 is that if we multiply $E[\tilde{s}^2]$ by $n/(n-1)$, we obtain the actual variance σ^2 of the distribution. This suggests that we might want to define a new quantity that is a slight modification of the $\tilde{s}^2$ in Eq. (3.60). We'll label it as s^2:

$$s^2 \equiv \left(\frac{n}{n-1}\right)\tilde{s}^2 = \frac{1}{n-1}\sum_1^n (x_i - \overline{x})^2 \qquad \text{(sample variance)} \qquad (3.73)$$

This quantity s^2 is called the *sample variance*. We'll discuss this terminology below. s^2 is a function of a particular set of n numbers, x_1 through x_n, just as $\tilde{s}^2$ is. But the expectation value of s^2 doesn't depend on n. The combination of Eqs. (3.72) and (3.73) tells us that the expectation value of s^2 is simply σ^2:

$$E[s^2] = \sigma^2 \qquad (3.74)$$

Our original quantity $\tilde{s}^2$ is a *biased estimator* of σ^2, in that its expectation value $E[\tilde{s}^2]$ depends on n and is smaller than σ^2 by the factor $(n-1)/n$. Our new quantity s^2 is an *unbiased estimator* of σ^2, in that its expectation value $E[s^2]$ is independent of n and equals σ^2. To summarize, the two quantities

$$\tilde{s}^2 \equiv \frac{1}{n}\sum_1^n (x_i - \overline{x})^2 \qquad \text{and} \qquad s^2 \equiv \frac{1}{n-1}\sum_1^n (x_i - \overline{x})^2 \qquad (3.75)$$

have expectation values of

$$E[\tilde{s}^2] = \frac{(n-1)\sigma^2}{n} \quad \text{and} \quad E[s^2] = \sigma^2. \tag{3.76}$$

A word on terminology: The quantity $\tilde{s}^2$ is called the "variance" of a particular set of n numbers, while the quantity s^2 is called the "sample variance" of the set. When talking about the sample variance, it is understood that you are concerned with producing an estimate of the actual variance of the underlying distribution. (This variance is often called the *population variance*, in view of the fact that it takes into account the entire population of possible outcomes, as opposed to just a sample of them.) The sample variance s^2 has the correct expectation value of σ^2. However, this terminology can get a little tricky. What if someone asks you to compute the variance of a sample of n numbers? Even though the word "sample" is used here, you should calculate $\tilde{s}^2$, because you are being asked to compute the *variance* of a set/sample of numbers, and the variance is defined via Eq. (3.37) or Eq. (3.60), with an n in the denominator. If someone actually wants you to compute the *sample variance*, then they should use this specific term, which is defined to be s^2, with an $(n-1)$ in the denominator. Of course, any ambiguity in terminology can be eliminated by simply using the appropriate symbol ($\tilde{s}^2$ or s^2) in addition to words.

Terminology aside, which of $\tilde{s}^2$ or s^2 *should* you be concerned with if you are given a set of n numbers? Well, if you are concerned only with these particular n numbers and nothing else (in particular, the underlying distribution, if the numbers came from one), then you should calculate $\tilde{s}^2$. This is the variance of these numbers.[2] But if the n numbers come from a distribution or a larger population, *and* if you are concerned with making a statement about this distribution or population, then you should calculate s^2, because this gives an unbiased estimate of σ^2. However, having said this, it is often the case that n is large enough so that the distinction between the n and the $n-1$ in the denominators of $\tilde{s}^2$ and s^2 doesn't matter. To summarize, the three related quantities we have encountered are:

σ^2: Distribution variance, or population variance.

$\tilde{s}^2$: Variance of a set of n numbers (a biased estimator of σ^2).

s^2: Sample variance of a set of n numbers (an unbiased estimator of σ^2).

Example ($n = 100$): We proved Theorem 3.5 mathematically, but let's now give some numerical evidence that $E[s^2]$ is in fact equal to σ^2. We'll arbitrarily choose $n = 100$. To demonstrate $E[s^2] = \sigma^2$, we'll numerically generate $N = 10^5$ sets of $n = 100$ values from a Gaussian (normal) distribution with $\mu = 0$ and $\sigma = 1$. (Eq. (3.74) holds for any type of distribution, so our Gaussian choice isn't important. We'll discuss the

---

[2]If you are concerned only with this set of numbers and nothing else, then you can rightly call the set a "population," in which case you can rightly call $\tilde{s}^2$ a "population" variance. But we'll just call it $\tilde{s}^2$.

Gaussian distribution in Section 4.8.) The μ value here is irrelevant, as we have noted. The $N = 10^5$ number is large enough so that we'll pretty much obtain the expectation value $E[s^2]$; see the second remark below.

The results of a numerical run are shown in Fig. 3.4. For each of the $N = 10^5$ sets of $n = 100$ values, we calculated the s^2 given by Eq. (3.73). The histogram gives the distribution of the N values of s^2. The average of these N values is 1.00062, which is very close to $\sigma^2 = 1$, consistent with Eq. (3.74).

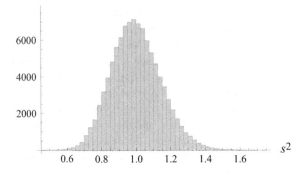

Figure 3.4: A histogram of the sample variances s^2 of $N = 10^5$ sets of numbers, with each set consisting of $n = 100$ numbers chosen from a Gaussian distribution with $\sigma = 1$.

REMARKS:

1. If you are interested in calculating the spread of the histogram, see Problem 3.12. A corollary to that problem is that if the underlying probability distribution is Gaussian (so now the Gaussian assumption matters), and if n is large, then $\text{Var}(s^2) \approx 2\sigma^4/n$. In the present setup with $\sigma = 1$ and $n = 100$, this gives $\text{Var}(s^2) \approx 0.02$. The standard deviation of the s^2 values is therefore about $\sqrt{0.02} \approx 0.14$. This is consistent with a visual inspection of the histogram.

2. If we make N larger (say, 10^6 or 10^7), the spread of the histogram remains the same. The standard deviation is still 0.14, because the variance $\text{Var}(s^2) \approx 2\sigma^4/n$ depends only on n, not on N. So the histogram will look the same. As far as the average value of s^2 (which is 1.00062 for the data in Fig. 3.4) goes, a larger N means that it is more likely to be very close to $\sigma^2 = 1$, due to the result in Eq. (3.53) for the standard deviation of the mean. (This effect is too small to see, so the histogram will still look the same.) Remember that the standard deviation of the average of N independent and identically distributed variables (the N values of s^2 here) is always smaller than the standard deviation of each of the variables (which is $\sqrt{2\sigma^4/n}$ here), by a factor of $1/\sqrt{N}$.

   In the present case, the σ in Eq. (3.53) is 0.14, and the n there is now N. If $N = 10^5$, then Eq. (3.53) says that the standard deviation of the average of the N values of s^2 is $(0.14)/\sqrt{10^5} = 4.4 \cdot 10^{-4}$. Our above numerical result of 1.00062 for the average of s^2 is therefore about one and a half standard deviations from the expected value ($\sigma^2 = 1$), which is quite reasonable. Larger values of N will cause the average of the N values of s^2 to be even closer to 1 (on average). ♣

If you want to produce an estimate of the *standard deviation* σ of a distribution, there are two things you might want to do. You can pick a value of n (say, 10) and then take the average of a large number N (say, a million) of values of the sample variance s^2 of n numbers. For very large N, this will essentially give you σ^2, by Eq. (3.74). You can then take the square root to obtain σ. This is a valid method for obtaining σ. Or, you can take the average of a large number N of values of the sample standard deviation s, each of which is the square root of the s^2 given in Eq. (3.73). However, this second method will *not* give you σ. Your calculated average will be *smaller* than σ; see Problem 3.11. Therefore, although Eq. (3.74) tells us that the sample variance s^2 is an unbiased estimator of the distribution variance σ^2, it is *not* true that the sample standard deviation s is an unbiased estimator of the distribution standard deviation σ.

3.6 Summary

- A *random variable* is a variable that can take on certain numerical values with certain probabilities. A random variable is denoted with an uppercase letter, such as X, while the actual values that the variable can take on are denoted with lowercase letters, such as x.

- The *expectation value* of a random variable X is the expected average value of the variable, over a large number of trials. It is given by

$$E(X) = p_1 x_1 + p_2 x_2 + \cdots + p_m x_m, \tag{3.77}$$

where the x's are the possible outcomes and the p's are the associated probabilities. The expectation value of the sum of two variables equals the sum of the individual expectation values:

$$E(X + Y) = E(X) + E(Y). \tag{3.78}$$

The expectation value of the product of two *independent* variables equals the product of the individual expectation values:

$$E(XY) = E(X) \cdot E(Y) \quad \text{(independent variables)} \tag{3.79}$$

- The *variance* of a random variable is related to the spread of the possible outcomes of the variable. It is given by

$$\text{Var}(X) \equiv E\left[(X - \mu)^2\right]. \tag{3.80}$$

It can also be written as $\text{Var}(X) = E(X^2) - \mu^2$. The variance of the sum of two *independent* variables equals the sum of the individual variances:

$$\text{Var}(X + Y) = \text{Var}(X) + \text{Var}(Y) \quad \text{(independent variables)} \tag{3.81}$$

The variance of the number of Heads in n tosses of a biased coin (involving probabilities p and $1 - p \equiv q$) is

$$\text{Var}(\text{Heads in } n \text{ flips}) = npq \quad \text{(biased coin)} \tag{3.82}$$

The variance of a set S of n numbers x_i is

$$\tilde{s}^2 \equiv \text{Var}(S) \equiv \frac{1}{n} \sum_1^n (x_i - \overline{x})^2. \tag{3.83}$$

- The *standard deviation* of a random variable gives a rough measure of the spread of the possible outcomes of the variable. It is defined as the square root of the variance, so we can write it in two ways:

$$\sigma_X \equiv \sqrt{\text{Var}(X)} = \sqrt{E[(X - \mu)^2]} = \sqrt{E(X^2) - \mu^2}. \tag{3.84}$$

If X and Y are two *independent* random variables, then

$$\sigma_{X+Y}^2 = \sigma_X^2 + \sigma_Y^2 \qquad \text{(independent variables)} \tag{3.85}$$

This is the statement that standard deviations "add in quadrature" for independent variables. The standard deviation of the number of Heads in n tosses of a biased coin (involving probabilities p and $1 - p \equiv q$) is

$$\sigma = \sqrt{npq} \qquad \text{(biased coin)} \tag{3.86}$$

- The *standard deviation of the mean* is the standard deviation of the average of a set of n random variables. If each of the random variables has the same standard deviation σ, then the standard deviation of their average equals

$$\sigma_{\overline{X}} = \frac{\sigma}{\sqrt{n}} \qquad \text{(standard deviation of the mean)} \tag{3.87}$$

- The *sample variance* s^2 of a set of n numbers x_i chosen from a given distribution is defined as

$$s^2 \equiv \frac{1}{n-1} \sum_1^n (x_i - \overline{x})^2. \tag{3.88}$$

The sample variance has the property that its expected value equals the actual variance σ^2 of the distribution.

3.7 Exercises

See **www.people.fas.harvard.edu/~djmorin/book.html** for a supply of problems without included solutions.

3.8 Problems

Section 3.1: Expectation value

3.1. **Flip until Heads** ∗

In Example 2 on page 136, we found that if you flip a coin until you get a Heads, the expectation value of the total number of coins is

$$\frac{1}{2} \cdot 1 + \frac{1}{4} \cdot 2 + \frac{1}{8} \cdot 3 + \frac{1}{16} \cdot 4 + \frac{1}{32} \cdot 5 + \cdots . \qquad (3.89)$$

We claimed that this sum equals 2. Demonstrate this by writing the sum as a geometric series starting with 1/2, plus another geometric series starting with 1/4, and so on. You can use the fact that the sum of a geometric series with first term a and ratio r is $a/(1-r)$.

3.2. **HT waiting time** ∗∗

We know from Example 2 on page 136 that the expected number of flips required to obtain a Heads is 2. What is the expected number of flips required to obtain a Heads and a Tails in succession (in that order)?

3.3. **Sum of dependent variables** ∗∗

Consider the example on page 137, but now let X and Y be dependent in the following manner: If $Y = 1$, then it is always the case that $X = 1$. If $Y = 2$, then it is always the case that $X = 2$. If $Y = 3$, then there are equal chances of X being 1 or 2. If we assume that Y takes on the values 1, 2, and 3 with equal probabilities of $1/3$, then you can quickly show that X takes on the values 1 and 2 with equal probabilities of $1/2$. So we have reproduced the probabilities in the original example. Show (by explicitly calculating the probabilities of the various outcomes) that in the present scenario where X and Y are dependent, the relation $E(X + Y) = E(X) + E(Y)$ still holds.

3.4. **Playing "unfair" games** ∗∗

(a) Assume that later on in life, things work out so that you have more than enough money in your retirement savings to take care of your needs and beyond, and that you truly don't have a need for any more money. Someone offers you the chance to play a one-time game where you have a 3/4 chance of doubling your money, and a 1/4 chance of losing it all. If you initially have N dollars, what is the expectation value of your resulting amount of money if you play the game? Would you want to play it?

(b) Assume that you are stranded somewhere, and that you have only $10 for a $20 bus ticket. Someone offers you the chance to play a one-time game where you have a 1/4 chance of doubling your money, and a 3/4 chance of losing it all. What is the expectation value of your resulting amount of money if you play the game? Would you want to play it?

3.5. Simpson's paradox **

During the baseball season in a particular year, player A has a higher batting average than player B. In the following year, A again has a higher average than B. But to your great surprise when you calculate the batting averages over the combined span of the two years, you find that A's average is *lower* than B's! Explain, by giving a concrete example, how this is possible.

Section 3.2: Variance

3.6. Variance of a product *

Let X and Y each be the result of independent (and fair) coin flips where we assign the value 1 to Heads and 0 to Tails. Show that $\text{Var}(XY)$ is not equal to $\text{Var}(X)\text{Var}(Y)$.

3.7. Variances *

For each of the three examples near the beginning of Section 3.2, show that the alternative $E(X^2) - \mu^2$ form of the variance given in Eq. (3.34) leads to the same results we obtained in the examples.

Section 3.3: Standard deviation

3.8. Random walk **

Consider the following one-dimensional random walk. A person starts at the origin and then takes n successive steps. Each step is equally likely to be to the right or to the left. All steps have the same length.

(a) What is the probability that the person is located back at the origin after the nth step?

(b) After n steps, what is the standard deviation of the person's position relative to the origin? (Assume that the length of each step is, say, one foot.)

Section 3.4: Standard deviation of the mean

3.9. Expected product, without replacement **

Consider a set of N given numbers, $a_1, a_2, \ldots, a_N$. Let the mean of these N numbers be μ, and let the standard deviation be σ. Draw two numbers X_1 and X_2 randomly *without replacement*. Show that the expectation value of their product is

$$E[X_1 X_2] = \mu^2 - \frac{\sigma^2}{N-1}. \tag{3.90}$$

Hint: All of the $a_i a_j$ possibilities (with $i \neq j$) are equally likely.

3.10. Standard deviation of the mean, without replacement ***

Consider a set of N given numbers, $a_1, a_2, \ldots, a_N$. Let the mean of these N numbers be μ, and let the standard deviation be σ. Draw a sample of n numbers X_i randomly *without replacement*, and calculate their sample mean,

$\sum X_i/n$. (n must be less than or equal to N, of course.) The variance of the sample mean is $E[(\sum X_i/n - \mu)^2]$. Show that this variance is given by

$$E\left[\left(\frac{\sum X_i}{n} - \mu\right)^2\right] = \frac{\sigma^2}{n}\left(1 - \frac{n-1}{N-1}\right). \tag{3.91}$$

The standard deviation of the sample mean is the square root of this. The result from Problem 3.9 will come in handy.

Section 3.5: Sample variance

3.11. **Biased sample standard deviation** ∗∗

We mentioned on page 163 that the sample standard deviation s is a *biased* estimator of the distribution standard deviation σ. The basic reason for this is that the square root operation is nonlinear, which means that the square root of the average of a set of numbers isn't equal to the average of their square roots. For example, the average of 1.1 and 0.9 is 1, but the average of $\sqrt{1.1}$ and $\sqrt{0.9}$ isn't 1. It is smaller than 1. Let's give a general proof that $E[s] \le \sigma$ (unlike $E[s^2] = \sigma^2$).

If we calculate the sample variances for a large number N of sets of n numbers, then the $E[s^2] = \sigma^2$ equality in Eq. (3.74) tells us that in the $N \to \infty$ limit, we have

$$\frac{s_1^2 + s_2^2 + \cdots + s_N^2}{N} = \sigma^2. \tag{3.92}$$

Our goal is to show that

$$\frac{s_1 + s_2 + \cdots + s_N}{N} \le \sigma, \tag{3.93}$$

in the $N \to \infty$ limit. To demonstrate this, square both sides of Eq. (3.93) and make copious use of the arithmetic-geometric-mean inequality, $\sqrt{ab} \le (a+b)/2$.

3.12. **Variance of the sample variance** ∗∗∗

Consider the sample variance s^2 (given in Eq. (3.73)) of a sample of n values, X_1 through X_n, chosen from a distribution with standard deviation σ and mean μ. We know from Eq. (3.74) that the expectation value of s^2 is σ^2, so the variance of s^2 (that is, the variance of the sample variance) is $\text{Var}(s^2) = E[(s^2 - \sigma^2)^2]$. The square root of this variance gives a measure of the spread of the results if you calculate s^2 for many different sets of n numbers (as we did in Fig. 3.4). Show that $\text{Var}(s^2)$ equals

$$\text{Var}(s^2) = \frac{1}{n}\left[\mu_4 - \sigma^4\left(\frac{n-3}{n-1}\right)\right], \tag{3.94}$$

where μ_4 is the distribution's fourth moment relative to the mean, that is, $\mu_4 \equiv E[(X-\mu)^4]$. The math here is extremely tedious, so you should attempt this problem only if you *really* enjoyed the proof of Theorem 3.5. Whatever adjective comes to mind for that proof, multiply it by 10 for this problem!

3.13. Sample variance for two dice rolls **

(a) We know from the first example in Section 3.2 that the variance of a single die roll is $\sigma^2 = 2.92$. If you use Eq. (3.73) to calculate the sample variance s^2 for $n = 2$ dice rolls, the expected value of s^2 should be $\sigma^2 = 2.92$, according to Eq. (3.74). By considering the 36 equally likely pairs of dice in Table 1.5, verify that this is indeed the case.

(b) Using the information you generated from Table 1.5, calculate $\mathrm{Var}(s^2)$. Then show that the result agrees with the expression for $\mathrm{Var}(s^2)$ in Eq. (3.94), with $n = 2$.

3.9 Solutions

3.1. Flip until Heads

The given sum equals

$$
\begin{aligned}
&\frac{1}{2} + \frac{1}{4} + \frac{1}{8} + \frac{1}{16} + \frac{1}{32} + \cdots \\
&\quad + \frac{1}{4} + \frac{1}{8} + \frac{1}{16} + \frac{1}{32} + \cdots \\
&\qquad + \frac{1}{8} + \frac{1}{16} + \frac{1}{32} + \cdots \\
&\qquad\quad + \frac{1}{16} + \frac{1}{32} + \cdots \\
&\qquad\qquad \vdots
\end{aligned}
\tag{3.95}
$$

This has the correct number of each type of term. For example, a "1/16" appears four times. The first line is a geometric series that sums to $a/(1-r) = (1/2)/(1-1/2) = 1$. The second line is also a geometric series, and it sums to $(1/4)/(1 - 1/2) = 1/2$. Likewise the third line sums to $(1/8)/(1 - 1/2) = 1/4$. And so on. The sum of the infinite number of lines in Eq. (3.95) therefore equals

$$
1 + \frac{1}{2} + \frac{1}{4} + \frac{1}{8} + \frac{1}{16} + \frac{1}{32} + \cdots .
\tag{3.96}
$$

But this itself is a geometric series, and it sums to $a/(1 - r) = 1/(1 - 1/2) = 2$, as desired.

3.2. HT waiting time

Our goal is to find the average number of flips to obtain an HT pair (including these two flips). We know that the average number of flips to obtain an H is 2. The important point to now realize is that once we obtain our first H, the game ends when we eventually obtain a T. This is true because if we obtain a T on the following flip, then we have obtained our HT, so we're done. If, on the other hand, we obtain a T, say, four flips later (that is, if we obtain three more H's and then a T), then our string looks like ...HHHHT, so we have obtained our HT pair. Basically, in any scenario, once we've obtained our first H, the first subsequent appearance of a T, whenever that may be, must necessarily follow an H, which means that we have obtained our HT pair.

We can therefore answer the original HT question if we can answer the question: How many flips on average does it take to obtain a T, following an H? Now, since H and T are interchangeable, this is exactly the same question as: How many flips on average does it take to obtain an H, starting at the beginning? (This is true because future flips can't depend on past flips. So we can imagine starting the process whenever we want. Starting after the first H is as valid a place to start as the actual beginning.) We already know that the answer to this question is 2. The average number of flips to obtain an HT string is therefore 2+2=4. It takes an average of two flips to obtain an H, and then an average of two more flips to obtain a T, at which point we necessarily have our HT sequence, as we noted above.

3.3. **Sum of dependent variables**

In the (X, Y) notation, the given information tells us that there is a 1/3 chance of obtaining $(1,1)$, a 1/3 chance of obtaining $(2,2)$, a 1/6 chance of obtaining $(1,3)$, and a 1/6 chance of obtaining $(2,3)$. Both $(2,2)$ and $(1,3)$ yield a sum of 4, so the probabilities of the various values of $X + Y$ are

$$P(2) = \frac{1}{3}, \quad P(3) = 0, \quad P(4) = \frac{1}{3} + \frac{1}{6} = \frac{1}{2}, \quad P(5) = \frac{1}{6}. \tag{3.97}$$

Eq. (3.4) then gives the expectation value of $X + Y$ as

$$E(X + Y) = \frac{1}{3} \cdot 2 + 0 \cdot 3 + \frac{1}{2} \cdot 4 + \frac{1}{6} \cdot 5 = \frac{21}{6} = 3.5. \tag{3.98}$$

This equals $E(X) + E(Y) = 1.5 + 2 = 3.5$, as Eq. (3.7) claims.

3.4. **Playing "unfair" games**

(a) The expectation value of your money after you play the game is $(3/4) \cdot 2N + (1/4) \cdot 0 = 3N/2$. So you will gain $N/2$ dollars, on average. It therefore seems like it would be a good idea to play the game. However, further thought shows that it would actually be a bad idea. There is basically no upside; you already have plenty of money, so twice the money won't help much. But there is a huge downside; you might lose all your money, and that would certainly be a bad thing.

The point here is that the important issue is your happiness, not the exact amount of money you have. On the happiness scale (from 0 to 1), you stand to gain nothing (or perhaps a tiny bit). Your happiness starts pretty much at 1, and even if you win the game, you can't climb any higher than 1. But you stand to lose a huge amount. This isn't to say that you can't be happy without money. But if you lose your entire savings, there's no doubt that it would put a damper on things. Let's assume that if you lose the game, your happiness decreases roughly to 0. Then if you play the game, the expectation value of your happiness is essentially $(3/4) \cdot 1 + (1/4) \cdot 0 = 3/4$. This is less than the starting value of 1, so it suggests that you shouldn't play the game. However, there is still another thing to consider; see the remark below.

(b) The expectation value of your money after you play the game is $(3/4) \cdot 0 + (1/4) \cdot 20 = 5$. So you will lose $5, on average. It therefore seems like it would be a bad idea to play the game. However, the $10 in your pocket is just as useless as $0, because either way, you're guaranteed to be stuck at the bus station. You therefore *should* play the game. That way, at least there's a 1/4 chance that you'll make it home. (We'll assume that the overall money you have back home washes out any effect of gaining or losing $10, in the long run.) The same argument we used above with the happiness level holds here. $0 and $10 yield the same level of happiness (or perhaps we should say misery), so there is basically no downside. But there is definitely an upside with the $20, because you can then buy a ticket. The expectation value of your happiness (on a scale from 0 to 1) is essentially $(3/4) \cdot 0 + (1/4) \cdot 1 = 1/4$. This is greater than the starting value of 0, so it suggests that you should play the game. But see the following remark.

REMARK: There is another consideration with these sorts of situations, in that they are *one-time events*. Even if we rig things so that the expectation value of your happiness level (or whatever measure you deem to be the important one) increases, it's still not obvious that you should play the game. Just as with any other probabilistic quantity, the expectation value has meaning only in the context of a *large number of identical trials*. You could imagine a situation where a group of many people play a particular game and the average happiness level increases. But *you* are only *one* person, and the increase in the overall happiness level of the group is of little comfort to you if you lose your shirt. Since you play the game only once, the expectation value is irrelevant to you. The decision mainly comes down to an assessment of the risk. Different people's reactions to risk are different, and you could imagine someone being very risk-averse and never playing a game with a significant downside, no matter what the upside is. ♣

3.5. Simpson's paradox

The two tables in Table 3.1 show an extreme scenario that gets to the heart of the matter. In the first year, player A has a small number of at-bats (6), while player B has a large number (600). In the second year, these numbers are reversed. You should examine these tables for a minute to see what's going on, before reading the next paragraph.

	First year	Second year
Player A	3/6 (.500)	150/600 (.250)
Player B	200/600 (.333)	1/6 (.167)

	Combined years
Player A	153/606 (.252)
Player B	201/606 (.332)

Table 3.1: Yearly and overall batting averages. The years with the large numbers of at-bats dominate the overall averages.

The main point to realize is that in the combined span of the two years, A's average is dominated by the .250 average coming from the large number of at-bats in the second year (yielding an overall average of .252, very close to .250), whereas B's average is dominated by the .333 average coming from the large number of at-bats in the first year (yielding an overall average of .332, very close to .333). B's .333 is lower than A's .500 in the first year, but that is irrelevant because A's very small number of at-bats that year hardly affects his overall average. Similarly, B's .167 is lower than A's .250 in the second year, but again, that is irrelevant because B's very small number of at-bats that year hardly affects his overall average. What matters is that B's .333 in the first year is higher than A's .250 in the second year. The large numbers of associated at-bats dominate the overall averages.

Fig. 3.5 shows a visual representation of the effect of the number of at-bats. The size of a data point in the figure gives a measure of the number of at-bats. So although B's average is lower than A's in each year, the large B data point in the first year is higher than the large A data point in the second year. These data points are what dominate the overall averages.

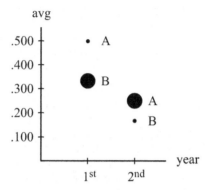

Figure 3.5: Visual representation of Simpson's paradox. The large data points dominate the overall averages.

REMARKS:

1. To generate the paradox where B's overall average surprisingly ends up being higher than A's overall average, the higher of B's two yearly averages must be higher than the lower of A's two yearly averages. If this weren't the case (that is, if the large B data point in Fig. 3.5 were lower than the large A data point), then A's overall average would necessarily be higher than B's overall average (as you can verify). So the paradox wouldn't be realized.

2. To generate the paradox, we must also have a disparity in the number of at-bats. If all four of the yearly at-bats in the first of the tables in Table 3.1 were the same, then A's overall average would necessarily be higher than B's overall average (as you can verify). The main point of the paradox is that when calculating the overall average for a given player, we can't just take the averages of the two averages. A year with more at-bats influences the average more than a year with fewer at-bats, as we saw above.

   The paradox can certainly be explained with at-bats that don't have values as extreme as 6 and 600, but we chose these in order to make the effect as clear as possible. Also, we chose the total number of at-bats in the above example to be the same for A and B over the two years, but this of course isn't necessary.

3. The paradox can also be phrased in terms of averages on exams, for example: For 10th graders taking a particular test, boys have a higher average than girls. For 11th graders taking the same test, boys again have a higher average than girls. But for the 10th and 11th graders combined, girls have a higher average than boys. Another real-life example deals with college admissions rates. The paradox can arise when looking at male/female acceptance rates to individual

departments, and then looking at the male/female acceptance rates to the college as a whole. (The departments are analogous to the different baseball years.)

4. One shouldn't get carried away with Simpson's paradox. There are plenty of scenarios where it doesn't apply, for example: In a particular school, the percentage of soccer players in the 10th grade is larger than the percentage of musicians. And the percentage of soccer players in the 11th grade is again larger than the percentage of musicians. Can the overall percentage of soccer players (in the combined grades) be smaller than the overall percentage of musicians? The answer to this question is a definite "No." One way to see why is to consider the *numbers* of soccer players and musicians, instead of the *percentages*. Since there are more soccer players than musicians in each grade, the total number (and hence percentage) of soccer players must be larger than the total number (and hence percentage) of musicians.

Another way to understand the "No" answer is to note that when calculating the percentages of soccer players and musicians in a given grade, we're dividing the number of students in each group by the *same* denominator (namely, the total number of students in the grade). We therefore can't take advantage of the effect in the baseball scenario above, where B's average was dominated by one year while A's was dominated by a different year, due to the different numbers of at-bats in a given year. Instead of the data points in Fig. 3.5, the present setup might yield something like the data points in Fig. 3.6. The critical feature here is that the dots in each year have the *same* size. The dots for the 11th grade happen to be larger because we're arbitrarily assuming that there are more students in that grade. The total percentage of soccer players in the two years is the weighted average of the two soccer dots (weighted by the size of the dots, or equivalently by the number of students in each grade). Likewise for the two music dots. The soccer weighted average is necessarily larger than the music weighted average. (This is fairly clear intuitively, but as an exercise you can prove it rigorously if you have your doubts.) ♣

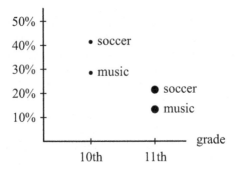

Figure 3.6: Simpson's paradox doesn't apply in this case. The overall percentage of soccer players is necessarily larger than the overall percentage of musicians.

3.6. **Variance of a product**

The random variable XY takes on the values of 0 and 1 with probabilities $1/4$ and $3/4$, because only HH yields an XY value of 1. The other three outcomes (HT, TH, TT) all yield 0. We therefore effectively have a single biased coin with probability $p = 1/4$ of obtaining a value of 1, and $q = 3/4$ of obtaining a value of 0. Eq. (3.33) then tells us that the variance of XY is $npq = 1 \cdot (1/4) \cdot (3/4) = 3/16$. And we know from Eq. (3.21) (or Eq. (3.33)) that the variance of each single (fair) coin flip is $\mathrm{Var}(X) = \mathrm{Var}(Y) = 1/4$. So $\mathrm{Var}(X)\mathrm{Var}(Y) = (1/4)(1/4) = 1/16$, which is not equal to $\mathrm{Var}(XY) = 3/16$.

3.7. **Variances**

For a die roll, we have

$$E(X^2) = \frac{1}{6}\left(1^2 + 2^2 + 3^2 + 4^2 + 5^2 + 6^2\right) = \frac{91}{6} = 15.17. \qquad (3.99)$$

And $\mu = 3.5$, so the variance is $E(X^2) - \mu^2 = 15.17 - 3.5^2 = 2.92$, as desired.
For a fair coin flip, we have

$$E(X^2) = \frac{1}{2}\left(1^2 + 0^2\right) = \frac{1}{2}. \qquad (3.100)$$

And $\mu = 1/2$, so the variance is $E(X^2) - \mu^2 = 1/2 - (1/2)^2 = 1/4$, as desired.
For a biased coin flip, we have

$$E(X^2) = p \cdot 1^2 + (1 - p) \cdot 0^2 = p. \qquad (3.101)$$

And $\mu = p$, so the variance is $E(X^2) - \mu^2 = p - p^2 = p(1 - p) \equiv pq$, as desired.

3.8. **Random walk**

(a) If the person ends up back at the origin after n steps, then it must be the case that $n/2$ of the steps were to the right and $n/2$ were to the left. (Note that this immediately tells us that n must be even, if there is to be any chance of ending up at the origin.) You can imagine the person flipping a coin, with Heads meaning a step to the right and Tails meaning a step to the left. So the given problem is equivalent to finding the probability that you obtain equal numbers $n/2$ of Heads and Tails in n coin flips. There is a total of 2^n possible outcomes (all equally likely) for the collection of n flips, and $\binom{n}{n/2}$ of these have exactly $n/2$ each of Heads and Tails. So the probability of obtaining exactly $n/2$ Heads is

$$p = \frac{1}{2^n}\binom{n}{n/2} = \frac{1}{2^n} \cdot \frac{n!}{((n/2)!)^2}. \qquad (3.102)$$

If n is even, this is the desired probability of ending up back at the origin. If n is odd, the probability is zero.

If n is large, then Eq. (2.66) gives an approximation to Eq. (3.102). The n in Eq. (2.66) corresponds to $n/2$ here, so the above result becomes $1/\sqrt{\pi(n/2)} = \sqrt{2/\pi n}$. For example, after $n = 100$ steps, the probability of being at the origin is about 8%.

(b) FIRST SOLUTION: Consider a single step, with the two possible outcomes of $+1$ and -1 (in feet). The mean displacement during the step is $\mu = 0$, so the first (or second) expression in Eq. (3.40) gives the standard deviation of a single step as

$$\sigma_1 = \sqrt{E[(X - \mu)^2]} = \sqrt{(1/2) \cdot (1)^2 + (1/2) \cdot (-1)^2} = 1. \qquad (3.103)$$

This makes sense, because the square of the length of a single step is guaranteed to be 1. The standard deviation of n independent steps (involving identical 50-50 processes) is then given by Eq. (3.45) as

$$\sigma_n = \sqrt{n} \cdot \sigma_1 = \sqrt{n}. \qquad (3.104)$$

REMARK: Since our random walk is basically the same as a series of coin flips (with Heads and Tails corresponding to 1 and -1 instead of the usual 1 and 0 that we have used in the past), the probability distribution for where the person is after n steps has the same basic shape as the binomial distribution in, say, the fourth plot in Fig. 3.1. In particular, we can make the same types of statements we made right before the example in Section 3.4. For example, assuming that the number of steps is large, there is a 99.7% chance that after n steps the person is within $3\sigma_n = 3\sqrt{n}$ of the origin. So for $n = 10,000$ steps, the person is 99.7% likely to be within $3\sqrt{n} = 300$ steps of the origin. ♣

SECOND SOLUTION: We can solve the problem from scratch, without invoking Eq. (3.45). This method allows us to see intuitively where the $\sigma_n = \sqrt{n}$ result comes from. Let the n steps be represented by the independent and identically distributed random variables X_i. Each X_i can take on the value of $+1$ or -1, with equal probabilities. Let Z be the sum of the X_i. So Z is the position after the n steps. We then have (see below for an explanation of these equations)

$$Z = X_1 + X_2 + X_3 + \cdots + X_n$$
$$\implies Z^2 = (X_1 + X_2 + X_3 + \cdots + X_n)^2$$
$$\implies Z^2 = (X_1^2 + X_2^2 + \cdots + X_n^2) + (\text{cross terms, like } 2X_1 X_2)$$
$$\implies Z^2 = (1 + 1 + \cdots + 1) + (\text{cross terms})$$
$$\implies E[Z^2] = n + E[\text{cross terms}]$$
$$\implies \sigma_n^2 = n + 0$$
$$\implies \sigma_n = \sqrt{n}. \qquad (3.105)$$

The second line is the square of the first. In the third line we expanded the square to obtain n "diagonal" terms X_i^2, along with $\binom{n}{2}$ cross terms $2X_i X_j$. In the fourth line we used the fact that since $X_i = \pm 1$, its square is always 1. The fifth line is the expectation value of the fourth line. To obtain the sixth line, we used the fact that since the mean value of Z is $\mu = 0$, Eq. (3.50) gives $E[Z^2] = \sigma_n^2$. And we also used the fact that the expectation value of the product $X_i X_j$ (with $i \neq j$) equals zero. This is true because X_i and X_j are independent variables, so Eq. (3.16) tells us that $E[X_i X_j] = E[X_i]E[X_j]$. And these individual expectation values are zero. The standard deviation is then $\sigma_n = \sqrt{n}$, as the seventh line states.

Whether we find σ_n in this manner or in the manner of the first solution above which used Eq. (3.45) (which can be traced back to Eq. (3.27) in Theorem 3.3), everything boils down to the fact that the cross terms in the squared expression have zero expectation value. So we are left with only the diagonal terms.

3.9. Expected product, without replacement

When drawing two numbers without replacing the first one, all of the $\binom{N}{2} = N(N - 1)/2$ possibilities for the product $a_i a_j$ are equally likely to be the value that $X_1 X_2$

takes. The expectation value $E[X_1X_2]$ is therefore simply the average of all the different $a_i a_j$ values. That is,[3]

$$E[X_1X_2] = \frac{\sum_{i<j} a_i a_j}{N(N-1)/2}. \tag{3.106}$$

Now, if we square the sum $a_1 + a_2 + \cdots + a_N$ and then subtract off the "diagonal" a_i^2 terms, we will be left with only the cross terms $2a_i a_j$. So we can rewrite the above numerator to obtain

$$E[X_1X_2] = \frac{[(\sum a_i)^2 - \sum a_i^2]/2}{N(N-1)/2}, \tag{3.107}$$

where the sums run from 1 to N. By the definition of the mean μ, we have $\mu \equiv \sum a_i/N \implies \sum a_i = N\mu$. And by the definition of $E[X^2]$, we have $E[X^2] \equiv \sum a_i^2/N \implies \sum a_i^2 = N \cdot E[X^2]$. (We are using X to denote a random draw from the complete set.) But $E[X^2] = \mu^2 + \sigma^2$ from Eq. (3.50), which is true for an arbitrary distribution, in particular the present one involving N equally likely outcomes. So $\sum a_i^2 = N(\mu^2 + \sigma^2)$, and Eq. (3.107) becomes

$$E[X_1X_2] = \frac{(N\mu)^2 - N(\mu^2 + \sigma^2)}{N(N-1)} = \frac{N\mu^2 - \mu^2 - \sigma^2}{N-1} = \mu^2 - \frac{\sigma^2}{N-1}, \tag{3.108}$$

as desired.

REMARKS:

1. This result for $E[X_1X_2]$ is *smaller* than the $E[X_1X_2] = E[X_1]E[X_2] = \mu^2$ result in the case where the X_i are *independent*, as they would be if we drew the numbers *with replacement*. This makes sense for the following reason. The expectation value of the product of two *independently* drawn numbers (which could be the same or different) is $E[X] \cdot E[X] = \mu^2$, whereas Eq. (3.50) tells us that the expectation value of the product of two *identical* numbers is $E[X^2] = \mu^2 + \sigma^2$, which is larger than μ^2. Therefore, if we remove these *identical* cases, then the expectation value of the product of two *different* numbers must be smaller than μ^2, so that the expectation value of *all* of the products is μ^2. This reasoning is basically just Eq. (3.108) described in words.

2. A quick corollary to Eq. (3.108) is the following. Consider a set of N given numbers, $a_1, a_2, \ldots, a_N$. Draw n numbers randomly without replacement. (So we must have $n \leq N$, of course.) Let X_i (with $1 \leq i \leq n$) be the random variables for these n draws. Then the expectation value of the product of *any two* of the X_i (that is, not just the first two, X_1 and X_2) is the same as in Eq. (3.108):

$$E[X_iX_j] = \mu^2 - \frac{\sigma^2}{N-1} \qquad (i \neq j). \tag{3.109}$$

This is true because the temporal ordering of the first draw through the nth draw is irrelevant. We will end up with the same setup if we imagine labeling n boxes 1 through n and then throwing (simultaneously, or in whatever temporal order we wish) n of the N given numbers into the n boxes, with one number in each. All of the $_NP_n$ ordered subgroups of n numbers are equally likely, so all of the expectation values $E[X_iX_j]$ (with $i \neq j$) are the same. They therefore all have the common value of, say, $E[X_1X_2]$. And this value is given in Eq. (3.108). ♣

---

[3]Instead of writing $i < j$ here, we can alternatively write $i \neq j$, as long as we divide by $N(N-1)$ instead of $N(N-1)/2$. The denominator is modified because the sum now includes, for example, both a_3a_5 and a_5a_3.

3.10. **Standard deviation of the mean, without replacement**

Note that we are calculating the variance here with respect to the known mean μ of all N numbers, as opposed to the sample mean $\bar{x}$ of the n numbers we draw (as we did for the sample variance in Section 3.5). The latter would make the lefthand side of Eq. (3.91) identically zero, of course.

If we expand the square on the lefthand side of Eq. (3.91) and also expand the square of $\sum X_i$, we obtain

$$E\left[\left(\frac{\sum X_i}{n} - \mu\right)^2\right] = E\left[\frac{\sum X_i^2 + 2\sum_{i<j} X_i X_j}{n^2} - 2\frac{\sum X_i}{n}\mu + \mu^2\right] \qquad (3.110)$$

$$= \frac{\sum E[X_i^2] + 2\sum_{i<j} E[X_i X_j]}{n^2} - 2\frac{\sum E[X_i]}{n}\mu + \mu^2,$$

where the sums run from 1 to n. All of the $E[X_i X_j]$ terms are equal, with their common value being $\mu^2 - \sigma^2/(N-1)$ from the second remark in Problem 3.9; there are $n(n-1)/2$ of these terms. Likewise, all of the $E[X_i^2]$ terms are equal, with their common value being $\mu^2 + \sigma^2$ from Eq. (3.50); there are n of these terms. (They are indeed all equal, by the same type of reasoning as in the second remark in Problem 3.9. The temporal ordering is irrelevant.) And $E[X_i] = \mu$, of course; there are n of these terms. (Again, the temporal ordering is irrelevant.) So we have

$$E\left[\left(\frac{\sum X_i}{n} - \mu\right)^2\right] = \frac{n(\mu^2 + \sigma^2)}{n^2} + \frac{n(n-1)}{2} \cdot \frac{2}{n^2}\left(\mu^2 - \frac{\sigma^2}{N-1}\right) - 2\frac{n\mu}{n}\mu + \mu^2$$

$$= \sigma^2\left(\frac{1}{n} - \frac{1}{n}\cdot\frac{n-1}{N-1}\right) + \mu^2\left(\frac{1}{n} + \frac{n-1}{n} - 2 + 1\right),$$

$$= \frac{\sigma^2}{n}\left(1 - \frac{n-1}{N-1}\right), \qquad (3.111)$$

as desired. The μ's all cancel here, leaving us with only σ's. It makes sense that the variance shouldn't depend on μ, because if we increase all of the N given numbers by a particular value b, then b will cancel out in the difference $(\sum X_i)/n - \mu$ on the lefthand side of Eq. (3.111), because both $(\sum X_i)/n$ and μ increase by b.

REMARK:

1. We can check some limiting cases of Eq. (3.111). If $n = 1$, then the variance reduces to σ^2. This is correct, because if $n = 1$ then we're drawing only one number X. The sample mean of one number is simply itself, so the variance on the lefthand side of Eq. (3.111) is $E[(X - \mu)^2]$, which is σ^2 by definition. In this $n = 1$ case, the "without replacement" qualifier is irrelevant. We're drawing only one number, so it doesn't matter if we replace it or not.

   If $n = N$, then the variance in Eq. (3.111) reduces to 0. This is correct, because we're drawing without replacement, so at the end of the $n = N$ drawings, we must have chosen all of the N given numbers exactly once. The sample mean $(\sum X_i)/N$ of the $n = N$ numbers is therefore guaranteed to be the mean μ of the entire set, so variance of the sample mean is zero.

2. If we instead draw n numbers *with* replacement, then all of the n draws are identical processes. At all stages (from the first draw to the nth draw) each of the N numbers in the complete set has a definite probability of being drawn; it doesn't matter what has already happened. (There is an equal probability of $1/N$

of picking any number on a given draw, but this equality isn't important here.) We therefore simply have a distribution consisting of N possible outcomes, in which case the result in Eq. (3.53) for the standard deviation of the mean is applicable. (Eq. (3.53) holds for independent and identical trials.) The variance of the sample mean is therefore σ^2/n.

Returning to the *without*-replacement case, we see that (except for $n = 1$) the variance in Eq. (3.111) is smaller than the with-replacement variance σ^2/n, due to the nonzero $(n-1)/(N-1)$ term that is subtracted off in Eq. (3.111). It makes intuitive sense that the without-replacement variance is smaller than the with-replacement variance, because the drawings are more constrained if there is no replacement; there are fewer possibilities for future draws. There is therefore less variance in the sample mean the larger n is, to the point where there is zero variance if $n = N$. ♣

3.11. Biased sample standard deviation

Let's label the lefthand side of Eq. (3.93) as K (in the $N \to \infty$ limit). If we square both sides of that equation, the numerator of K^2 contains N terms of the form s_i^2, along with $\binom{N}{2} = N(N-1)/2$ cross terms of the form $2s_is_j$. That is,

$$K^2 = \frac{(s_1^2 + s_2^2 + \cdots + s_N^2) + (2s_1s_2 + 2s_1s_3 + \cdots + 2s_{N-1}s_N)}{N^2}. \tag{3.112}$$

If we let $a \equiv s_i^2$ and $b \equiv s_j^2$ in the $\sqrt{ab} \leq (a + b)/2$ arithmetic-geometric-mean inequality, we obtain

$$\sqrt{s_i^2 s_j^2} \leq \frac{s_i^2 + s_j^2}{2} \implies 2s_is_j \leq s_i^2 + s_j^2. \tag{3.113}$$

Therefore, in the above expression for K^2, if we replace each of the $\binom{N}{2}$ cross terms $2s_is_j$ with $s_i^2 + s_j^2$, we obtain a result that is larger than (or equal to) K^2. In this modified expression for K^2, a particular s_i^2 term such as s_1^2 appears $N - 1$ times (once with each of the other s_i^2 terms). Hence,

$$\begin{aligned}
K^2 &\leq \frac{(s_1^2 + s_2^2 + \cdots + s_N^2) + (N-1)(s_1^2 + s_2^2 + \cdots + s_N^2)}{N^2} \\
&= \frac{N(s_1^2 + s_2^2 + \cdots + s_N^2)}{N^2} \\
&= \frac{s_1^2 + s_2^2 + \cdots + s_N^2}{N} \\
&= \sigma^2,
\end{aligned} \tag{3.114}$$

in the $N \to \infty$ limit. Therefore, $K \leq \sigma$, and we have demonstrated Eq. (3.93), as desired.

REMARKS: The arithmetic-geometric-mean inequality, $\sqrt{ab} \leq (a + b)/2$, is very easy to prove. In fact, the ratio of its usefulness to proof-length is perhaps the largest of any mathematical result! Since the square of a number is necessarily nonnegative, we have

$$\left(\sqrt{a} - \sqrt{b}\right)^2 \geq 0 \implies a - 2\sqrt{ab} + b \geq 0 \implies \frac{a+b}{2} \geq \sqrt{ab}, \tag{3.115}$$

as desired.

How much smaller is $E[s]$ than σ? Consider the $n = 100$ case in the example near the end of Section 3.5. Fig. 3.4 showed the histogram of $N = 10^5$ values of s^2. We can take the square root of each of these N values of s^2 to obtain N values of s. And then we can average these N values to obtain the average value of s. The result is 0.9975, give or take a little, depending on the numerical run. So $E[s]$ must be about 0.9975, which is 0.0025 smaller than $\sigma = 1$. This 0.9975 result is reasonable, based on the following (extremely hand-wavy!) argument. We're just trying to get the correct order of magnitude here, so we won't be concerned with factors of order 1.

If two values of s^2 take the form of $1 + a$ and $1 - a$, then their average equals 1, of course. But the average value of s, which is $\sqrt{1 + a} + \sqrt{1 - a}$, is *not* equal to 1. It is smaller than 1, and this is the basic idea behind the fact that $E[s]$ is smaller than σ. To produce some actual numbers, let's pretend that the whole right half of the histogram in Fig. 3.4 is lumped together at the one-standard-deviation mark. (This is the hand-wavy part!) We found in the discussion of Fig. 3.4 that the standard deviation of s^2 is $\sqrt{2\sigma^4/n} = 0.14$. So we'll lump the whole right half of the histogram at the 1.14 mark. Similarly, we'll lump the whole left half at the 0.86 mark. We then have just two values of s^2, so the average value of s is $(\sqrt{1.14} + \sqrt{0.86})/2 = 0.9975$. This result agrees with the above numerical 0.9975 result a little *too* well. We had no right to expect such good agreement. But in any case, it is clear that the $\sigma - E[s]$ difference decreases with n, because the above 0.14 standard-deviation value came from $\sqrt{2\sigma^4/n}$, which decreases with n. ♣

3.12. Variance of the sample variance

Starting with $\mathrm{Var}(s^2) = E[(s^2 - \sigma^2)^2]$, we can rewrite this variance in the same manner as in Eq. (3.35):

$$
\begin{aligned}
\mathrm{Var}(s^2) &= E[(s^2 - \sigma^2)^2] \\
&= E[s^4] - 2E[s^2] \cdot \sigma^2 + \sigma^4 \\
&= E[s^4] - 2\sigma^2 \cdot \sigma^2 + \sigma^4 \\
&= E[s^4] - \sigma^4,
\end{aligned}
\tag{3.116}
$$

where we have used the fact that $E[s^2] = \sigma^2$. Our task is therefore to calculate $E[s^4]$. Let's rewrite the sample variance in Eq. (3.73), again in the manner of Eq. (3.35):

$$
\begin{aligned}
s^2 &= \frac{1}{n-1} \sum_{1}^{n} (x_i - \bar{x})^2 \\
&= \frac{1}{n-1}\left(\left(\sum x_i^2\right) - 2\left(\sum x_i\right)\bar{x} + n\bar{x}^2 \right) \\
&= \frac{1}{n-1}\left(\left(\sum x_i^2\right) - n\bar{x}^2 \right),
\end{aligned}
\tag{3.117}
$$

where the sum runs from 1 to n. We have used the fact that $\sum x_i = n\bar{x}$, by the definition of $\bar{x}$. Squaring the above s^2 and plugging the resulting expression for s^4 into Eq. (3.116), we find that the variance of s^2 is (switching from definite values x to random variables X)

$$\text{Var}(s^2) = E[s^4] - \sigma^4$$

$$= \frac{1}{(n-1)^2} E\left[\left(\left(\sum X_i^2\right) - n\overline{X}^2\right)^2\right] - \sigma^4 \tag{3.118}$$

$$= \frac{1}{(n-1)^2}\left(E\left[\left(\sum X_i^2\right)^2\right] - 2nE\left[\left(\sum X_i^2\right)\overline{X}^2\right] + n^2 E\left[\overline{X}^4\right]\right) - \sigma^4.$$

Note that we can't combine the second and third terms of the expansion of the square here, as we did in Eq. (3.117), because the expression analogous to $\sum X_i = n\overline{X}$ isn't valid when dealing with the square of X. That is, $\sum X_i^2 \neq n\overline{X}^2$. We therefore have to treat the second and third terms separately.

When calculating the three expectation values that appear in Eq. (3.118), it is *much* easier to work with random variables that are measured relative to the distribution's mean μ. So let's define the random variable Z by $Z \equiv X - \mu$. That is, $Z_i \equiv X_i - \mu$. etc. The Z_i's then all have the property that $E[Z_i] = 0$. We're effectively just shifting the distribution so that its mean is zero. This will *greatly* simplify the following calculations of the expectation values in Eq. (3.118).[4] Since the s^2 in Eq. (3.73) is independent of μ, the $\text{Var}(s^2)$ in Eq. (3.118) is also independent of μ. We are therefore free to replace all the X_i's with Z_i's in Eq. (3.118), without changing the value of $\text{Var}(s^2)$.

Look at the first term in Eq. (3.118). With $X_i \to Z_i$, this term is $E\left[\left(\sum Z_i^2\right)^2\right]$. When $(Z_1^2 + \cdots + Z_n^2)^2$ is multiplied out, there will be n terms of the form Z_i^4, which all have the same expectation value; call it $E[Z^4]$. And there will be $\binom{n}{2} = n(n-1)/2$ terms of the form $2Z_i^2 Z_j^2$, which again all have the same expectation value; call it $2E[Z_1^2 Z_2^2]$. So we obtain

$$E\left[\left(\sum Z_i^2\right)^2\right] = nE[Z^4] + \frac{n(n-1)}{2} \cdot 2E[Z_1^2 Z_2^2]$$

$$= nE[Z^4] + n(n-1)E[Z_1^2]E[Z_2^2]$$

$$= nE[Z^4] + n(n-1)\sigma^4, \tag{3.119}$$

where we have used the fact that $E[Z_i^2] = \sigma^2$ for any Z_i, which is just Eq. (3.50) with $\mu = 0$. We have also used Eq. (3.16), which holds here because the Z_i are independent variables.

Now look at the second term in Eq. (3.118), with $X_i \to Z_i$. When $\overline{Z}^2 = (1/n^2)(Z_1 + \cdots + Z_n)^2$ is expanded, there will be terms of the form Z_i^2 and $2Z_i Z_j$. When the latter is multiplied by $(Z_1^2 + \cdots + Z_n^2)$, it will produce terms of the form $Z_i Z_j Z_k^2$ and $Z_i Z_j^3$. Both of these contain a Z_i raised to the first power, so from Eq. (3.16) the expectation value will involve a factor of $E[Z_i]$, which is zero. We therefore need concern ourselves only with the Z_i^2 terms in $\overline{Z}^2$. The second term in the parentheses in Eq. (3.118) then becomes

$$-2nE\left[\left(\sum Z_i^2\right) \cdot \frac{1}{n^2}\left(\sum Z_i^2\right)\right] = -\frac{2}{n}E\left[\left(\sum Z_i^2\right)^2\right]$$

$$= -2E[Z^4] - 2(n-1)\sigma^4, \tag{3.120}$$

---

[4]We could have used this strategy in the proof of Theorem 3.5, but it wouldn't have saved a huge amount of time. The μ's that appeared in Eqs. (3.70) and (3.71) didn't cause much of a headache. But in the present solution they definitely would.

where we have used the fact that we ended up with the same form as the first term in Eq. (3.118), which we already calculated in Eq. (3.119).

Now for the third term in Eq. (3.118), with $X_i \to Z_i$. When we multiply out $\overline{Z}^4 = (1/n^4)(Z_1 + \cdots + Z_n)^4$, we obtain five different types of terms, as you can verify. They are Z_i^4, $Z_i^3 Z_j$, $Z_i^2 Z_j^2$, $Z_i^2 Z_j Z_k$, $Z_i Z_j Z_k Z_l$. The second, fourth, and fifth of these terms involve a single power of at least one Z_i, so their expectation values are zero. We therefore care only about the Z_i^4 and $Z_i^2 Z_j^2$ terms. There are n of the former type (all with the same expectation value). And there are $\binom{n}{2}\binom{4}{2} = 3n(n-1)$ of the latter type (again all with the same expectation value). This is true because there are $\binom{n}{2}$ ways to pick a particular pair of (i, j) indices, and for each of these pairs there are $\binom{4}{2} = 6$ ways to pick the two Z_i's from the four factors of $(Z_1 + \cdots + Z_n)$ in $\overline{Z}^4$. The third term in the parentheses in Eq. (3.118) is therefore

$$n^2 E\left[\overline{Z}^4\right] = n^2 \cdot \frac{1}{n^4}\left(nE[Z^4] + 3n(n-1)E[Z_1^2 Z_2^2]\right)$$

$$= \frac{1}{n}E[Z^4] + \frac{3(n-1)}{n}E[Z_1^2]E[Z_2^2]$$

$$= \frac{1}{n}E[Z^4] + \frac{3(n-1)}{n}\sigma^4, \tag{3.121}$$

where we have used the fact that $E[Z^2] = \sigma^2$. Plugging the results from Eqs. (3.119), (3.120), and (3.121) into Eq. (3.118), and grouping the $E[Z^4]$ and σ^4 terms together, gives

$$\text{Var}(s^2) = \frac{1}{(n-1)^2}\left[E[Z^4]\left(n - 2 + \frac{1}{n}\right) + \sigma^4(n-1)\left(n - 2 + \frac{3}{n}\right)\right] - \sigma^4. \tag{3.122}$$

If we factor out a $1/n$, the coefficient of $E[Z^4]$ in the parentheses becomes $(n-1)^2$, so we obtain

$$\text{Var}(s^2) = \frac{1}{n}\left[E[Z^4] + \sigma^4\left(\frac{n^2 - 2n + 3}{n-1} - n\right)\right]$$

$$= \frac{1}{n}\left[E[Z^4] - \sigma^4\left(\frac{n-3}{n-1}\right)\right], \tag{3.123}$$

which agrees with Eq. (3.94) because $\mu_4 \equiv E[(X - \mu)^4] = E[Z^4]$.

In the case where Z is a Gaussian distribution, you can use Eq. (4.123) in Problem 4.23 to show that $\mu_4 = 3\sigma^4$. $\text{Var}(s^2)$ then simplifies to $\text{Var}(s^2) = 2\sigma^4/(n-1)$. If n is large, then this is essentially equal to $2\sigma^4/n$, as we claimed in the example near the end of Section 3.5. Remember that this $2\sigma^4/n$ result holds only in the case of a Gaussian distribution and large n. In contrast, the result in Eq. (3.123) is valid for any distribution and for any n.

3.13. **Sample variance for two dice rolls**

(a) When $n = 2$, the $n - 1$ factor in the denominator of Eq. (3.73) equals 1, so the sample variance of two given dice rolls with values x_1 and x_2 is

$$s^2 = (x_1 - \overline{x})^2 + (x_2 - \overline{x})^2, \tag{3.124}$$

where $\overline{x} \equiv (x_1 + x_2)/2$. Let's determine the values of s^2 that Table 1.5 produces. If x_1 equals x_2, then $s^2 = 0$. In the table, there are six such pairs (along the main diagonal). If x_1 and x_2 differ by 1, then they each differ from $\bar{x}$ by $\pm 1/2$, so $s^2 = (1/2)^2 + (1/2)^2 = 1/2$. There are ten such pairs (along the two diagonals adjacent to the main diagonal). Continuing in this manner, if x_1 and x_2 differ by 2, then $s^2 = 1^2 + 1^2 = 2$; there are eight such pairs. If x_1 and x_2 differ by 3, then $s^2 = (3/2)^2 + (3/2)^2 = 9/2$; there are six such pairs. If x_1 and x_2 differ by 4, then $s^2 = 2^2 + 2^2 = 8$; there are four such pairs. Finally, if x_1 and x_2 differ by 5, then $s^2 = (5/2)^2 + (5/2)^2 = 25/2$; there are two such pairs. The expectation value of s^2 is therefore

$$E[s^2] = \frac{1}{36}\left(6 \cdot 0 + 10 \cdot \frac{1}{2} + 8 \cdot 2 + 6 \cdot \frac{9}{2} + 4 \cdot 8 + 2 \cdot \frac{25}{2}\right) = 2.92, \quad (3.125)$$

which correctly equals σ^2, as Eq. (3.74) states. If we want to instead calculate $\tilde{s}^2$, we simply need to tack on a factor of $n = 2$ in the denominator. We then end up with $E[\tilde{s}^2] = \sigma^2/2 = 1.46$, in agreement with Eq. (3.65) for the $n = 2$ case.

(b) Using the above results, the variance of s^2 is

$$\begin{aligned}
\mathrm{Var}(s^2) &= E[(s^2 - \sigma^2)^2] \\
&= \frac{1}{36}\left(6 \cdot (0 - 2.92)^2 + 10 \cdot (0.5 - 2.92)^2 + 8 \cdot (2 - 2.92)^2 \right. \\
&\qquad \left. + 6 \cdot (4.5 - 2.92)^2 + 4 \cdot (8 - 2.92)^2 + 2 \cdot (12.5 - 2.92)^2\right) \\
&= 11.6. \quad (3.126)
\end{aligned}$$

We'll now show that this agrees with Eq. (3.94) when $n = 2$. The calculation of $\mu_4 \equiv E[(X - \mu)^4]$ is similar to the calculation of $\mathrm{Var}(X)$ in Eq. (3.20). The only difference is that we now have fourth powers instead of squares. So

$$\begin{aligned}
\mu_4 &= E[(X - 3.5)^4] \\
&= \frac{1}{6}\left[(1 - 3.5)^4 + (2 - 3.5)^4 + (3 - 3.5)^4 \right. \\
&\qquad \left. + (4 - 3.5)^4 + (5 - 3.5)^4 + (6 - 3.5)^4\right] \\
&= 14.73. \quad (3.127)
\end{aligned}$$

When $n = 2$, the $(n - 3)/(n - 1)$ factor in Eq. (3.94) equals -1, so Eq. (3.94) gives

$$\mathrm{Var}(s^2) = \frac{1}{2}\left(\mu_4 + \sigma^4\right) = \frac{1}{2}\left(14.73 + 2.92^2\right) = 11.6, \quad (3.128)$$

in a agreement with Eq. (3.126). The standard deviation of s^2 is then $\sqrt{11.6} = 3.4$, which seems reasonable, considering that the six possible values of s^2 range from 0 to 12.5.

Chapter 4

Distributions

At the beginning of Section 3.1, we introduced the concepts of *random variables* and *probability distributions*. A random variable is a variable that can take on certain numerical values with certain probabilities. The collection of these probabilities is called the *probability distribution* for the random variable. A probability distribution specifies how the total probability (which is always 1) is distributed among the various possible outcomes.

In this chapter, we will discuss probability distributions in detail. In Section 4.1 we warm up with some examples of discrete distributions, and then in Section 4.2 we discuss continuous distributions. These involve the *probability density*, which is the main new concept in this chapter. It takes some getting used to, but we'll have plenty of practice with it. In Sections 4.3–4.8 we derive and discuss a number of the more common and important distributions. They are, respectively, the uniform, Bernoulli, binomial, exponential, Poisson, and Gaussian (or normal) distributions.

Parts of this chapter are a bit mathematical, but there's no way around this if we want to do things properly. However, we've relegated some of the more technical issues to Appendices B and C. If you want to skip those and just accept the results that we derive there, that's fine. But you are strongly encouraged to at least take a look at Appendix B, where we derive many properties of the number e, which is the most important number in probability and statistics.

4.1 Discrete distributions

In this section we'll give a few simple examples of discrete distributions. To start off, consider the results from Example 3 in Section 2.3.4, where we calculated the probabilities of obtaining the various possible numbers of Heads in five coin flips. We found:

$$P(0) = \frac{1}{32}, \quad P(1) = \frac{5}{32}, \quad P(2) = \frac{10}{32},$$
$$P(3) = \frac{10}{32}, \quad P(4) = \frac{5}{32}, \quad P(5) = \frac{1}{32}. \tag{4.1}$$

These probabilities add up to 1, as they should. Fig. 4.1 shows a plot of $P(n)$ versus n. The random variable here is the number of Heads, and it can take on the values of 0 through 5, with the above probabilities.

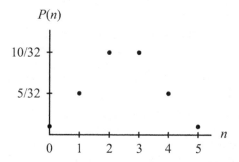

Figure 4.1: The probability distribution for the number of Heads in five coin flips.

As we've done in Fig. 4.1, the convention is to plot the random variable on the horizontal axis and the probability on the vertical axis. The collective information, given either visually in Fig. 4.1 or explicitly in Eq. (4.1), is the probability distribution. A probability distribution simply tells you what all the probabilities are for the values that the random variable can take. Note that $P(n)$ in the present example is nonzero only if n takes on one of the *discrete* values, 0, 1, 2, 3, 4, or 5. It's a silly question to ask for the probability of getting 4.27 Heads, because n must of course be an integer. The probability of getting 4.27 Heads is trivially zero. Hence the word "discrete" in the title of this section.

Another simple example of a discrete probability distribution is the one for the six possible outcomes of the roll of one die. The random variable in this setup is the number on the top face of the die. If the die is fair, then all six numbers have equal probabilities, so the probability for each is $1/6$, as shown in Fig. 4.2.

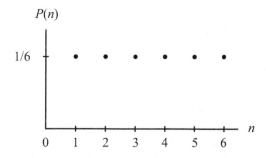

Figure 4.2: The probability distribution for the roll of one die.

What if the die isn't fair? For example, what if we make the "1" face heavier than the others by embedding a small piece of lead in the center of that face, just below the surface? The die is then more likely to land with the "1" face pointing down. The "6" face is opposite the "1," so the die is more likely to land with the "6" pointing up. Fig. 4.2 will therefore be modified by raising the "6" dot and lowering the other five dots; the sum of the probabilities must still be 1, of course. P_2 through P_5 are all equal, by symmetry. The exact values of all the probabilities depend in a complicated way on how the mass of the lead weight compares with the mass of the die, and also on the nature of both the die and the table on which the die is rolled (how much friction, how bouncy, etc.).

As mentioned at the beginning of Section 3.1, a random variable is assumed to take on *numerical* values, by definition. So the outcomes of Heads and Tails for a single coin flip technically aren't random variables. But it still makes sense to plot the probabilities as shown in Fig. 4.3, even though the outcomes on the horizontal axis aren't associated with a random variable. Of course, if we define a random variable to be the number of Heads, then the "Heads" in the figure turns into a 1, and the "Tails" turns into a 0. In most situations, however, the outcomes take on numerical values right from the start, so we can officially label them as random variables. But even if they don't, we'll often take the liberty of still referring to the thing being plotted on the horizontal axis of a probability distribution as a random variable.

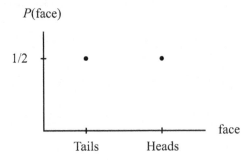

Figure 4.3: The probability distribution for a single coin flip.

4.2   Continuous distributions

4.2.1   Motivation

Probability distributions are fairly straightforward when the random variable is discrete. You just list (or plot) the probabilities for each of the possible values of the random variable. These probabilities will always add up to 1. However, not everything comes in discrete quantities. For example, the temperature outside your house takes on a continuous set of values, as does the amount of water in a glass. (We'll ignore the atomic nature of matter!)

In finding the probability distribution for a continuous random variable, you might think that the procedure should be exactly the same as in the discrete case. That is, if our random variable is the temperature at a particular location at noon tomorrow, then you might think that you simply have to answer questions of the form: What is the probability that the temperature at noon tomorrow will be 70° Fahrenheit?

Unfortunately, there is something wrong with this question, because it is too easy to answer. The answer is that the probability is *zero*, because there is simply no chance that the temperature at a specific time (and a specific location) will be *exactly* 70°. If it's 70.1°, that's not good enough. And neither is 70.01°, nor even 70.00000001°. Basically, since the temperature takes on a continuous set of values (and hence an infinite number of possible values), the probability of a specific value occurring is $1/\infty$, which is zero.[1]

However, even though the above question ("What is the probability that the temperature at noon tomorrow will be 70°?") is a poor one, that doesn't mean we should throw in the towel and conclude that probability distributions don't exist for continuous random variables. They do in fact exist, because there *are* some useful questions we can ask. These useful questions take the general form of: What is the probability that the temperature at a particular location at noon tomorrow lies somewhere between 69° and 71°? This question has a nontrivial answer, in the sense that it isn't automatically zero. And depending on what the forecast is for tomorrow, the answer might be something like 20%.

We can also ask: What is the probability that the temperature at noon lies somewhere between 69.5° and 70.5°? The answer to this question is smaller than the answer to the previous one, because it involves a range of only one degree instead of two degrees. If we assume that inside the range of 69° to 71° the temperature is equally likely to be found anywhere (which is a reasonable approximation although undoubtedly not exactly correct), and if the previous answer was 20%, then the present answer is (roughly) 10%, because the range is half the size.

The point here is that the smaller the range, the smaller the chance that the temperature lies in that range. Conversely, the larger the range, the larger the chance that the temperature lies in that range. Taken to an extreme, if we ask for the probability that the temperature at noon lies somewhere between $-100°$ and 200°, then the answer is exactly equal to 1 (ignoring liquid nitrogen spills, forest fires, and such things!).

In addition to depending on the size of the range, the probability also of course depends on where the range is located on the temperature scale. For example, the probability that the temperature at noon lies somewhere between 69° and 71° is undoubtedly different from the probability that it lies somewhere between 11° and 13°. Both ranges have a span of two degrees, but if the given day happens to be in late summer, the temperature is much more likely to be around 70° than to be sub-freezing (let's assume we're in, say, Boston). To actually figure out the probabilities, many different pieces of data would have to be considered. In the present temperature example, the data would be of the meteorological type. But if we were interested in the probability that a random person is between 69 and 71 inches tall, then we'd need to consider a whole different set of data.

The lesson to take away from all this is that if we're looking at a random variable that can take on a continuous set of values, the probability that this random variable falls into a given range depends on three things. It depends on:

---

[1]Of course, if you're using a digital thermometer that measures the temperature to the nearest tenth of a degree, then it *does* make sense to ask for the probability that the thermometer reads, say, 70.0 degrees. This probability is generally nonzero. This is due to the fact that the *reading* on the digital thermometer is a *discrete* random variable, whereas the *actual temperature* is a *continuous* random variable.

1. the location of the range,

2. the size of the range,

3. the specifics of the situation we're dealing with.

(The third of these is what determines the *probability density*, which is a function whose argument is the location of the range. We'll now discuss probability densities.)

4.2.2 Probability density

Consider the plot in Fig. 4.4, which gives a hypothetical probability distribution for the temperature example we've been discussing. This plot shows the probability distribution on the vertical axis, as a function of the temperature T (the random variable) on the horizontal axis. We have chosen to measure the temperature in Fahrenheit. We're denoting the probability distribution by[2] $\rho(T)$ instead of $P(T)$, to distinguish it from the type of probability distribution we've been talking about for discrete variables. The reason for this new notation is that $\rho(T)$ is a probability *density* and not an actual probability. We'll talk about this below. When writing the functional form of a probability distribution, we'll denote probability *densities* with lowercase letters, like the ρ in $\rho(T)$ or the f in $f(x)$. And we'll denote actual *probabilities* with uppercase letters, like the P in $P(n)$.

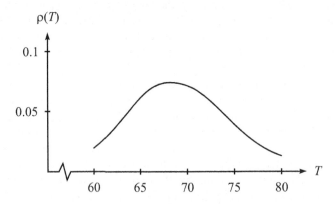

Figure 4.4: A hypothetical probability distribution for the temperature.

We haven't yet said exactly what we mean by $\rho(T)$. But in any case, it's clear from Fig. 4.4 that the temperature is more likely to be near $70°$ than near $60°$. The following definition of $\rho(T)$ allows us to be precise about what we mean by this.

---

[2]As mentioned at the beginning of Section 3.1, a random variable is usually denoted with an uppercase letter, while the actual values are denoted with lowercase letters. So we should technically be writing $\rho(t)$ here. But since an uppercase T is the accepted notation for temperature, we'll use T for the actual value.

- Definition of the probability density function, $\rho(T)$:

  $\rho(T)$ *is the function of T that, when multiplied by a small interval ΔT, gives the probability that the temperature lies between T and $T + \Delta T$. That is,*

  $$P(\text{temp lies between } T \text{ and } T + \Delta T) = \rho(T) \cdot \Delta T. \qquad (4.2)$$

Note that the lefthand side contains an actual probability P, whereas the righthand side contains a probability *density*, $\rho(T)$. The latter needs to be multiplied by a range of T (or whatever quantity we're dealing with) in order to obtain an actual probability. The above definition is relevant to any continuous random variable, of course, not just temperature.

Eq. (4.2) might look a little scary, but a few examples should clear things up. From Fig. 4.4, it looks like $\rho(70°)$ is about 0.07. So if we pick $\Delta T = 1°$, we find that the probability of the temperature lying between 70° and 71° is about

$$\rho(T) \cdot \Delta T = (0.07)(1) = 0.07 = 7\%. \qquad (4.3)$$

If we instead pick a smaller ΔT, say 0.5°, we find that the probability of the temperature lying between 70° and 70.5° is about $(0.07)(0.5) = 3.5\%$. And if we pick an even smaller ΔT, say 0.1°, we find that the probability of the temperature lying between 70° and 70.1° is about $(0.07)(0.1) = 0.7\%$.

Similarly, we can apply Eq. (4.2) to any other value of T. For example, it looks like $\rho(60°)$ is about 0.02. So if we pick $\Delta T = 1°$, we find that the probability of the temperature lying between 60° and 61° is about $(0.02)(1) = 2\%$. And as above, we can pick other values of ΔT too.

Note that, in accordance with Eq. (4.2), we have been using the value of ρ at the *lower* end of the given temperature interval. That is, when the interval was 70° to 71°, we used $\rho(70°)$ and then multiplied this by ΔT. But couldn't we just as well use the value of ρ at the *upper* end of the interval? That is, couldn't the righthand side of Eq. (4.2) just as well be $\rho(T + \Delta T) \cdot \Delta T$? Indeed it could. But as long as ΔT is small, it doesn't matter much which value of ρ we use. They will both give essentially the same answer. See the second remark below.

Remember that *three* inputs are necessary when finding the probability that the temperature lies in a specified range. As we noted at the end of Section 4.2.1, the first input is the value of T we're concerned with, the second is the range ΔT, and the third is the information encapsulated in the probability density function, $\rho(T)$, evaluated at the given value of T. The latter two of these three quantities are the two quantities that are multiplied together on the righthand side of Eq. (4.2). Knowing only one of these isn't enough to give you a probability.

To recap, there is a very important difference between the probability distribution for a continuous random variable and that for a discrete random variable. For a continuous variable, the probability distribution consists of a *probability density*. But for a discrete variable, it consists of *actual probabilities*. We plot a *density* for a continuous distribution, because it wouldn't make sense to plot actual probabilities, since they're all zero. This is true because the probability of obtaining *exactly* a particular value is zero, since there is an infinite number of possible values.

Conversely, we plot *actual probabilities* for a discrete distribution, because it wouldn't make sense to plot a density, since it consists of a collection of infinite spikes. This is true because on a die roll, for example, there is a 1/6 chance of obtaining a number between, say, 4.9999999 and 5.0000001. The probability density at the outcome of 5, which from Eq. (4.2) equals the probability divided by the interval length, is then (1/6)/(0.0000002), which is huge. And the interval can be made arbitrarily small, which means that the density is arbitrarily large. To sum up, the term "probability distribution" applies to both continuous and discrete variables, whereas the term "probability density" applies only to continuous variables.

REMARKS:

1. $\rho(T)$ is a function of T, so it depends on what units we're using to measure T. We used Fahrenheit above, but what if we instead want to use Celsius? Problem 4.1 addresses this issue (but you will need to read Section 4.2.3 first).

2. Note the inclusion of the word "small" in the definition of the probability density in Eq. (4.2). The reason for this word is that we want $\rho(T)$ to be (roughly) constant over the specified range. If ΔT is small enough, then this is approximately true. If $\rho(T)$ varied greatly over the range of ΔT, then it wouldn't be clear which value of $\rho(T)$ we should multiply by ΔT to obtain the probability. The point is that if ΔT is small enough, then all of the $\rho(T)$ values are roughly the same, so it doesn't matter which one we pick.

   An alternative definition of the density $\rho(T)$ is

   $$P(\text{temp lies between } T - (\Delta T)/2 \text{ and } T + (\Delta T)/2) = \rho(T) \cdot \Delta T. \qquad (4.4)$$

   The only difference between this definition and the one in Eq. (4.2) is that we're now using the value of $\rho(T)$ at the midpoint of the temperature range, instead of the left-end value we used in Eq. (4.2). Both definitions are equally valid, because they give essentially the same result for $\rho(T)$, provided that ΔT is small. Similarly, we could use the value of $\rho(T)$ at the right end of the temperature range.

   How small do we need ΔT to be? The answer to this will be evident when we talk about probability in terms of area in Section 4.2.3. In short, we need the change in $\rho(T)$ over the span of ΔT to be small compared with the values of $\rho(T)$ in that span.

3. The probability density function involves only (1) the value of T (or whatever) we're concerned with, and (2) the specifics of the situation at hand (meteorological data in the above temperature example, etc.). The density is completely independent of the arbitrary value of ΔT that we choose. This is how things work with any kind of density.

   For example, consider the mass density of gold. This mass density is a property of the gold itself. More precisely, it is a function of each point in the gold. For pure gold, the density is constant throughout the volume, but we could imagine impurities that would make the mass density be a varying function of position, just as the above probability density is a varying function of temperature. Let's call the mass density $\rho(\mathbf{r})$, where $\mathbf{r}$ signifies the possible dependence of ρ on the location of a given point within the volume. (The position of a given point can be described by the vector pointing from the origin to the point. And vectors are generally denoted by boldface letters like $\mathbf{r}$.) Let's call the small volume we're concerned with ΔV. Then the mass in the small volume ΔV is given by the product of the density and the volume, that is, $\rho(\mathbf{r}) \cdot \Delta V$. This is directly analogous to the fact that the probability in the above temperature example is given by the product of the probability density and the temperature span, that is, $\rho(T) \cdot \Delta T$. The correspondence among the various quantities is

$$\text{Mass in } \Delta V \text{ around location } \mathbf{r} \iff \text{Prob that temp lies in } \Delta T \text{ around } T$$

$$\rho(\mathbf{r}) \iff \rho(T)$$

$$\Delta V \iff \Delta T. \; \clubsuit \tag{4.5}$$

4.2.3 Probability equals area

The graphical interpretation of the product $\rho(T) \cdot \Delta T$ in Eq. (4.2) is that it is the area of the rectangle shown in Fig. 4.5. This is true because ΔT is the base of the rectangle, and $\rho(T)$ is the height.

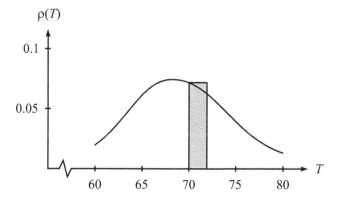

Figure 4.5: Interpretation of the product $\rho(T) \cdot \Delta T$ as an area.

We have chosen ΔT to be $2°$ in the figure. With this choice, the area of the rectangle, which equals $\rho(70°) \cdot (2°)$, gives a reasonably good approximation to the probability that the temperature lies between $70°$ and $72°$. But it isn't exact, because $\rho(T)$ isn't constant over the $2°$ interval. A better approximation to the probability that the temperature lies between $70°$ and $72°$ is achieved by splitting the $2°$ interval into two intervals of $1°$ each, and then adding up the probabilities of lying in each of these two intervals. These two probabilities are approximately equal to $\rho(70°) \cdot (1°)$ and $\rho(71°) \cdot (1°)$, and the two corresponding rectangles are shown in Fig. 4.6.

But again, the sum of the areas of these two rectangles is still only an approximate result for the true probability that the temperature lies between $70°$ and $72°$, because $\rho(T)$ isn't constant over the $1°$ intervals either. A better approximation is achieved by splitting the $1°$ intervals into smaller intervals, and then again into even smaller ones. And so on. When we get to the point of having 100 or 1000 extremely thin rectangles, the sum of their areas will essentially be the area shown in Fig. 4.7. This area is the correct probability that the temperature lies between $70°$ and $72°$. So in retrospect, we see that the rectangular area in Fig. 4.5 exceeds the true probability by the area of the tiny triangular-ish region in the upper righthand corner of the rectangle.

We therefore arrive at a more precise definition (compared with Eq. (4.2)) of the probability density, $\rho(T)$:

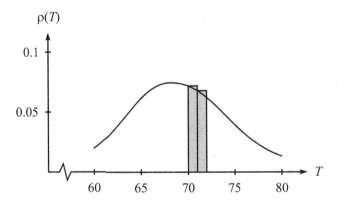

Figure 4.6: Subdividing the area, to produce a better approximation to the probability.

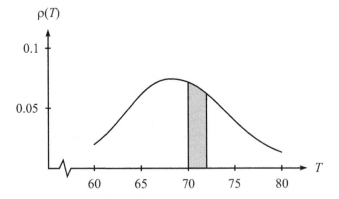

Figure 4.7: The area below the curve between $70°$ and $72°$ equals the probability that the temperature lies between $70°$ and $72°$.

- Improved definition of the probability density function, $\rho(T)$:

  $\rho(T)$ is the function of T for which the area under the $\rho(T)$ curve between T and $T + \Delta T$ gives the probability that the temperature (or whatever quantity we're dealing with) lies between T and $T + \Delta T$.

This is an exact definition, and there is no need for ΔT to be small, as there was in the definition in Eq. (4.2). The difference is that the present definition involves the exact area, whereas Eq. (4.2) involved the area of a rectangle (via simple multiplication by ΔT), which was only an approximation. But technically the only thing we need to add to Eq. (4.2) is the requirement that we take the $\Delta T \to 0$ limit. That makes the definition rigorous.

The total area under any probability density curve must be 1, because this area equals the probability that the temperature (or whatever) takes on some value between $-\infty$ and $+\infty$, and because every possible result is included in the $-\infty$ to $+\infty$ range. However, in any realistic case, the density is essentially zero outside a specific finite region. So there is essentially no contribution to the area from the parts of the plot outside that region. There is therefore no need to go to $\pm\infty$. The total area under each of the curves in the above figures, including the tails on either side which we haven't bothered to draw, is indeed equal to 1 (at least roughly; the curves were drawn by hand).

Given a probability density function $f(x)$, the *cumulative distribution function* $F(x)$ is defined to be the probability that X takes on a value that is less than or equal to x. That is, $F(x) = P(X \le x)$. For a continuous distribution, this definition implies that $F(x)$ equals the area under the $f(x)$ curve from $-\infty$ up to the given x value. A quick corollary is that the probability $P(a < x \le b)$ that x lies between two given values a and b is equal to $F(b) - F(a)$. For a discrete distribution, the definition $F(x) = P(X \le x)$ still applies, but we now calculate $P(X \le x)$ by forming a discrete sum instead of finding an area. Although the cumulative distribution function can be very useful in probability and statistics, we won't use it much in this book.

We'll now spend a fair amount of time in Sections 4.3–4.8 discussing some common types of probability distributions. There is technically an infinite number of possible distributions, although only a hundred or so come up frequently enough to have names. And even many of these are rather obscure. A handful, however, come up again and again in a variety of settings, so we'll concentrate on these. They are the uniform, Bernoulli, binomial, exponential, Poisson, and Gaussian (or normal) distributions.

4.3 Uniform distribution

We'll start with a very simple continuous probability distribution, one that is uniform over a given interval, and zero otherwise. Such a distribution might look like the one shown in Fig. 4.8. If the distribution extends from x_1 to x_2, then the value of $\rho(x)$ in that region must be $1/(x_2 - x_1)$, so that the total area is 1.

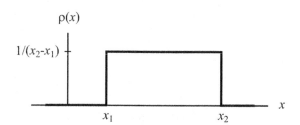

Figure 4.8: A uniform distribution.

This type of distribution could arise, for example, from a setup where a rubber ball bounces around in an empty rectangular room. When it finally comes to rest, we measure its distance x from a particular one of the walls. If you initially throw the ball hard enough, then it's a pretty good approximation to say that x is equally likely to take on any value between 0 and L, where L is the length of the room in the relevant direction. In this setup, the x_1 in Fig. 4.8 equals 0 (so we would need to shift the rectangle to the left), and the x_2 equals L.

The random variable here is X, and the value it takes is denoted by x. So x is what we plot on the horizontal axis. Since we're dealing with a continuous distribution, we plot the probability *density* (not the probability!) on the vertical axis. If L equals 10 feet, then outside the region $0 < x < 10$, the probability density $\rho(x)$ equals zero. Inside this region, the density equals the total probability divided by the total interval, which gives 1 per 10 feet, or equivalently $1/10$ per foot. If we want to find the actual probability that the ball ends up between, say, $x = 6$ and $x = 8$, then we just multiply $\rho(x)$ by the interval length, which is 2 feet. The result is ($1/10$ per foot)(2 feet), which equals $2/10 = 1/5$. This makes sense, of course, because the 2-foot interval is $1/5$ of the total distance.

A uniform density is easy to deal with, because the area under a given part of the curve (which equals the probability) is simply a rectangle. And the area of a rectangle is just the base times the height, which is the interval length times the density. This is exactly the product we formed above. When the density isn't uniform, it can be very difficult sometimes to find the area under a given part of the curve.

Note that the larger the region of nonzero $\rho(x)$ in a uniform distribution, the smaller the value of $\rho(x)$. This follows from the fact that the total area under the density "curve" (which is just a straight line segment in this case) must equal 1. So if the base becomes longer, the height must become shorter.

4.4 Bernoulli distribution

We'll now consider a very simple discrete distribution, called the Bernoulli distribution. This is the distribution for a process in which only two possible outcomes, 1 and 0, can occur, with probabilities p and $1 - p$, respectively. (They must add up to 1, of course.) The plot of this probability distribution is shown in Fig. 4.9. It is common to call the outcome of 1 a success and the outcome of 0 a failure. A special case of a Bernoulli distribution is the distribution for a coin toss, where the probabilities for Heads and Tails (which we can assign the values of 1 and 0, respectively) are both equal to $1/2$.

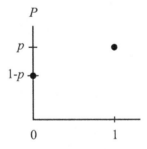

Figure 4.9: A Bernoulli distribution takes on the values 1 and 0 with probabilities p and $1 - p$.

The Bernoulli distribution is the simplest of all distributions, with the exception of the trivial case where only one possible outcome can occur, which therefore has a probability of 1. The uniform and Bernoulli distributions are simple enough that there isn't much to say. In contrast, the distributions in the following four sections (binomial, exponential, Poisson, and Gaussian) are a bit more interesting, so we'll have plenty to say about them.

4.5 Binomial distribution

The binomial distribution, which is discrete, is an extension of the Bernoulli distribution. The binomial distribution is defined to be the probability distribution for the total number of successes that arise in an arbitrary number of independent and identically distributed Bernoulli processes. An example of a binomial distribution is the probability distribution for the number of Heads in, say, five coin tosses, which we discussed in Section 4.1. We could just as well pick any other number of tosses.

In the case of five coin tosses, each coin toss is a Bernoulli process. When we put all five tosses together and look at the total number of successes (Heads), we get a binomial distribution. Let's label the total number of successes as k. In this specific example, there are $n = 5$ Bernoulli processes, with each one having a $p = 1/2$ probability of success. The probability distribution $P(k)$ is simply the one we plotted earlier in Fig. 4.1, where we counted the number of Heads.

Let's now find the binomial distribution associated with a general number n of independent Bernoulli trials, each with the same probability of success, p. So our goal is to find the value of $P(k)$ for all of the different possible values of the total number of successes, k. The possible values of k range from 0 up to the number of trials, n.

To calculate the binomial distribution (for given n and p), we first note that p^k is the probability that a *specific set* of k of the n Bernoulli processes all yield success, because each of the k processes has a p probability of yielding success. We then need the other $n - k$ processes to *not* yield success, because we want *exactly* k successes. This happens with probability $(1 - p)^{n-k}$, because each of the $n - k$ processes has a $1 - p$ probability of yielding failure. The probability that a specific set of k processes (and no others) all yield success is therefore $p^k \cdot (1 - p)^{n-k}$. Finally, since there are $\binom{n}{k}$ ways to pick a specific set of k processes, we see that the probability that exactly k of the n processes yield success is

$$\boxed{P(k) = \binom{n}{k} p^k (1 - p)^{n-k}} \qquad \text{(binomial distribution)} \qquad (4.6)$$

This is the desired binomial distribution. Note that this distribution depends on two parameters – the number n of Bernoulli trials and the probability p of success in each trial. If you want to make these parameters explicit, you can write the Binomial distribution $P(k)$ as $B_{n,p}(k)$. That is,

$$B_{n,p}(k) = \binom{n}{k} p^k (1 - p)^{n-k}. \qquad (4.7)$$

But we'll generally just use the simple $P(k)$ notation.

In the special case of a binomial distribution generated from n coin tosses, we have $p = 1/2$. So Eq. (4.6) gives the probability of obtaining k Heads as

$$P(k) = \frac{1}{2^n}\binom{n}{k}. \tag{4.8}$$

To recap: In Eq. (4.6), n is the total number of Bernoulli processes, p is the probability of success in each Bernoulli process, and k is the total number of successes in the n processes. (So k can be anything from 0 to n.) Fig. 4.10 shows the binomial distribution for the cases of $n = 30$ and $p = 1/2$ (which arises from 30 coin tosses), and $n = 30$ and $p = 1/6$ (which arises from 30 die rolls, with a particular one of the six numbers representing success).

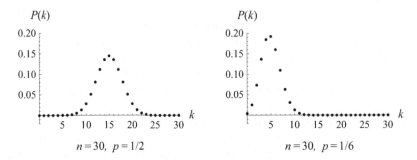

Figure 4.10: Two binomial distributions with $n = 30$ but different values of p.

Example (Equal probabilities): Given n, for what value of p is the probability of zero successes equal to the probability of one success?

Solution: In Eq. (4.6) we want $P(0)$ to equal $P(1)$. This gives

$$\binom{n}{0}p^0(1-p)^{n-0} = \binom{n}{1}p^1(1-p)^{n-1}$$

$$\implies 1 \cdot 1 \cdot (1-p)^n = n \cdot p \cdot (1-p)^{n-1}$$

$$\implies 1 - p = np \implies p = \frac{1}{n+1}. \tag{4.9}$$

This $p = 1/(n+1)$ value is the special value of p for which various competing effects cancel. On one hand, $P(1)$ contains an extra factor of n from the $\binom{n}{1}$ coefficient, which arises from the fact that there are n different ways for one success to happen. But on the other hand, $P(1)$ also contains a factor of p, which arises from the fact that one success *does* happen. The first of these effects makes $P(1)$ larger than $P(0)$, while the second makes it smaller.[3] The effects cancel when $p = 1/(n+1)$. Fig. 4.11 shows the plot for $n = 10$ and $p = 1/11$.

The $p = 1/(n+1)$ case is the cutoff between the maximum of $P(k)$ occurring when k is zero or nonzero. If p is larger than $1/(n+1)$, as it is in both plots in Fig. 4.10

---

[3]Another effect is that $P(1)$ is larger because it contains one fewer factor of $(1-p)$. But this effect is minor when p is small, which is the case if n is large, due to the $p = 1/(n+1)$ form of the answer.

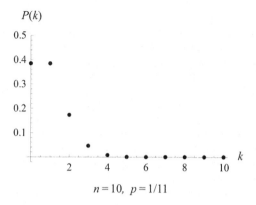

$P(k)$

$n = 10, \ p = 1/11$

Figure 4.11: $P(0)$ equals $P(1)$ if $p = 1/(n + 1)$.

above, then the maximum occurs at a nonzero value of k. That is, the distribution has a bump. On the other hand, if p is smaller than $1/(n + 1)$, then the maximum occurs at $k = 0$. That is, the distribution has its peak at $k = 0$ and falls off from there.

Having derived the binomial distribution in Eq. (4.6), there is a simple double check that we can perform on the result. Since the number of successes, k, can take on any integer value from 0 to n, the sum of the $P(k)$ probabilities from $k = 0$ to $k = n$ must equal 1. The $P(k)$ expression in Eq. (4.6) does indeed satisfy this requirement, due to the binomial expansion, which tells us that

$$\left(p + (1 - p)\right)^n = \sum_{k=0}^{n} \binom{n}{k} p^k (1 - p)^{n-k}. \tag{4.10}$$

This is just Eq. (1.21) from Section 1.8.3, with $a = p$ and $b = 1 - p$. The lefthand side of Eq. (4.10) is simply $1^n = 1$. And each term in the sum on the righthand side is a $P(k)$ term from Eq. (4.6). So Eq. (4.10) becomes

$$1 = \sum_{k=0}^{n} P(k), \tag{4.11}$$

as we wanted to show. You are encouraged to verify this result for the probabilities in, say, the left plot in Fig. 4.10. Feel free to make rough estimates of the probabilities when reading them off the plot. You will find that the sum is indeed 1, up to the rough estimates you make.

The task of Problem 4.4 is to use Eq. (3.4) to explicitly demonstrate that the expectation value of the binomial distribution in Eq. (4.6) equals pn. In other words, if our binomial distribution is derived from n Bernoulli trials, each having a probability p of success, then we should expect a total of pn successes (on average, if we do a large number of sets of n trials). This must be true, of course, because a fraction p of the n trials yield success, on average, by the definition of p for the given Bernoulli process.

REMARK: (We should emphasize what is meant by a probability distribution.) Let's say that you want to experimentally verify that the left plot in Fig. 4.10 is the correct probability distribution for the total number of Heads that show up in 30 coin flips. You of course can't do this by flipping a coin just once. And you can't even do it by flipping a coin 30 times, because all you'll get from that is just one number for the total number of Heads. For example, you might obtain 17 Heads. In order to experimentally verify the distribution, you need to perform *a large number of sets of 30 coin flips,* and you need to record the total number of Heads you get in each 30-flip set. The result will be a long string of numbers such as $13, 16, 15, 16, 18, 14, 11, 17, \ldots$. If you then calculate the fractions of the time that each number appears, these fractions should (roughly) agree with the probabilities shown in Fig. 4.10. The longer the string of numbers, the better the agreement, in general. (The main point here is that the distribution does't say much about *one particular* set of 30 flips. Rather, it says what the expected distribution of outcomes is for a *large number* of sets of 30 flips.) ♣

4.6   Exponential distribution

In Sections 4.6–4.8 we'll look at three probability distributions (exponential, Poisson, and Gaussian) that are a bit more involved than the three we've just discussed (uniform, Bernoulli, and binomial). We'll start with the exponential distribution, which takes the general form,

$$\rho(t) = Ae^{-bt}, \tag{4.12}$$

where A and b are quantities that depend on the specific situation at hand. We will find below in Eq. (4.26) that these quantities must be related in a certain way in order for the total probability to be 1. The parameter t corresponds to whatever the random variable is. The exponential distribution is a continuous one, so $\rho(t)$ is a probability density. The most common type of situation where this distribution arises is the following.

Consider a repeating event that happens completely randomly in time. By "completely randomly" we mean that there is a uniform probability that the event happens at any given instant (or more precisely, in any small time interval of a given length), independent of what has already happened. That is, the process has no "memory." The exponential distribution that we'll eventually arrive at (after a lot of work!) in Eq. (4.26) gives the probability distribution for the *waiting time* until the next event occurs. (Since the time t is a continuous quantity, we'll need to develop some formalism to analyze the distribution. To ease into it, let's start with the slightly easier case where time is assumed to be discrete.)

4.6.1   Discrete case

(Consider a process where we roll a hypothetical 10-sided die once every second. So time is discretized into 1-second intervals.) It's actually not necessary to introduce time here at all. We could simply talk about the number of iterations of the process. But it's easier to talk about things like the "waiting time" than the "number of iterations you need to wait for." So for convenience, we'll discuss things in the context of time.

If the die shows a "1," we'll consider that a success. The other nine numbers represent failure. There are two reasonable questions we can ask: What is the average waiting time (that is, the expectation value of the waiting time) between successes? And what is the probability distribution of the waiting times between successes?

Average waiting time

It is fairly easy to determine the average waiting time. There are 10 possible numbers on the die, so on average we can expect 1/10 of them to be 1's. If we run the process for a long time, say, an hour (which consists of 3600 seconds), then we can expect about 360 1's. The average waiting time between successes is therefore (3600 seconds)/360 = 10 seconds.

More generally, if the probability of success in each trial is p, then the average waiting time is $1/p$ (assuming that the trials happen at 1-second intervals). This can be seen by the same reasoning as above. If we perform n trials of the process, then pn of them will yield success, on average. The average waiting time between successes is the total time (n) divided by the number of successes (pn):

$$\text{Average waiting time} = \frac{n}{pn} = \frac{1}{p}. \tag{4.13}$$

Note that the preceding reasoning gives us the average waiting time, without requiring any knowledge of the actual probability distribution of the waiting times (which we will calculate below). Of course, once we *do* know what the probability distribution is, we should be able to calculate the average (the expectation value) of the waiting times. This is the task of Problem 4.7.

Distribution of waiting times

Finding the probability distribution of the waiting times requires a little more work than finding the average waiting time. For the 10-sided die example, the question we're trying to answer is: What is the probability that if we consider two successive 1's, the time between them will be 6 seconds? Or 30 seconds? Or 1 second? And so on. Although the *average* waiting time is 10 seconds, this certainly doesn't mean that the waiting time will always be 10 seconds. In fact, we will find below that the probability that the waiting time is exactly 10 seconds is quite small.

Let's be general and say that the probability of success in each trial is p (so $p = 1/10$ in our present setup). Then the question is: What is the probability, $P(k)$, that we will have to wait exactly k iterations (each of which is 1 second here) to obtain the next success?

To answer this, note that in order for the next success to happen on the kth iteration, there must be failure (which happens with probability $1 - p$) on the first $k - 1$ iterations, and then success on the kth one. The probability of this happening is

$$\boxed{P(k) = (1 - p)^{k-1}p} \qquad \text{(geometric distribution)} \tag{4.14}$$

This is the desired (discrete) probability distribution for the waiting time. This distribution goes by the name of the *geometric distribution*, because the probabilities form a geometric progression, due to the increasing power of the $(1 - p)$ factor. The geometric distribution is the discrete version of the exponential distribution that we'll arrive at in Eq. (4.26) below.

Eq. (4.14) tells us that the probability that the next success comes on the very next iteration is p, the probability that it comes on the second iteration is $(1-p)p$, the probability that it comes on the third iteration is $(1-p)^2 p$, and so on. Each probability is smaller than the previous one by the factor $(1-p)$. A plot of the distribution for $p = 1/10$ is shown in Fig. 4.12. The distribution is maximum at $k = 1$ and falls off from that value. Even though $k = 10$ is the average waiting time, the probability of the waiting time being *exactly* $k = 10$ is only $P(10) = (0.9)^9(0.1) \approx 0.04 = 4\%$.

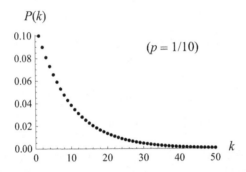

Figure 4.12: The geometric distribution with $p = 1/10$.

If p is large (close to 1), the plot of $P(k)$ starts high (at p, which is close to 1) and then falls off quickly, because the factor $(1-p)$ is close to 0. On the other hand, if p is small (close to 0), the plot of $P(k)$ starts low (at p, which is close to 0) and then falls off slowly, because the factor $(1-p)$ is close to 1.

As a double check on the result in Eq. (4.14), we know that the next success has to eventually happen *sometime*, so the sum of all the $P(k)$ probabilities must be 1. These $P(k)$ probabilities form a geometric series whose first term is p and whose ratio is $1-p$. The general formula for the sum of a geometric series with first term a and ratio r is $a/(1-r)$, so we have

$$
\begin{aligned}
P(1) + P(2) + P(3) + \cdots &= p + p(1-p) + p(1-p)^2 + \cdots \\
&= \frac{p}{1-(1-p)} \\
&= 1,
\end{aligned}
\tag{4.15}
$$

as desired. As another check, we can verify that the expectation value (the average) of the waiting time for the geometric distribution in Eq. (4.14) equals $1/p$, as we already found above; see Problem 4.7.

You are encouraged to use a coin to experimentally "verify" Eq. (4.14) (or equivalently, the plot analogous to Fig. 4.12) for the case of $p = 1/2$. Just flip a coin as many times as you can in ten minutes, each time writing down a 1 if you get Heads and a 0 if you get Tails. Then make a long list of the waiting times between the 1's. Then count up the number of one-toss waits, the number of two-toss waits, and so on. Then divide each of these numbers by the total number of waits (not the total number of tosses!) to find the probability of each waiting length. The results should be (roughly) consistent with Eq. (4.14) for $p = 1/2$. In this case, the probabilities in Eq. (4.14) for $k = 1, 2, 3, 4, \ldots$ are $1/2, 1/4, 1/8, 1/16, \ldots$.

4.6.2 Rates, expectation values, and probabilities

Let's now consider the case where time is a continuous quantity. That is, let's assume that we can have a "successful" event at *any* instant, not just at the evenly-spaced 1-second marks as above. (A continuous process whose probability is uniform in time can be completely described by just *one* number – the average rate of success, which we'll call λ.) We generally won't bother writing the word "average," so we'll just call λ the "rate." Before getting into the derivation of the continuous exponential distribution in Section 4.6.3, we'll need to talk a little about rates.

The rate λ can be determined by counting the number of successful events that occur during a long time interval, and then dividing by this time. For example, if 300 (successful) events happen during 100 minutes, then the rate λ is 3 events per minute. Of course, if you count the number of events in a different span of 100 minutes, you will most likely get a slightly different number, perhaps 313 or 281. But in the limit of a very long time interval, you will find essentially the same rate, independent of which specific long interval you use.

(If the rate λ is 3 events per minute, you can alternatively write this as 1 event per 20 seconds, or $1/20$ of an event per second. There is an infinite number of ways to write λ, and it's personal preference which one you pick. Just remember that you have to state the "per time" interval you're using. If you just say that the rate is 3, that doesn't mean anything.)

What is the expectation value of the number of events that happen during a time t? This expected number simply equals the product λt, from the definition of λ. If the expected number were anything other than λt, then if we divided it by t to obtain the rate, we wouldn't get λ. If you want to be a little more rigorous, consider a very large number n of intervals with length t. The total time in these intervals is nt. This total time is very large, so the number of events that happen during this time is (approximately) equal to $(nt)\lambda$, by the definition of λ. The expected number of events in each of the n intervals with length t is therefore $nt\lambda/n = \lambda t$, as above. So we can write

$$\boxed{\text{(Expected number of events in time } t) = \lambda t} \qquad (4.16)$$

In the above setup where λ equals 3 events per minute, the expected number of events that happen in, say, 5 minutes is

$$\lambda t = (3 \text{ events per minute})(5 \text{ minutes}) = 15 \text{ events}. \qquad (4.17)$$

Does this mean that we are guaranteed to have exactly 15 events during a particular 5-minute span? Absolutely not. We can theoretically have any number of events, although there is essentially zero chance that the number will differ significantly from 15. (The probability of obtaining the various numbers of events is governed by the Poisson distribution, which we'll discuss in Section 4.7.) But the *expectation value is* 15. That is, if we perform a large number of 5-minute trials and then calculate the average number of events that occur in each trial, the result will be close to 15.

A trickier question to ask is: What is the probability that *exactly one* event happens during a time t? Since λ is the rate, you might think that you can just multiply λ by t, as we did above, to say that the probability is λt. But this certainly can't be correct, because it would imply a probability of 15 for a 5-minute interval in the above setup. This is nonsense, because probabilities can't be larger than 1. If we instead pick a time interval of 20 seconds (1/3 of a minute), we obtain a λt value of 1. This doesn't have the fatal flaw of being larger than 1, but it has another issue, in that it says that exactly one event is *guaranteed* to happen during a 20-second interval. This can't be correct either, because it's certainly possible for zero (or two or three, etc.) events to occur. We'll figure out the exact probabilities of these numbers in Section 4.7.

The strategy of multiplying λ by t to obtain a probability doesn't seem to work. However, there is one special case where it *does* work. If the time interval is extremely small (let's call it ϵ, which is a standard letter to use for something that is very small), then it *is* true that the probability of *exactly one* event occurring during the ϵ time interval is *essentially* equal to $\lambda \epsilon$. We're using the word "essentially" because, although this statement is technically not true, it becomes arbitrarily close to being true in the limit where ϵ approaches zero. In the above example with $\lambda = 1/20$ events per second, the statement, "λt is the probability that exactly one event happens during a time t," is a lousy approximation if $t = 20$ seconds, a decent approximation if $t = 2$ seconds, and a very good approximation if $t = 0.2$ seconds. And it only gets better as the time interval gets smaller. We'll explain why in the first remark below.

We can therefore say that if $P_\epsilon(1)$ stands for the probability that exactly one event happens during a small time interval ϵ, then

$$\boxed{P_\epsilon(1) \approx \lambda \epsilon} \qquad \text{(if } \epsilon \text{ is very small)} \qquad (4.18)$$

The smaller ϵ is, the better this approximation is. Technically, the condition in Eq. (4.18) is really "if $\lambda \epsilon$ is very small." But we'll generally be dealing with "normal" sized λ's, so $\lambda \epsilon$ being small is equivalent to ϵ being small. When we deal with continuous time below, we'll actually be taking the $\epsilon \to 0$ limit. In this mathematical limit, the "$\approx$" sign in Eq. (4.18) becomes an exact "=" sign. To sum up:

- If t is very small, then λt is both the expected number of events that happen during the time t *and* (essentially) the probability that exactly one event happens during the time t.

- If t *isn't* very small, then λt is only the expected number of events.

REMARKS:

1. We claimed above that λt equals the probability of *exactly one* event occurring, only if t is very small. The reason for this restriction is that if t *isn't* small, then there is the possibility of *multiple* events occurring during the time t. We can be explicit about this as follows. Since we know from Eq. (4.16) that the expected number of events during

any time t is λt, we can use the expression for the expectation value in Eq. (3.4) to write

$$\lambda t = P_t(0) \cdot 0 + P_t(1) \cdot 1 + P_t(2) \cdot 2 + P_t(3) \cdot 3 + \cdots, \qquad (4.19)$$

where $P_t(k)$ is the probability of obtaining exactly k events during the time t. Solving for $P_t(1)$ gives

$$P_t(1) = \lambda t - P_t(2) \cdot 2 - P_t(3) \cdot 3 + \cdots. \qquad (4.20)$$

We see that $P_t(1)$ is smaller than λt due to the $P_t(2)$ and $P_t(3)$, etc., probabilities. So $P_t(1)$ isn't equal to λt. However, if all of the probabilities of multiple events occurring ($P_t(2)$, $P_t(3)$, etc.) are very small, then $P_t(1)$ is *essentially* equal to λt. And this is exactly what happens if the time interval is very small. For small times, there is hardly any chance of the event even occurring *once*. So it is even less likely that it will occur *twice*, and even less likely for three times, etc.

We can be a little more precise about this. The following argument isn't completely rigorous, but it should convince you that if t is very small, then $P_t(1)$ is essentially equal to λt. If t is very small, then assuming we don't know yet that $P_t(1)$ *equals* λt, we can still say that it should be roughly *proportional* to λt. This is true because if an event has only a tiny chance of occurring, then if you cut λ in half, the probability is essentially cut in half. Likewise if you cut t in half. This proportionality then implies that the probability that exactly two events occur is essentially proportional to $(\lambda t)^2$. We'll see in Section 4.7 that there is actually a factor of $1/2$ involved here, but that is irrelevant in the present argument. The important point is the *quadratic* nature of $(\lambda t)^2$. If λt is sufficiently small, then $(\lambda t)^2$ is negligible compared with λt. Likewise for $P_t(3) \propto (\lambda t)^3$, etc. We can therefore ignore the scenarios where multiple events occur. So with $t \to \epsilon$, Eq. (4.20) becomes

$$P_\epsilon(1) \approx \lambda \epsilon - \cancel{P_\epsilon(2)} \cdot 2 - \cancel{P_\epsilon(3)} \cdot 3 + \cdots, \qquad (4.21)$$

in agreement with Eq. (4.18). As mentioned above, if $\lambda \epsilon$ is small, it is because ϵ is small, at least in the situations we'll be dealing with.

2. Imagine drawing the λ vs. t "curve." We have put "curve" in quotes because the curve is actually just a straight horizontal line, since we're assuming a constant λ. If we consider a time interval Δt, the associated area under the curve equals $\lambda \Delta t$, because we have a simple rectangular region. So from Eq. (4.18), this area gives the probability that an event occurs during a time Δt, *provided* that Δt is very small. This might make you think that λ can be interpreted as a probability distribution, because we found in Section 4.2.3 that the area under a distribution curve gives the probability. However, the λ "curve" *cannot* be interpreted as a probability distribution, because this area-equals-probability result holds *only for very small* Δt. The area under a distribution curve has to give the probability for *any* interval on the horizontal axis. The λ "curve" doesn't satisfy this property. The total area under the λ "curve" is infinite (because the straight horizontal line extends for all time), whereas actual probability distributions must have a total area of 1.

3. Since only *one* quantity, λ, is needed to describe everything about a random process whose probability is uniform in time, any other quantity we might want to determine must be able to be written in terms of λ. This will become evident below. ♣

4.6.3 Continuous case

In the case of discrete time in Section 4.6.1, we asked two questions: What is the average waiting time between successes? And what is the probability distribution of the waiting times between successes? We'll now answer these two questions in the case where time is a continuous quantity.

Average waiting time

As in the discrete case, the first of the two questions is fairly easy to answer. Let the average rate of success be λ, and consider a large time t. We know from Eq. (4.16) that the average total number of events that occur during the time t is λt. The average waiting time (which we'll call τ) is the total time divided by the total number of events, λt. That is,

$$\tau = \frac{t}{\lambda t} \implies \boxed{\tau = \frac{1}{\lambda}} \qquad \text{(average waiting time)} \qquad (4.22)$$

We see that the average waiting time is simply the reciprocal of the rate at which the events occur. For example, if the rate is 5 events per second, then the average waiting time is $1/5$ of a second, which makes sense. This would of course be true in the nonrandom case where the events occur at exactly equally spaced intervals of $1/5$ second. But the nice thing is that Eq. (4.22) holds even for the random process we're discussing, where the intervals aren't equally spaced.

It makes sense that the rate λ is in the denominator in Eq. (4.22), because if λ is small, the average waiting time is large. And if λ is large, the average waiting time is small. And as promised in the third remark above, τ depends on λ.

Distribution of waiting times

Now let's answer the second (more difficult) question: What is the probability distribution of the waiting times between successes? Equivalently, what is the probability that the waiting time from a given event to the next event is between t and $t + \Delta t$, where Δt is small? To answer this, we'll use the same general strategy that we used in the discrete case in Section 4.6.1, except that now the time interval between iterations will be a very small time ϵ instead of 1 second. We will then take the $\epsilon \to 0$ limit, which will make time continuous.

The division of time into little intervals is summarized in Fig. 4.13. From time zero (which is when we'll assume the initial event happens) to time t, we'll break up time into a very large number of very small intervals with length ϵ (which means that there are t/ϵ of these intervals). And then the interval of Δt sits at the end. Both ϵ and Δt are assumed to be very small, but they need not have anything to do with each other. ϵ exists as a calculational tool only, while Δt is the arbitrarily-chosen small time interval that appears in Eq. (4.2).

In order for the next success (event) to happen between t and $t + \Delta t$, there must be failure during every one of the t/ϵ intervals of length ϵ shown in Fig. 4.13, and then there must be success between t and $t + \Delta t$. From Eq. (4.18), the latter happens with probability $\lambda \Delta t$, because Δt is assumed to be very small. Also, Eq. (4.18) says

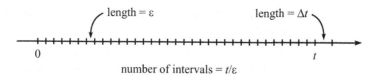

number of intervals = t/ε

Figure 4.13: Dividing time into little intervals.

that the probability of success in any given small interval of length ϵ is $\lambda\epsilon$, which means that the probability of failure is $1 - \lambda\epsilon$. And since there are t/ϵ of these intervals, the probability of failure in all of them is $(1 - \lambda\epsilon)^{t/\epsilon}$. The probability that the next success happens between t and $t + \Delta t$, which we'll label as $P(t,\Delta t)$, is therefore

$$P(t,\Delta t) = \left((1 - \lambda\epsilon)^{t/\epsilon}\right)(\lambda\,\Delta t). \tag{4.23}$$

The reasoning that led to this equation is in the same spirit as the reasoning that led to Eq. (4.14). See the first remark below.

It's now time to use one of the results from Appendix C, namely the approximation given in Eq. (7.14), which says that for small a we can write[4]

$$(1 + a)^n \approx e^{na}. \tag{4.24}$$

This works for negative a as well as positive a. Here e is Euler's number, which has the value of $e \approx 2.71828$. (If you want to know more about e, there's plenty of information in Appendix B!) For the case at hand, a comparison of Eqs. (4.23) and (4.24) shows that we want to define $a \equiv -\lambda\epsilon$ and $n \equiv t/\epsilon$, which yields $na = (t/\epsilon)(-\lambda\epsilon) = -\lambda t$. Eq. (4.24) then gives $(1 - \lambda\epsilon)^{t/\epsilon} \approx e^{-\lambda t}$, so Eq. (4.23) becomes

$$P(t,\Delta t) = e^{-\lambda t}\lambda\,\Delta t. \tag{4.25}$$

The probability distribution (or density) is obtained by simply erasing the Δt, because Eq. (4.2) says that the density is obtained by dividing the probability by the interval length. We therefore see that the desired probability distribution for the waiting time between successes is

$$\boxed{\rho(t) = \lambda e^{-\lambda t}} \quad \text{(exponential distribution)} \tag{4.26}$$

This is known as the *exponential distribution*. This name is appropriate, of course, because the distribution decreases exponentially with t. As promised in the third remark on page 201, the distribution depends on λ (along with t, of course). In the present setup involving waiting times, it is often more natural to work in terms of the average waiting time τ than the rate λ, in which case the preceding result becomes (using $\lambda = 1/\tau$ from Eq. (4.22))

$$\boxed{\rho(t) = \frac{e^{-t/\tau}}{\tau}} \quad \text{(exponential distribution)} \tag{4.27}$$

---

[4]You are strongly encouraged to read Appendix C at this point, if you haven't already. But if you want to take Eq. (4.24) on faith, that's fine too. However, you should at least verify with a calculator that it works fairly well for, say, $a = 0.01$ and $n = 200$.

In the notation of Eq. (4.12), both A and b are equal to $1/\tau$ (or λ). So they are in fact related, as we noted right after Eq. (4.12).

Fig. 4.14 shows plots of $\rho(t)$ for a few different values of the average waiting time, τ. The two main properties of each of these curves are the starting value at $t = 0$ and the rate of decay as t increases. From Eq. (4.27), the starting value at $t = 0$ is $e^0/\tau = 1/\tau$. So the bigger τ is, the smaller the starting value. This makes sense, because if the average waiting time τ is large (equivalently, if the rate λ is small), then there is only a small chance that the next event will happen right away.

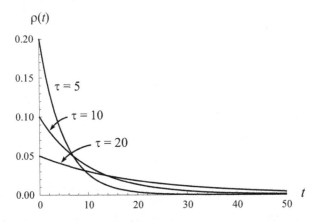

Figure 4.14: Examples of exponential distributions with different values of the average waiting time τ.

How fast do the curves decay? This is governed by the denominator of the exponent in Eq. (4.27). For every τ units that t increases by, $\rho(t)$ decreases by a factor of $1/e$. This can be seen by plugging a time of $t + \tau$ into Eq. (4.27), which gives

$$\rho(t + \tau) = \frac{e^{-(t+\tau)/\tau}}{\tau} = \frac{(e^{-t/\tau} \cdot e^{-1})}{\tau} = \frac{1}{e} \cdot \frac{e^{-t/\tau}}{\tau} = \frac{1}{e}\rho(t). \qquad (4.28)$$

So $\rho(t + \tau)$ is $1/e$ times as large as $\rho(t)$, and this holds for any value of t. A few particular values of $\rho(t)$ are

$$\rho(0) = \frac{1}{\tau}, \quad \rho(\tau) = \frac{1}{e\tau}, \quad \rho(2\tau) = \frac{1}{e^2\tau}, \quad \rho(3\tau) = \frac{1}{e^3\tau}, \qquad (4.29)$$

and so on. If τ is large, the curve takes longer to decrease by a factor of $1/e$. This is consistent with Fig. 4.14, where the large-τ curve falls off slowly, and the small-τ curve falls off quickly. To sum up, if τ is large, the $\rho(t)$ curve starts off low and then decays slowly. And if τ is small, the curve starts off high and then decays quickly.

Example (Same density): Person A measures a very large number of waiting times for a process with $\tau = 5$. Person B does the same for a process with $\tau = 20$. To their surprise, they find that for a special value of t, they both observe (roughly) the same number of waiting times that fall into a given small interval around t. What is this special value of t?

Solution: The given information tells us that the probability densities for the two processes are equal at the special value of t. Plugging the τ values of 5 and 20 into Eq. (4.27) and setting the results equal to each other gives

$$\frac{e^{-t/5}}{5} = \frac{e^{-t/20}}{20} \implies \frac{20}{5} = e^{t/5-t/20} \implies \ln\left(\frac{20}{5}\right) = t\left(\frac{1}{5} - \frac{1}{20}\right)$$

$$\implies \ln 4 = t\left(\frac{15}{100}\right) \implies t = 9.24. \tag{4.30}$$

This result agrees (at least to the accuracy of a visual inspection) with the value of t where the $\tau = 5$ and $\tau = 20$ curves intersect in Fig. 4.14.

Although it might seem surprising that there exists a value of t for which the densities associated with two different values of τ are equal, it is actually fairly clear, due to the following continuity argument. For small values of t, the $\tau = 5$ process has a larger density (because the events happen closer together), while for large values of t, the $\tau = 20$ process has a larger density (because the events happen farther apart). Therefore, by continuity, there must exist a particular value of t for which the densities are equal. But it takes the above calculation to find the exact value.

REMARKS:

1. In comparing Eq. (4.23) with Eq. (4.14), we see in retrospect that we could have obtained Eq. (4.23) by simply replacing the first p in Eq. (4.14) with $\lambda\epsilon$ (because $\lambda\epsilon$ is the probability of success at each intermediate step), the second p with $\lambda\,\Delta t$ (this is the probability of success at the last step), and $k - 1$ with t/ϵ (this is the number of intermediate steps). But you might find these replacements a bit mysterious without the benefit of the reasoning preceding Eq. (4.23).

2. The area under each of the curves in Fig. 4.14 must be 1. The waiting time has to be *something*, so the sum of all the probabilities must be 1. The proof of this fact is very quick, but it requires calculus, so we'll relegate it to Problem 4.8(a). (But note that we did demonstrate this for the discrete case in Eq. (4.15).) Likewise, the expectation value of the waiting time must be τ, because that's how τ was defined. Again, the proof is quick but requires calculus; see Problem 4.8(c). (The demonstration for the discrete case is the task of Problem 4.7.)

3. We've been referring to $\rho(t)$ as the probability distribution of the waiting times from one event to the next. However, $\rho(t)$ is actually the distribution of the waiting times from *any point in time* to the occurrence of the next event. That is, you can start your stopwatch at any time, not just at the occurrence of an event. If you go back through the above discussion, you will see that nowhere did we use the fact that an event actually occurred at $t = 0$.

   However, beware of the following incorrect reasoning. Let's say that an event happens at $t = 0$, but that you don't start your stopwatch until, say, $t = 1$. The fact that the next event after $t = 1$ doesn't happen (on average) until $t = 1 + \tau$ (from the previous paragraph) seems to imply that the average waiting time from $t = 0$ is $1 + \tau$. But it better not be, because we know from above that it's just τ. The error here is that we forgot about the scenarios where the next event after $t = 0$ happens *between* $t = 0$ and $t = 1$. When these events are included, the average waiting time, starting at $t = 0$, ends up correctly being τ. (The demonstration of this fact requires calculus.) In short, the waiting time from $t = 1$ is indeed τ, but the next event (after the $t = 0$ event) might have already happened before $t = 1$.

4. In a sense, the curves for all of the different values of τ in Fig. 4.14 are really the same curve. They're just stretched or squashed in the horizontal and vertical directions. The general form of the curve described by the expression in Eq. (4.27) is shown in Fig. 4.15.

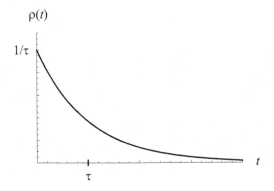

Figure 4.15: The general form of the exponential distribution.

As long as we change the scales on the axes so that τ and $1/\tau$ are always located at the same positions, then the curves will look the same for any τ. For example, as we saw in Eq. (4.29), no matter what the value of τ is, the value of the curve at $t = \tau$ is always $1/e$ times the value at $t = 0$. Of course, when we plot things, we usually keep the scales fixed, in which case the τ and $1/\tau$ positions move along the axes, as shown in Fig. 4.16 (these are the same curves as in Fig. 4.14). But by suitable uniform stretching/squashing of the axes, the curve in Fig. 4.15 can be turned into any of the curves in Fig. 4.16.

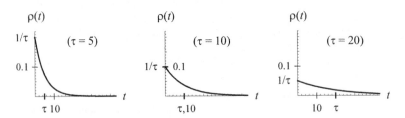

Figure 4.16: These curves can be obtained from the curve in Fig. 4.15 by suitable stretching/squashing of the axes.

5. The fact that any of the curves in Fig. 4.16 can be obtained from any of the other curves by stretching and squashing the two directions by inverse (as you can verify) factors

implies that the areas under all of the curves are the same. (This is consistent with the fact that all of the areas must be 1.) To see how these inverse factors work together to keep the area constant, imagine the area being broken up into a large number of thin vertical rectangles, stacked side by side under the curve. The stretching and squashing of the curve does the same thing to each rectangle. All of the widths get stretched by a factor of f, and all of the heights get squashed by the same factor of f (or $1/f$, depending on your terminology). So the area of each rectangle remains the same. The same thing must then be true for the area under the whole curve.

6. Note that the distribution for the waiting time is a discrete distribution in the case of discrete time (see Eq. (4.14)), and a continuous distribution in the case of continuous time (see Eq. (4.27)). Although these facts make perfect sense, one should be careful about extrapolating to a general conclusion. In the Poisson discussion in the following section, we'll encounter a discrete distribution in the case of continuous time. ♣

4.7 Poisson distribution

The goal of this section is to derive the Poisson probability distribution,

$$\boxed{P(k) = \frac{a^k e^{-a}}{k!}} \qquad \text{(Poisson distribution)} \qquad (4.31)$$

The parameter a depends on the situation at hand, and k is the value of the random variable, which is the number of events that occur in a certain region of time (or space, or whatever), as we'll discuss below. Since k is an integer (because it is the number of events that occur), the Poisson distribution is a discrete one. A common type of situation where this distribution arises is the following.

As with the exponential distribution in the previous section, consider a repeating event that happens completely randomly in time. We will show that the probability distribution of the *number of events that occur during a given time interval* takes the form of the above Poisson distribution. Whereas the exponential distribution deals with the *waiting time* until the next event, the Poisson distribution deals with the *number of events* in a given time interval. As in the case of the exponential distribution, our strategy for deriving the Poisson distribution will be to first consider the case of discrete time, and then the case of continuous time.

4.7.1 Discrete case

Consider a process that is repeated each second (so time is discretized into 1-second intervals), and let the probability of success in each trial be p (the same for all trials). For example, as in Section 4.6.1, we can roll a hypothetical 10-sided die once every second, and if the die shows a "1," then we consider that a success. The other nine numbers represent failure. As in Section 4.6.1, it isn't actually necessary to introduce time here. We could simply talk about the number of iterations of the process, as we will in the balls-in-boxes example below.

The question we will answer here is: What is the probability distribution of the number of successes that occur in a time interval of n seconds? In other words, what is the probability, $P(k)$, that exactly k events happen during a time span of *n* seconds? It turns out that this is *exactly* the same question that we answered in Section 4.5 when we derived the binomial distribution in Eq. (4.6). So we can just copy over the reasoning here. We'll formulate things in the language of rolls of a die, with a "1" being a success. But the setup could be anything with a probability *p* of success.

The probability that a *specific set* of *k* of the *n* rolls all yield a 1 equals p^k, because each of the *k* rolls has a *p* probability of yielding a 1. We then need the other $n - k$ rolls to *not* yield a 1, because we want *exactly k* 1's. This happens with probability $(1 - p)^{n-k}$, because each of the $n - k$ rolls has a $1 - p$ probability of being something other than a 1. The probability that a specific set of *k* rolls (and no others) all yield success is therefore $p^k \cdot (1 - p)^{n-k}$. Finally, since there are $\binom{n}{k}$ ways to pick a specific set of *k* rolls, we see that the probability that exactly *k* of the *n* rolls yield a 1 is

$$P(k) = \binom{n}{k} p^k (1 - p)^{n-k} \tag{4.32}$$

This distribution is exactly the same as the binomial distribution in Eq. (4.6), so there's nothing new here. But there will indeed be something new when we discuss the continuous case in Section 4.7.2.

Example (Balls in boxes): Let *n* balls be thrown randomly into *b* boxes. What is the probability, $P(k)$, that a given box has exactly *k* balls in it?

Solution: This is a restatement of the problem we just solved. Imagine randomly throwing one ball each second into the boxes, and consider a particular box. (As mentioned above, the time interval of one second is irrelevant. All that matters is that we perform *n* iterations of the process, sooner or later.) If a given ball ends up in that box, we'll call that a success. For each ball, this happens with probability $1/b$, because there are *b* boxes. So the *p* in the above discussion equals $1/b$. Since we're throwing *n* balls into the boxes, we're simply performing *n* iterations of a process that has a probability $p = 1/b$ of success. Eq. (4.32) is therefore applicable, and with $p = 1/b$ it gives the probability of obtaining exactly *k* successes (that is, exactly *k* balls in a particular box) as

$$P(k) = \binom{n}{k} \left(\frac{1}{b}\right)^k \left(1 - \frac{1}{b}\right)^{n-k}. \tag{4.33}$$

We've solved the problem, but let's now see if our answer makes sense. As a concrete example, consider the case where we have $n = 1000$ balls and $b = 100$ boxes. On average, we expect to have $n/b = 10$ balls in each box. But many (in fact, most) of the boxes will have other numbers of balls. In theory, the number *k* of balls in a particular box can take on any value from 0 to $n = 1000$. But intuitively we expect most of the boxes to have *roughly* 10 balls (say, between 5 and 15 balls). We certainly don't expect many boxes to have 2 or 50 balls.

Fig. 4.17 shows a plot of the $P(k)$ in Eq. (4.33), for the case where $n = 1000$ and $b = 100$. As expected, it is peaked near the average value, $n/b = 10$, and it becomes negligible a moderate distance away from $k = 10$. There is very little chance of having fewer than 3 or more than 20 balls in a given box; Eq. (4.33) gives $P(2) \approx 0.2\%$ and $P(21) \approx 0.1\%$. We've arbitrarily chopped off the plot at $k = 30$ because the probabilities between $k = 30$ (or even earlier) and $k = 1000$ are indistinguishable from zero. But technically all of these probabilities are nonzero. For example, $P(1000) = (1/100)^{1000}$, because if $k = 1000$ then all of the 1000 balls need to end up in the given box, and each one ends up there with probability $1/100$. The resulting probability of 10^{-2000} is utterly negligible.

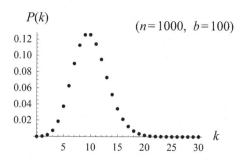

Figure 4.17: The probability distribution for the number of balls in a given box, if $n = 1000$ balls are thrown into $b = 100$ boxes.

4.7.2 Continuous case

As with the exponential distribution in Section 4.6.3, we'll now consider the case where time is continuous. That is, we'll assume that we can have a successful event at *any* instant, not just at the evenly-spaced 1-second marks, as we assumed above. As in Section 4.6.3, such a process can be completely described by just *one* number – the average rate of events, which we'll again call λ. Eq. (4.18) tells us that $\lambda \epsilon$ is the probability that exactly one event occurs in a very small time interval ϵ. The smaller the ϵ, the smaller the probability that the event occurs. We're assuming that λ is constant in time, that is, the event is just as likely to occur at one time as any other.

Our goal here is to answer the question: What is the probability, $P(k)$, that exactly k events occur during a given time span of t? To answer this, we'll use the same general strategy that we used above in the discrete case, except that now the time interval between iterations will be a very small time ϵ instead of 1 second. We will then take the $\epsilon \to 0$ limit, which will make time continuous. The division of time into little intervals is summarized in Fig. 4.18. We're dividing the time interval t into a very large number of very small intervals with length ϵ. There are t/ϵ of these intervals, which we'll label as n. There is no need to stick a Δt interval on the end, as there was in Fig. 4.13.

Compared with the discrete case we addressed above, Eq. (4.18) tells us that the probability of exactly one event occurring in a given small interval of length ϵ is now $\lambda \epsilon$ instead of p. So we can basically just repeat the derivation preceding Eq. (4.32), which itself was a repetition of the derivation preceding Eq. (4.6). You're probably getting tired of it by now!

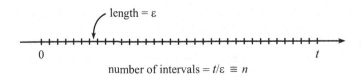

Figure 4.18: Dividing time into little intervals.

The probability that a *specific set* of k of the n little intervals all yield exactly one event each equals $(\lambda\epsilon)^k$, because each of the k intervals has a $\lambda\epsilon$ probability of yielding one event. We then need the other $n - k$ intervals to *not* yield an event, because we want *exactly* k events. This happens with probability $(1 - \lambda\epsilon)^{n-k}$, because each of the $n - k$ intervals has a $1 - \lambda\epsilon$ chance of yielding zero events. The probability that a specific set of k intervals (and no others) all yield an event is therefore $(\lambda\epsilon)^k \cdot (1 - \lambda\epsilon)^{n-k}$. Finally, since there are $\binom{n}{k}$ ways to pick a specific set of k intervals, we see that the probability that exactly k of the n intervals yield an event is

$$P(k) = \binom{n}{k}(\lambda\epsilon)^k(1 - \lambda\epsilon)^{n-k}. \tag{4.34}$$

This is simply Eq. (4.32) with p replaced by $\lambda\epsilon$.

Now it's time to have some mathematical fun. Let's see what Eq. (4.34) reduces to in the $\epsilon \to 0$ limit, which will give us the desired continuous-time limit. Note that $\epsilon \to 0$ implies that $n \equiv t/\epsilon \to \infty$. The math here will be a little more involved than the math that led to the exponential distribution in Eq. (4.26).

If we write out the binomial coefficient and expand things a bit, Eq. (4.34) becomes

$$P(k) = \frac{n!}{(n - k)!\,k!}(\lambda\epsilon)^k(1 - \lambda\epsilon)^n(1 - \lambda\epsilon)^{-k}. \tag{4.35}$$

Of the various letters in this equation, n is huge, ϵ is tiny, and λ and k are "normal," not assumed to be huge or tiny. λ is determined by the setup, and k is the number of events we're concerned with. (We'll see below that the relevant k's are roughly the size of the product $\lambda t = \lambda n\epsilon$.) In the $\epsilon \to 0$ limit (and hence $n \to \infty$ limit), we can make three approximations to Eq. (4.35):

- First, in the $n \to \infty$ limit, we can say that

$$\frac{n!}{(n - k)!} \approx n^k, \tag{4.36}$$

  at least in a multiplicative sense (we don't care about an additive sense). This follows from the fact that $n!/(n - k)!$ is the product of the k numbers from n down to $n - k + 1$. And if n is large compared with k, then all of these k numbers are essentially equal to n (multiplicatively). Therefore, since there are k of them, we simply get n^k. You can verify this for, say, the case of $n = 1,000,000$ and $k = 10$. The product of the 10 numbers from $1,000,000$ down to $999,991$ equals $1,000,000^{10}$ to within an error of 0.005%

- Second, we can apply the $(1 + a)^n \approx e^{na}$ approximation from Eq. (7.14) in Appendix C, which we already used once in the derivation of the exponential distribution; see the discussion following Eq. (4.24). We can use this approximation to simplify the $(1 - \lambda\epsilon)^n$ term. With $a \equiv -\lambda\epsilon$, Eq. (7.14) gives

$$(1 - \lambda\epsilon)^n \approx e^{-n\lambda\epsilon}. \tag{4.37}$$

- Third, in the $\epsilon \to 0$ limit, we can use the $(1+a)^n \approx e^{na}$ approximation again, this time to simplify the $(1 - \lambda\epsilon)^{-k}$ term. The result is

$$(1 - \lambda\epsilon)^{-k} \approx e^{k\lambda\epsilon} \approx e^0 = 1, \tag{4.38}$$

because for any fixed values of k and λ, the $k\lambda\epsilon$ exponent becomes infinitesimally small as $\epsilon \to 0$. Basically, in $(1 - \lambda\epsilon)^{-k}$ we're forming a finite power of a number that is essentially equal to 1. Note that this reasoning doesn't apply to the $(1 - \lambda\epsilon)^n$ term in Eq. (4.37), because n isn't a fixed number. It changes with ϵ, in that it becomes large as ϵ becomes small.

In the $\epsilon \to 0$ and $n \to \infty$ limits, the "$\approx$" signs in the approximations in the preceding three equations turn into exact "=" signs. Applying these three approximations to Eq. (4.35) gives

$$\begin{aligned}
P(k) &= \frac{n!}{(n-k)!\,k!}(\lambda\epsilon)^k(1 - \lambda\epsilon)^n(1 - \lambda\epsilon)^{-k} \\
&= \frac{n^k}{k!}(\lambda\epsilon)^k e^{-n\lambda\epsilon} \cdot 1 \\
&= \frac{1}{k!}(\lambda \cdot n\epsilon)^k e^{-\lambda \cdot n\epsilon} \\
&= \frac{1}{k!}(\lambda t)^k e^{-\lambda t},
\end{aligned} \tag{4.39}$$

where we have used $n \equiv t/\epsilon \implies n\epsilon = t$ to obtain the last line. Now, from Eq. (4.16) λt is the average number of events that are expected to occur in the time t. Let's label this average number of events as $a \equiv \lambda t$. We can then write Eq. (4.39) as

$$\boxed{P(k) = \frac{a^k e^{-a}}{k!}} \qquad \text{(Poisson distribution)} \tag{4.40}$$

where a is the average number of events in the time interval under consideration. If you want, you can indicate the a value by writing $P(k)$ as $P_a(k)$.

Since a is the only parameter left on the righthand side of Eq. (4.40), the distribution is completely specified by a. The individual values of λ and t don't matter. All that matters is their product $a \equiv \lambda t$. This means that if we, say, double the time interval t under consideration and also cut the rate λ in half, then a remains unchanged; so we have exactly the same distribution $P(k)$. Although it is clear that doubling t and halving λ yields the same *average* number of events (since the average equals the product λt), it might not be intuitively obvious that the entire $P(k)$ *distribution* is the same. But the result in Eq. (4.40) shows that this is indeed the case.

The Poisson distribution in Eq. (4.40) gives the probability of obtaining exactly k events during a period of time for which the expected number is a. Since k is a discrete variable (being the integer number of times that an event occurs), the Poisson distribution is a discrete distribution. Although the Poisson distribution is derived from a *continuous* process (in that the time t is continuous, which means that an event can happen at any time), the distribution itself is a *discrete* distribution, because k must be an integer. Note that while the *observed* number of events k must be an integer, the *average* number of events a need not be.

REMARK: Let's discuss this continuous/discrete issue a little further. In the last remark in Section 4.6.3, we noted that the exponential distribution for the waiting time, t, is a discrete distribution in the case of discrete time, and a continuous distribution in the case of continuous time. This seems reasonable. But for the Poisson distribution, the distribution for the number of events, k, is a discrete distribution in the case of discrete time, and also (as we just noted) a discrete distribution in the case of continuous time. It is simply always a discrete distribution, because the random variable is the number of events, k, which is discrete. The fact that time might be continuous is irrelevant, as far as the discreteness of k goes. The difference in the case of the exponential distribution is that *time itself* is the random variable (because we're considering waiting times). So if we make time continuous, then by definition we're also making the random variable continuous, which means that we have a continuous distribution. ♣

Example (Number of shoppers): On average, one shopper enters a given store every 15 seconds. What is the probability that in a given time interval of one minute, zero shoppers enter the store? Four shoppers? Eight shoppers?

Solution: The given average time interval of 15 seconds tells us that the average number of shoppers who enter the store in one minute is $a = 4$. Having determined a, we simply need to plug the various values of k into Eq. (4.40). For $k = 0, 4,$ and 8 we have

$$P(0) = \frac{4^0 e^{-4}}{0!} = 1 \cdot e^{-4} \approx 0.018 \approx 2\%,$$

$$P(4) = \frac{4^4 e^{-4}}{4!} = \frac{32}{3} \cdot e^{-4} \approx 0.195 \approx 20\%,$$

$$P(8) = \frac{4^8 e^{-4}}{8!} = \frac{512}{315} \cdot e^{-4} \approx 0.030 = 3\%. \tag{4.41}$$

We see that the probability that four shoppers enter the store in a given minute is about 10 times the probability that zero shoppers enter. The probabilities quickly die off as k gets larger. For example, $P(12) \approx 0.06\%$.

The above results are a subset of the information contained in the plot of $P(k)$ shown in Fig. 4.19. Note that $P(3) = P(4)$. This is evident from the above expression for $P(4)$, because if we cancel a factor of 4 in the numerator and denominator, we end up with $4^3 e^{-4}/3!$ which equals $P(3)$. See Problem 4.10 for more on this equality.

Remember that when finding $P(k)$, the only parameter that matters is a. If we modify the problem by saying that on average one shopper enters the store every 15 *minutes*, and if we change the time interval to one *hour* (in which case a again equals 4), then all of the $P(k)$ values are exactly the same as above.

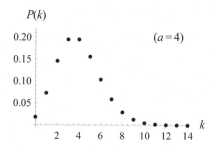

Figure 4.19: The Poisson distribution with $a = 4$.

Example (Balls in boxes, again): Although Eq. (4.40) technically holds only in the limit of a continuous process, it still provides a very good approximation for discrete processes, as long as the numbers involved are fairly large. Consider the balls-in-boxes example in Section 4.7.1. With $n = 1000$ and $b = 100$, the average number of balls in a box is $a = n/b = 10$. Since b is fairly large, we expect that the Poisson distribution in Eq. (4.40) with $a = 10$ will provide a good approximation to the exact binomial distribution in Eq. (4.33) with $n = 1000$ and $b = 100$, or equivalently Eq. (4.32) with $n = 1000$ and $p = 1/b = 1/100$.

Let's see how good the approximation is. Fig. 4.20 shows plots for two different sets of n and b values: $n = 100$, $b = 10$; and $n = 1000$, $b = 100$. With these values, both plots have $a = 10$. The dots in the second plot are a copy of the dots in Fig. 4.17. In both plots we have superimposed the exact discrete binomial distribution (the dots) and the Poisson distribution (the curves).[5] Since the plots have the same value of a, they have the same Poisson curve. In the right plot, the points pretty much lie on the curve, so the approximate Poisson probabilities in Eq. (4.40) are essentially the same as the exact binomial probabilities in Eq. (4.33). In other words, the approximation is a very good one.

However, in the left plot, the points lie slightly off the curve. The average $a = n/b$ still equals 10, so the Poisson curve is exactly the same as in the right plot. But the exact binomial probabilities in Eq. (4.33) are changed from the $n = 1000$ and $b = 100$ case. The Poisson approximation doesn't work as well here, although it's still reasonably good. The condition under which the Poisson approximation is a good one turns out to be the very simple relation, $p \equiv 1/b \ll 1$. See Problem 4.14.

The Poisson distribution in Eq. (4.40) works perfectly well for small a, even $a < 1$. It's just that in this case, the plot of $P(k)$ doesn't have a bump, as it does in Figs. 4.19 and 4.20. Instead, it starts high and then falls off as k increases. Fig. 4.21 shows the plot of $P(k)$ for various values of a. We've arbitrarily decided to cut off the plots at $k = 20$, even though they technically go on forever. Since we are assuming that time is continuous, we can theoretically have an arbitrarily large number of

---

[5]We've drawn the Poisson distribution as a continuous curve (the $k!$ in Eq. (4.40) can be extrapolated to non-integer values of k), because it would be difficult to tell what's going on in the figure if we plotted two sets of points nearly on top of each other. But you should remember that we're really only concerned with integer values of k, since the k in Eq. (4.40) is the number of times something occurs. We've plotted the whole curve for visual convenience only.

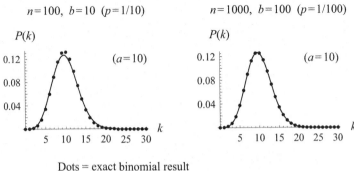

Dots = exact binomial result

Curves = Poisson approximation (both plots have $a=10$)

Figure 4.20: Comparison between the exact binomial result and the Poisson approximation.

events in any given time interval, although the probability will be negligibly small. In the plots, the probabilities are effectively zero by $k = 20$, except in the $a = 15$ case.

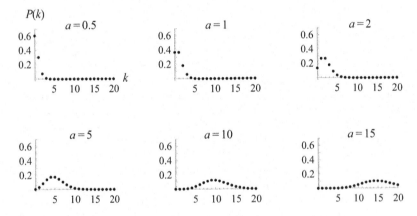

Figure 4.21: The Poisson distribution for various values of a.

As a increases, the bump in the plots (once it actually becomes a bump) does three things: (1) it shifts to the right, because it is centered near $k = a$, due to the result in Problem 4.10, (2) it decreases in height, due to the result in Problem 4.11, and (3) it becomes wider, due to the result in Problem 4.13. The last two of these properties are consistent with each other, in view of the fact that the sum of all the probabilities must equal 1, for any value of a.

Eq. (4.40) gives the probability of obtaining zero events as $P(0) = e^{-a}$. If $a = 0.5$ then $P(0) = e^{-0.5} \approx 0.61$. This agrees with the first plot in Fig. 4.21. Likewise, if $a = 1$ then $P(0) = e^{-1} \approx 0.37$, in agreement with the second plot. If a is large then the $P(0) = e^{-a}$ probability goes to zero, in agreement with the bottom three plots. This makes sense; if the average number of events is *large*, then it is very *unlikely* that we will obtain zero events. In the opposite extreme, if a is very small (for example, $a = 0.01$), then the $P(0) = e^{-a}$ probability is very close to 1.

This again makes sense; if the average number of events is very *small*, then it is very *likely* that we will obtain zero events.

To make it easier to compare the six plots in Fig. 4.21, we have superimposed them in Fig. 4.22. Although we have drawn these Poisson distributions as continuous curves to make things clearer, remember that the distribution applies only to integer values of k.

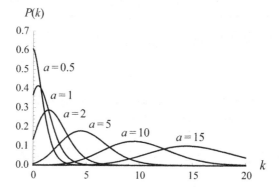

Figure 4.22: Superimposing the plots in Fig. 4.21, drawn as continuous curves.

Problems 4.9 through 4.13 cover various aspects of the Poisson distribution, namely: the fact that the total probability is 1, the location of the maximum, the value of the maximum, the expectation value, and the variance.

4.8 Gaussian distribution

The Gaussian probability distribution (also known as the "normal distribution" or the "bell curve") is the most important of all the probability distributions. The reason, as we will see in Chapter 5, is that in the limit of large numbers, many other distributions reduce to a Gaussian. But for now, we'll just examine the mathematical properties of the Gaussian distribution. The distribution is commonly written in either of the following forms:

$$f(x) = \sqrt{\frac{b}{\pi}}\, e^{-b(x-\mu)^2} \quad \text{or} \quad \sqrt{\frac{1}{2\pi\sigma^2}}\, e^{-(x-\mu)^2/2\sigma^2} \qquad (4.42)$$

If you want to explicitly indicate the parameters that appear, you can write the distribution as $f_{\mu,b}(x)$ or $f_{\mu,\sigma}(x)$. The Gaussian distribution is a *continuous* one. That is, x can take on a continuum of values, like t in the exponential distribution, but unlike k in the binomial and Poisson distributions. The Gaussian probability distribution is therefore a probability *density*. As mentioned at the beginning of Section 4.2.2, the standard practice is to use lowercase letters (like the f in $f(x)$) for probability densities, and to use uppercase letters (like the P in $P(k)$) for actual probabilities.

The second expression in Eq. (4.42) is obtained from the first by letting $b = 1/2\sigma^2$. The first expression is simpler, but the second one is more common. This is due to the fact that the standard deviation, which we introduced in Section 3.3, turns out simply to be σ. Hence our use of the letter σ here. Note that b (or σ) appears twice in the distribution – in the exponent and in the prefactor. These two appearances conspire to make the total area under the distribution equal to 1. See Problem 4.22 for a proof of this fact.

The quantities μ and b (or μ and σ) depend on the specific situation at hand. Let's look at how these quantities affect the shape and location of the curve. We'll work mainly with the first form in Eq. (4.42) here, but any statements we make about b can be converted into statements about σ by replacing b with $1/2\sigma^2$.

Mean

Let's consider μ first. Fig. 4.23 shows the plots of two Gaussian distributions, one with $b = 2$ and $\mu = 6$, and the other with $b = 2$ and $\mu = 10$. The two functions are

$$f(x) = \sqrt{\frac{2}{\pi}}\, e^{-2(x-6)^2} \quad \text{and} \quad f(x) = \sqrt{\frac{2}{\pi}}\, e^{-2(x-10)^2}. \tag{4.43}$$

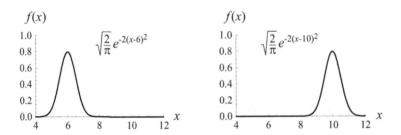

Figure 4.23: Gaussian distributions with different means.

It is clear from the plots that μ is the location of the maximum of the curve. Mathematically, this is true because the $e^{-b(x-\mu)^2}$ exponential factor has an exponent that is either zero or negative (because a square is always zero or positive). So this exponential factor is always less than or equal to 1. Its maximum value occurs when the exponent is zero, that is, when $x = \mu$. The peak is therefore located at $x = \mu$. If we increase μ (while keeping b the same), the whole curve just shifts to the right, keeping the same shape. This is evident from the figure.

Because the curve is symmetric around the maximum, μ is also the mean (or expectation value) of the distribution:

$$\text{Mean} = \mu. \tag{4.44}$$

Since we used the letter μ for the mean throughout Chapter 3, it was a natural choice to use μ the way we did in Eq. (4.42). Of course, for the same reason, it would also have been natural to use μ for the mean of the exponential and Poisson distributions. But we chose to label those means as τ and a, so that there wouldn't be too many μ's floating around in this chapter.

Height

Now let's consider b. Fig. 4.24 shows the plots of two Gaussian distributions, one with $b = 2$ and $\mu = 6$, and the other with $b = 8$ and $\mu = 6$. The two functions are

$$f(x) = \sqrt{\frac{2}{\pi}}\, e^{-2(x-6)^2} \qquad \text{and} \qquad f(x) = \sqrt{\frac{8}{\pi}}\, e^{-8(x-6)^2}. \qquad (4.45)$$

Note that the scales on both the x and y axes in Fig. 4.24 are different from those in Fig. 4.23. The first function here is the same as the first function in Fig. 4.23.

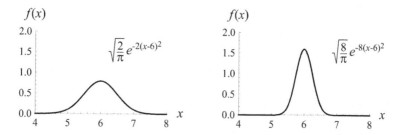

Figure 4.24: Gaussian distributions with different values of b. Both the heights and the widths differ.

It is clear from the plots that b affects both the height and width of the curve. Let's see how these two effects come about. The effect on the height is easy to understand, because the height of the curve (the maximum value of the function) is simply $\sqrt{b/\pi}$. This is true because when x equals μ (which is the location of the maximum), the $e^{-b(x-\mu)^2}$ factor equals 1, in which case the value of $\sqrt{b/\pi}\, e^{-b(x-\mu)^2}$ is just $\sqrt{b/\pi}$. (By the same reasoning, the second expression in Eq. (4.42) gives the height in terms of σ as $1/\sqrt{2\pi\sigma^2}$.) Looking at the two functions in Eq. (4.45), we see that the ratio of the heights is $\sqrt{8/2} = 2$. And this is indeed the ratio we observe in Fig. 4.24. To summarize:

$$\text{Height} = \sqrt{\frac{b}{\pi}} = \sqrt{\frac{1}{2\pi\sigma^2}}. \qquad (4.46)$$

Width in terms of b

Now for the width. We see that the second function in Fig. 4.24 is both taller and narrower than the first. (But it has the same midpoint, because we haven't changed μ.) The factor by which it is shrunk in the horizontal direction appears to be about $1/2$. And in fact, it is exactly $1/2$. It turns out that the width of a Gaussian curve is proportional to $1/\sqrt{b}$. This means that since we increased b by a factor of 4 in constructing the second function, we decreased the width by a factor of $1/\sqrt{4} = 1/2$. Let's now show that the width is in fact proportional to $1/\sqrt{b}$.

But first, what do we mean by "width"? A vertical rectangle has a definite width, but a Gaussian curve doesn't, because the "sides" are tilted. We could arbitrarily define the width to be how wide the curve is at a height equal to half the maximum height. Or instead of half, we could say a third. Or a tenth. We can define it however we want, but the nice thing is that however we choose to define it, the above "proportional to $1/\sqrt{b}$" result will still hold, as long as we pick one definition and stick with it for whatever curves we're looking at. Similarly, if we want to work with the second expression in Eq. (4.42), then since $1/\sqrt{b} \propto \sigma$, the width will be proportional to σ, independent of the specifics of our arbitrary definition.

The definition we'll choose here is: The width of a curve is the width at the height equal to $1/e$ (which happens to be about 0.37) times the maximum height (which is $\sqrt{b/\pi}$). This $1/e$ choice is a natural one, because the x values that correspond to this height are easy to find. They are simply $\mu \pm 1/\sqrt{b}$, because the first expression in Eq. (4.42) gives

$$
\begin{aligned}
f(\mu \pm 1/\sqrt{b}) &= \sqrt{b/\pi}\, e^{-b[(\mu \pm 1/\sqrt{b}) - \mu]^2} \\
&= \sqrt{b/\pi}\, e^{-b(\pm 1/\sqrt{b})^2} \\
&= \sqrt{b/\pi}\, e^{-b/b} \\
&= \sqrt{\frac{b}{\pi}} \cdot \frac{1}{e} ,
\end{aligned}
\tag{4.47}
$$

as desired. Since the difference between $\mu + 1/\sqrt{b}$ and $\mu - 1/\sqrt{b}$ equals $2/\sqrt{b}$, the width of the Gaussian curve (by our arbitrary definition) is $2/\sqrt{b}$. So $1/\sqrt{b}$ is half of the width, which we'll call the "half-width". (The term "half-width" can also refer to the full width of the curve at half of the maximum height. We won't use that meaning here.) Again, any other definition of the width would also yield the $\sqrt{b}$ in the denominator. That's the important part. The 2 in the numerator doesn't have much significance. The half-width is shown below in Fig. 4.25, following the discussion of the width in terms of σ.

Width in terms of σ

When working with the second form in Eq. (4.42) (which is the more common of the two), the default definition of the width is the width at the height equal to $1/\sqrt{e}$ times the maximum height. This definition (which is different from the above $1/e$ definition) is used because the values of x that correspond to this height are simply $x \pm \sigma$. This is true because if we plug $x = \mu \pm \sigma$ into the second expression in Eq. (4.42), we obtain

$$
\begin{aligned}
f(\mu \pm \sigma) &= \sqrt{1/2\pi\sigma^2}\, e^{-[(\mu \pm \sigma) - \mu]^2/2\sigma^2} \\
&= \sqrt{1/2\pi\sigma^2}\, e^{-(\pm \sigma)^2/2\sigma^2} \\
&= \sqrt{1/2\pi\sigma^2}\, e^{-1/2} \\
&= \sqrt{\frac{1}{2\pi\sigma^2}} \cdot \frac{1}{\sqrt{e}} .
\end{aligned}
\tag{4.48}
$$

The factor of $1/\sqrt{e}$ here equals $1/\sqrt{2.718} \approx 0.61$, which is larger than the $1/e \approx$ 0.37 factor in our earlier definition. This is consistent with the fact that the $x = \mu \pm \sigma$ points (where the height is $1/\sqrt{e} \approx 0.61$ times the maximum) are closer to the center than the $x = \mu \pm 1/\sqrt{b} = \mu \pm \sqrt{2}\,\sigma$ points (where the height is $1/e \approx 0.37$ times the maximum). This is summarized in Fig. 4.25; we have chosen $\mu = 0$ for convenience.

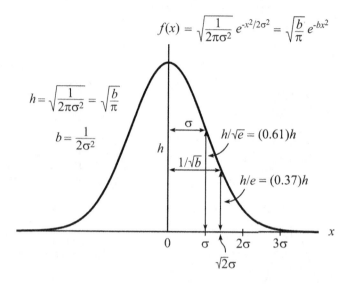

$$f(x) = \sqrt{\frac{1}{2\pi\sigma^2}}\, e^{-x^2/2\sigma^2} = \sqrt{\frac{b}{\pi}}\, e^{-bx^2}$$

$$h = \sqrt{\frac{1}{2\pi\sigma^2}} = \sqrt{\frac{b}{\pi}}$$

$$b = \frac{1}{2\sigma^2}$$

$h/\sqrt{e} = (0.61)h$

$h/e = (0.37)h$

Figure 4.25: Different definitions of the half-width, in terms of b and σ.

Although the $x = \mu \pm \sigma$ points yield a nice value of the Gaussian distribution ($1/\sqrt{e}$ times the maximum), the *really* nice thing about the $x = \mu \pm \sigma$ points is that they are one *standard deviation* from the mean μ. It can be shown (with calculus, see Problem 4.23) that the standard deviation (defined in Eq. (3.40)) of the Gaussian distribution given by the second expression in Eq. (4.42) is simply σ. This is why the second form in Eq. (4.42) is more widely used than the first. And for the same reason, people usually choose to (arbitrarily) define the half-width of the Gaussian curve to be σ instead of the $1/\sqrt{b} = \sqrt{2}\,\sigma$ half-width that we found earlier. That is, they're defining the width by looking at where the function is $1/\sqrt{e}$ times the maximum, instead of $1/e$ times the maximum. As we noted earlier, any such definition is perfectly fine; it's a matter of person preference. The critical point is that the width is proportional to σ (or $1/\sqrt{b}$). The exact numerical factor involved is just a matter of definition.

As mentioned on page 153, it can be shown numerically that about 68% of the total area (probability) under the Gaussian curve lies between the points $\mu \pm \sigma$. In other words, you have a 68% chance of obtaining a value of x that is within one standard deviation from the mean μ. We used the word "numerically" above, because although the areas under the curves (or the discrete sums) for all of the other distributions we've dealt with in the chapter can be calculated in closed form, this isn't true for the Gaussian distribution. So when finding the area under the Gaussian curve, you always need to specify the numerical endpoints of your interval, and then you can use a computer to calculate the area (numerically, to whatever accuracy you want). It can likewise be shown that the percentage of the total area that is within two standard deviations from μ (that is, between the points $\mu \pm 2\sigma$) is about

95%. And the percentage within three standard deviations from μ is about 99.7%. These percentages are consistent with a visual inspection of the shaded areas in Fig. 4.26. The percentages rapidly approach 100%. The percentage within five standard deviations from μ is about 99.99994%.

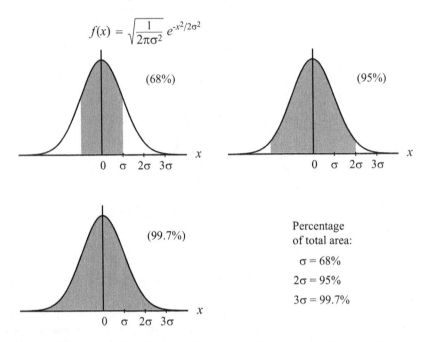

$$f(x) = \sqrt{\frac{1}{2\pi\sigma^2}}\, e^{-x^2/2\sigma^2}$$

(68%)

(95%)

(99.7%)

Percentage of total area:

$\sigma = 68\%$

$2\sigma = 95\%$

$3\sigma = 99.7\%$

Figure 4.26: Areas under a Gaussian distribution within σ, 2σ, and 3σ from the mean.

REMARKS:

1. The Gaussian distribution is a continuous one, because x can take on any value. The distribution applies (either exactly or approximately) to a nearly endless list of processes with continuous random variables such as length, time, light intensity, affinity for butternut squash, etc.

   We'll find in Sections 5.1 and 5.3 that the Gaussian distribution is a good approximation to the binomial and Poisson distributions if the numbers involved are large. In these cases, only integer values of x are relevant, so the distribution is effectively discrete. You can still draw the continuous curve described by Eq. (4.42), but it is relevant only for integer values of x.

2. We mentioned near the beginning of this section that the value of the prefactor in the expressions in Eq. (4.42) makes the total area under the distribution curve be equal to 1. Problem 4.22 gives a proof of this, but for now we can at least present an argument that explains why the prefactor must be proportional to $1/\sigma$ (or equivalently, to $\sqrt{b}$). Basically, since the width of the curve is proportional to σ (as we showed above), the height must be proportional to $1/\sigma$. This is true because if you increase σ by a factor of, say, 10 and thereby stretch the curve by a factor of 10 in the horizontal direction, then you also have to squash the curve by a factor of 10 in the vertical direction, if you want to keep the area the same. (See the fifth remark on page 206.) A factor of $1/\sigma$ in the prefactor accomplishes this. But note that this reasoning tells us only that

the prefactor is *proportional* to $1/\sigma$, and not what the constant of proportionality is. It happens to be $1/\sqrt{2\pi}$.

3. Two parameters are needed to describe the Gaussian distribution: μ and σ (or μ and b). This should be contrasted with the Poisson distribution, where only one parameter, a, is needed. Similarly, the exponential distribution depends on only the one parameter λ (or τ). In the Poisson case, not only does the width determine the height, but it also determines the location of the bump. In contrast, the Gaussian mean μ need not have anything to do with σ (or b). ♣

4.9 Summary

In this chapter we learned about probability distributions. In particular, we learned:

- A *probability distribution* is the collective information about how the total probability (which is always 1) is distributed among the various possible outcomes of the random variable.

- A probability distribution for a continuous random variable is given in terms of a *probability density*. To obtain an actual probability, the density must be multiplied by an interval of the random variable. More generally, the probability equals the area under the density curve.

We discussed six specific probability distributions:

- 1. *Uniform:* (Continuous) The probability density is uniform over a given span of random-variable values, and zero otherwise. The uniform distribution can be described by two parameters: the mean and the width, or alternatively the endpoints of the nonzero region. These two parameters then determine the height.

- 2. *Bernoulli:* (Discrete) The random variable can take on only two values, 1 and 0, with probabilities p and $1 - p$. An example with $p = 1/2$ is a coin toss with Heads = 1 and Tails = 0. The Bernoulli distribution is described by one parameter: p.

- 3. *Binomial:* (Discrete) The random variable is the number k of successes in a collection of n Bernoulli processes. An example is the total number of Heads in n coin tosses. The distribution takes the form,

$$P(k) = \binom{n}{k} p^k (1 - p)^{n-k}. \tag{4.49}$$

The number k of successes must be an integer, of course. The binomial distribution is described by two parameters: n and p.

- 4. *Exponential:* (Continuous) This is the probability distribution for the waiting time t until the next event, for a completely random process. We derived this by taking the continuum limit of the analogous discrete result, which was the *geometric distribution* given in Eq. (4.14). The exponential distribution takes the form,

$$\rho(t) = \frac{e^{-t/\tau}}{\tau},\tag{4.50}$$

where τ is the average waiting time. Equivalently, $\rho(t) = \lambda e^{-\lambda t}$, where $\lambda = 1/\tau$ is the average rate at which the events happen. The exponential distribution is described by one parameter: τ (or λ).

- 5. *Poisson:* (Discrete) This is the probability distribution for the number of events that happen in a given region (of time, space, etc.), for a completely random process. We derived this by taking the continuum limit of the analogous discrete result, which was simply the binomial distribution. The Poisson distribution takes the form,

$$P(k) = \frac{a^k e^{-a}}{k!},\tag{4.51}$$

where a is the expected number of events in the given region. The number k of observed events must be an integer, of course. But a need not be. The Poisson distribution is described by one parameter: a.

- 6. *Gaussian:* (Continuous) This distribution takes the form,

$$f(x) = \sqrt{\frac{b}{\pi}}\, e^{-b(x-\mu)^2} \quad \text{or} \quad \sqrt{\frac{1}{2\pi\sigma^2}}\, e^{-(x-\mu)^2/2\sigma^2}.\tag{4.52}$$

σ is the *standard deviation* of the distribution. About 68% of the probability is contained in the range from $\mu - \sigma$ to $\mu + \sigma$. The width of the distribution is proportional to σ (and to $1/\sqrt{b}$). The Gaussian distribution is described by two parameters: μ and σ (or μ and b).

4.10   Exercises

See **www.people.fas.harvard.edu/~djmorin/book.html** for a supply of problems without included solutions.

4.11   Problems

Section 4.2: Continuous distributions

4.1. **Fahrenheit and Celsius** *

Fig. 4.4 shows the probability density for the temperature, with the temperature measured in Fahrenheit. Draw a reasonably accurate plot of the same probability density, but with the temperature measured in Celsius. (The conversion from Fahrenheit temperature F to Celsius temperature C is $C = (5/9)(F - 32)$. So it takes a ΔF of $9/5 = 1.8$ degrees to create a ΔC of 1 degree.)

4.2. **Expectation of a continuous distribution** ∗ *(calculus)*

The expectation value of a *discrete* random variable is given in Eq. (3.4). Given a *continuous* random variable with probability density $\rho(x)$, explain why the expectation value is given by the integral $\int x\rho(x)\,dx$.

Section 4.3: Uniform distribution

4.3. **Variance of the uniform distribution** ∗ *(calculus)*

Using the general idea from Problem 4.2, find the variance of a uniform distribution that extends from $x = 0$ to $x = a$.

Section 4.5: Binomial distribution

4.4. **Expectation of the binomial distribution** ∗∗

Use Eq. (3.4) to explicitly demonstrate that the expectation value of the binomial distribution in Eq. (4.6) equals pn. This must be true, of course, because a fraction p of the n trials yield success, on average, by the definition of p. *Hint*: The goal is to produce the result of pn, so try to factor a pn out of the sum in Eq. (3.4). You will eventually need to use an expression analogous to Eq. (4.10).

4.5. **Variance of the binomial distribution** ∗∗∗

As we saw in Problem 4.4, the expectation value of the binomial distribution is $\mu = pn$. Use the technique in either of the solutions to that problem to show that the variance of the binomial distribution is $np(1 - p) \equiv npq$ (in agreement with Eq. (3.33)). *Hint*: The form of the variance in Eq. (3.34) works best. When finding the expectation value of k^2 (or really K^2, where K is the random variable whose value is k), it is easiest to find the expectation value of $k(k - 1)$ and then add on the expectation value of k.

4.6. **Hypergeometric distribution** ∗∗∗

(a) A box contains N balls. K of them are red, and the other $N - K$ are blue. (K here is just a given number, not a random variable.) If you draw n balls *without replacement*, what is the probability of obtaining exactly k red balls? The resulting probability distribution is called the *hypergeometric distribution*.

(b) In the limit where N and K are very large, explain in words why the hypergeometric distribution reduces to the binomial distribution given in Eq. (4.6), with $p = K/N$. Then demonstrate this fact mathematically. What exactly is meant by "N and K are very large"?

Section 4.6: Exponential distribution

4.7. **Expectation of the geometric distribution** ∗∗

Verify that the expectation value of the geometric distribution in Eq. (4.14) equals $1/p$. (This is the waiting time we found by an easier method in Eq. (4.13).) The calculation involves a math trick, so you should do Problem 3.1 before solving this one.

4.8. Properties of the exponential distribution ** *(calculus)*

(a) By integrating the exponential distribution in Eq. (4.27) from $t = 0$ to $t = \infty$, show that the total probability is 1.

(b) What is the *median* value t? That is, for what value t_{med} are you equally likely to obtain a t value larger or smaller than t_{med}?

(c) By using the result from Problem 4.2, show that the expectation value is τ, as we know it must be.

(d) Again by using Problem 4.2, find the variance.

Section 4.7: Poisson distribution

4.9. Total probability *

Show that the sum of all of the probabilities in the Poisson distribution given in Eq. (4.40) equals 1, as we know it must. *Hint:* You will need to use Eq. (7.7) in Appendix B.

4.10. Location of the maximum **

For what (integer) value of k is the Poisson distribution $P(k)$ maximum?

4.11. Value of the maximum *

For large a, what approximately is the height of the bump in the Poisson $P(k)$ plot? You will need the result from the previous problem. *Hint:* You will also need to use Stirling's formula, given in Eq. (2.64) in Section 2.6.

4.12. Expectation of the Poisson distribution **

Use Eq. (3.4) to verify that the expectation value of the Poisson distribution equals a. This must be the case, of course, because a is defined to be the expected number of events in the given interval.

4.13. Variance of the Poisson distribution **

As we saw in Problem 4.12, the expectation value of the Poisson distribution is $\mu = a$. Use the technique in the solution to that problem to show that the variance of the Poisson distribution is a (which means that the standard deviation is $\sqrt{a}$). *Hint:* When finding the expectation value of k^2, it is easiest to find the expectation value of $k(k-1)$ and then add on the expectation value of k.

4.14. Poisson accuracy **

In the "balls in boxes, again" example on page 213, we saw that in the right plot in Fig. 4.20, the Poisson distribution is an excellent approximation to the exact binomial distribution. But in the left plot, it is only a so-so approximation. What parameter(s) determine how good the approximation is?

To answer this, we'll define the "goodness" of the approximation to be the ratio of the Poisson expression $P_P(k)$ in Eq. (4.40) to the exact binomial expression $P_B(k)$ in Eq. (4.32), with both functions evaluated at the expected value of k, namely $a = pn$, which we'll assume is an integer. (We're using Eq. (4.32) instead of Eq. (4.33), just because it's easier to work with. The expressions are equivalent, with $p \leftrightarrow 1/b$.) The closer the ratio $P_P(pn)/P_B(pn)$ is to 1, the better the Poisson approximation is. Calculate this ratio. You will need to use Stirling's formula, given in Eq. (2.64). You may assume that n is large (because otherwise there wouldn't be a need to use the Poisson approximation).

4.15. **Bump or no bump** *

In Fig. 4.21, we saw that $P(0) = P(1)$ when $a = 1$. (This is the cutoff between the distribution having or not having a bump.) Explain why this is consistent with what we noted about the binomial distribution (namely, that $P(0) = P(1)$ when $p = 1/(n+1)$) in the example in Section 4.5.

4.16. **Typos** *

A hypothetical writer has an average of one typo per 50 pages of work. (Wishful thinking, perhaps!) What is the probability that there are no typos in a 350-page book?

4.17. **Boxes with zero balls** *

You randomly throw n balls into 1000 boxes and note the number of boxes that end up with zero balls in them. If you repeat this process a large number of times and observe that the average number of boxes with zero balls is 20, what is n?

4.18. **Twice the events** **

(a) Assume that on average, the events in a random process happen a times, where a is large, in a given time interval t. With the notation $P_a(k)$ representing the Poisson distribution, use Stirling's formula (given in Eq. (2.64)) to produce an approximate expression for the probability $P_a(a)$ that exactly a events happen during the time t.

(b) Consider the probability that exactly *twice* the number of events, $2a$, happen during *twice* the time, $2t$. What is the ratio of this probability to $P_a(a)$?

(c) Consider the probability that exactly *twice* the number of events, $2a$, happen during the *same* time, t. What is the ratio of this probability to $P_a(a)$?

4.19. **P(0) the hard way** ***

For a Poisson process with a expected events, Eq. (4.40) gives the probability of having zero events as

$$P(0) = \frac{a^0 e^{-a}}{0!} = e^{-a} = 1 - \left(a - \frac{a^2}{2!} + \frac{a^3}{3!} - \cdots \right), \qquad (4.53)$$

where we have used the Taylor series for e^x given in Eq. (7.7). With the above grouping of the terms, the sum in parentheses must be the probability of having *at least one* event, because when this is subtracted from 1, we obtain the probability of zero events. Explain why this is the case, by accounting for the various multiple events that can occur. You will want to look at the remark in the solution to Problem 2.3 first. The task here is to carry over that reasoning to a continuous Poisson process.

4.20. Probability of at least 1 **

A million balls are thrown at random into a billion boxes. Consider a particular one of the boxes. What (approximately) is the probability that *at least one* ball ends up in that box? Solve this by:

(a) using the Poisson distribution in Eq. (4.40); you will need to use the approximation in Eq. (7.9),

(b) working with probabilities from scratch; you will need to use the approximation in Eq. (7.14).

Note that since the probability you found is very small, it is also approximately the probability of obtaining *exactly one* ball in the given box, because multiple events are extremely rare; see the discussion in the first remark in Section 4.6.2.

4.21. Comparing probabilities ***

(a) A hypothetical 1000-sided die is rolled three times. What is the probability that a given number (say, 1) shows up all three times?

(b) A million balls are thrown at random into a billion boxes. (So from the result in Problem 4.20, the probability that exactly one ball ends up in a given box is approximately $1/1000$.) If this process (of throwing a million balls into a billion boxes) is performed three times, what (approximately) is the probability that exactly one ball lands in a given box all three times? (It can be a different ball each time.)

(c) A million balls are thrown at random into a billion boxes. This process is performed a *single* time. What (approximately) is the probability that exactly three balls end up in a given box? Solve this from scratch by using a counting argument.

(d) Solve part (c) by using the Poisson distribution.

(e) The setups in parts (b) and (c) might seem basically the same, because both setups involve three balls ending up in the given box, and there is a $1/b = 1/10^9$ probability that any given ball ends up in the given box. Give an intuitive explanation for why the answers differ.

Section 4.8: Gaussian distribution

4.22. Area under a Gaussian curve ** *(calculus)*

Show that the area (from $-\infty$ to ∞) under the Gaussian distribution, $f(x) = \sqrt{b/\pi}\, e^{-bx^2}$, equals 1. That is, show that the total probability equals 1. (We have set $\mu = 0$ for convenience, since μ doesn't affect the total area.) There is a very sneaky way to do this. But since it's completely out of the blue, we'll give a hint: Calculate the *square* of the desired integral by multiplying it by the integral of $\sqrt{b/\pi}\,e^{-by^2}$. Then make use of a change of variables from Cartesian to polar coordinates, to convert the Cartesian double integral into a polar double integral.

4.23. Variance of the Gaussian distribution ∗∗ *(calculus)*

Show that the variance of the second Gaussian expression in Eq. (4.42) equals σ^2 (which means that the standard deviation is σ). You may assume that $\mu = 0$ (because μ doesn't affect the variance), in which case the expression for the variance in Eq. (3.19) becomes $E(X^2)$. And then by the reasoning in Problem 4.2, this expectation value is $\int x^2 f(x)\,dx$. So the task of this problem is to evaluate this integral. The straightforward method is to use integration by parts.

4.12 Solutions

4.1. Fahrenheit and Celsius

A density is always given in terms of "something per something else." In the temperature example in Section 4.2, the "units" of probability density were probability per Fahrenheit degree. These units are equivalent to saying that we need to multiply the density by a certain number of Fahrenheit degrees (the ΔT) to obtain a probability; see Eq. (4.2). Analogously, we need to multiply a mass density (mass per volume) by a volume to obtain a mass.

If we want to instead write the probability density in terms of probability per *Celsius* degree, we can't simply use the same function $\rho(T)$ that appears in Fig. 4.4. Since there are 1.8 Fahrenheit degrees in each Celsius degree, the correct plot of $\rho(T)$ is shown in Fig. 4.27.

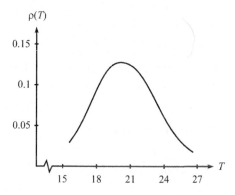

Figure 4.27: Expressing Fig. 4.4 in terms of Celsius instead of Fahrenheit.

This plot differs from Fig. 4.4 in three ways. First, since the peak of the curve in Fig. 4.4 was at about 68 degrees Fahrenheit, it is now *shifted* and located at about $(5/9)(68 - 32) = 20$ degrees Celsius in Fig. 4.27.

Second, compared with Fig. 4.4, the curve in Fig. 4.27 is *contracted* by a factor of 1.8 in the horizontal direction due to the conversion from Fahrenheit to Celsius. The span is only about 11 Celsius degrees, compared with a span of about 20 Fahrenheit degrees in Fig. 4.4. This follows from the fact that each Celsius degree is worth 1.8 Fahrenheit degrees.

Third, since the area under the entire curve in Fig. 4.27 must still be 1, the curve must also be *expanded* by a factor of 1.8 in the vertical direction. So the maximum value is about 0.13, compared with the maximum value of about 0.07 in Fig. 4.4.

REMARK: These contraction and expansion countereffects cause the probabilities we calculate here to be consistent with ones we calculated in Section 4.2. For example, we found in Eq. (4.3) that the probability of the temperature falling between 70 °F and 71 °F is about 7%. Now, 70 °F and 71 °F correspond to 21.11 °C and 21.67 °C, as you can show using $C = (5/9)(F - 32)$. So the probability of the temperature falling between 21.11 °C and 21.67 °C had better also be 7%. It's the same temperature interval; we're just describing it in a different way. And indeed, from the Celsius plot, the value of the density near 21° is about 0.12. Therefore, the probability of falling between 21.11 °C and 21.67 °C, which equals the density times the interval, is $(0.12)(21.67 - 21.11) = 0.067 \approx 7\%$, in agreement with the Fahrenheit calculation (up to the rough readings we made from the plots). If we had forgotten to expand the height of the curve by the factor of 1.8 in Fig. 4.27, we would have obtained only about half of this probability, and therefore a different answer to exactly the same question (asked in a different language). That wouldn't be good. ♣

4.2. Expectation of a continuous distribution

For a general probability density $\rho(x)$, the probability associated with a span dx around a given value of x is $\rho(x)\,dx$; this is true by the definition of the probability density. Now, the expectation value of a *discrete* random variable is given in Eq. (3.4). To extract from this expression the expectation value of a *continuous* random variable, we can imagine dividing up the x axis into a very large number of little intervals dx. The probabilities p_i in Eq. (3.4) get replaced with the various $\rho(x)\,dx$ probabilities. And the outcomes x_i in Eq. (3.4) get replaced with the various values of x.

In making these replacements, we're pretending that all of the x values in a tiny interval dx are equal to the value at, say, the midpoint (call it x_i). This x_i then occurs with probability $p_i = \rho(x_i)\,dx$. We therefore have a discrete distribution that in the $dx \to 0$ limit is the same as the original continuous distribution. The discreteness of our approximate distribution allows us to apply Eq. (3.4) and say that the expectation value equals

$$\text{Expectation value} = \sum p_i x_i = \sum (\rho(x_i)\,dx)x_i. \qquad (4.54)$$

In the $dx \to 0$ limit, this discrete sum turns into the integral,

$$\text{Expectation value} = \int \left(\rho(x)\,dx\right)x = \int x\rho(x)\,dx, \qquad (4.55)$$

as desired. This is the general expression for the expectation value of a continuous random variable. The limits of the integral are technically $-\infty$ to ∞, although it is often the case that $\rho(x) = 0$ everywhere except in a finite region. For example, the density $\rho(t)$ for the exponential distribution is zero for $t < 0$, and it becomes negligibly small for $t \gg \tau$, where τ is the average waiting time.

The above result generalizes to the expectation value of things other than x. For example, the same reasoning shows that the expectation value of x^2 (which is relevant when calculating the variance) equals $\int x^2 \rho(x)\, dx$. And the expectation value of x^7 (if you ever happened to be interested in such a quantity) equals $\int x^7 \rho(x)\, dx$.

4.3. Variance of the uniform distribution

FIRST SOLUTION: Since the nonzero part of the distribution has length a on the x axis, the value of the distribution in that region must be $1/a$, so that the total area is 1. We'll use the $E(X^2) - \mu^2$ form of the variance in Eq. (3.34), with $\mu = a/2$ here. Our task is therefore to calculate $E(X^2)$. From the last comment in the solution to Problem 4.2, this equals

$$E(X^2) = \int_0^a x^2 \rho(x)\, dx = \int_0^a x^2 \cdot \frac{1}{a}\, dx = \frac{1}{a} \cdot \frac{x^3}{3}\Big|_0^a = \frac{a^2}{3}. \tag{4.56}$$

The variance is then

$$\mathrm{Var}(X) = E(X^2) - \mu^2 = \frac{a^2}{3} - \left(\frac{a}{2}\right)^2 = \frac{a^2}{12}. \tag{4.57}$$

The standard deviation is therefore $a/(2\sqrt{3}) \approx (0.29)a$.

SECOND SOLUTION: Let's shift the distribution so that it is nonzero from $x = -a/2$ to $x = a/2$. This shift doesn't affect the variance, which is now simply $E(X^2)$, because $\mu = 0$. So

$$\mathrm{Var}(X) = E(X^2) = \int_{-a/2}^{a/2} x^2 \rho(x)\, dx = \frac{1}{a} \cdot \frac{x^3}{3}\Big|_{-a/2}^{a/2}$$

$$= \frac{1}{3a}\left(\left(\frac{a}{2}\right)^3 - \left(-\frac{a}{2}\right)^3\right) = \frac{a^2}{12}. \tag{4.58}$$

THIRD SOLUTION: We can use the $E[(X - \mu)^2]$ form of the variance in Eq. (3.19), with the original $0 < x < a$ span. This gives

$$\mathrm{Var}(X) \equiv E[(X - a/2)^2] = \int_0^a (x - a/2)^2 \rho(x)\, dx$$

$$= \frac{1}{a}\int_0^a (x^2 - ax + a^2/4)\, dx$$

$$= \frac{1}{a}\left(\frac{a^3}{3} - a\frac{a^2}{2} + \frac{a^2}{4}a\right) = \frac{a^2}{12}. \tag{4.59}$$

4.4. Expectation of the binomial distribution

FIRST SOLUTION: The $k = 0$ term doesn't contribute anything to the sum in Eq. (3.4), so we can start with the $k = 1$ term. The sum goes up to $k = n$. Plugging the probabilities from Eq. (4.6) into Eq. (3.4), we obtain an expectation value of

$$\sum_{k=1}^n k \cdot P(k) = \sum_{k=1}^n k \cdot \binom{n}{k} p^k (1 - p)^{n-k}. \tag{4.60}$$

If the factor of k weren't on the righthand side, we would know how to evaluate this sum; see Eq. (4.10). So let's get rid of the k and create a sum that looks like Eq. (4.10).

The steps are the following.

$$\sum_{k=1}^{n} k \cdot P(k)$$

$$= \sum_{k=1}^{n} k \cdot \frac{n!}{k!(n-k)!} p^k (1-p)^{n-k} \quad \text{(expanding the binomial coeff.)}$$

$$= pn \sum_{k=1}^{n} k \cdot \frac{(n-1)!}{k!(n-k)!} p^{k-1} (1-p)^{n-k} \quad \text{(factoring out } pn)$$

$$= pn \sum_{k=1}^{n} \frac{(n-1)!}{(k-1)!(n-k)!} p^{k-1} (1-p)^{n-k} \quad \text{(canceling the } k)$$

$$= pn \sum_{k=1}^{n} \binom{n-1}{k-1} p^{k-1} (1-p)^{(n-1)-(k-1)} \quad \text{(rewriting)}$$

$$= pn \sum_{j=0}^{n-1} \binom{n-1}{j} p^j (1-p)^{(n-1)-j} \quad \text{(defining } j \equiv k-1)$$

$$= pn(p + (1-p))^{n-1} \quad \text{(using the binomial expansion)}$$

$$= pn \cdot 1, \tag{4.61}$$

as desired. Note that in the sixth line, the sum over j goes from 0 to $n-1$, because the sum over k went from 1 to n.

Even though we know that the expectation value has to be pn (as mentioned in the statement of the problem), it's nice to see that the math does in fact work out.

SECOND SOLUTION: Here is another (sneaky) way to obtain the expectation value. This method uses calculus. The binomial expansion tells us that

$$(p + q)^n = \sum_{k=0}^{n} \binom{n}{k} p^k q^{n-k}. \tag{4.62}$$

This relation is identically true for arbitrary values of p (and q), so we can take the derivative with respect to p to obtain another valid relation:

$$n(p + q)^{n-1} = \sum_{k=1}^{n} k \binom{n}{k} p^{k-1} q^{n-k}. \tag{4.63}$$

If we now multiply both sides by p and then set q to equal $1 - p$ (the relation is true for all values of q, in particular this specific one), we obtain

$$np(1)^{n-1} = \sum_{k=1}^{n} k \binom{n}{k} p^k (1-p)^{n-k} \implies np = \sum_{k=1}^{n} k \cdot P(k), \tag{4.64}$$

as desired.

4.5. Variance of the binomial distribution

FIRST SOLUTION: As suggested in the statement of the problem, let's find the expectation value of $k(k-1)$. Since we've already done a calculation like this in Problem 4.4, we won't list out every step here as we did in Eq. (4.61). The $k = 0$ and $k - 1$ terms don't contribute anything to the expectation value of $k(k-1)$, so we can start the sum with the $k = 2$ term. We have (with $j \equiv k - 2$ in the 5th line)

$$\sum_{k=2}^{n} k(k-1) \cdot P(k)$$

$$= \sum_{k=2}^{n} k(k-1) \cdot \frac{n!}{k!(n-k)!} p^k (1-p)^{n-k}$$

$$= p^2 n(n-1) \sum_{k=2}^{n} \frac{(n-2)!}{(k-2)!(n-k)!} p^{k-2} (1-p)^{n-k}$$

$$= p^2 n(n-1) \sum_{k=2}^{n} \binom{n-2}{k-2} p^{k-2} (1-p)^{(n-2)-(k-2)}$$

$$= p^2 n(n-1) \sum_{j=0}^{n-2} \binom{n-2}{j} p^j (1-p)^{(n-2)-j}$$

$$= p^2 n(n-1)(p + (1-p))^{n-2}$$

$$= p^2 n(n-1) \cdot 1. \tag{4.65}$$

The expectation value of k^2 equals the expectation value of $k(k-1)$ plus the expectation value of k. The latter is just pn, from Problem 4.4. So the expectation value of k^2 is

$$p^2 n(n-1) + pn. \tag{4.66}$$

To obtain the variance, Eq. (3.34) tells us that we need to subtract off $\mu^2 = (pn)^2$ from this result. The variance is therefore

$$\left(p^2 n(n-1) + pn\right) - p^2 n^2 = \left(p^2 n^2 - p^2 n + pn\right) - p^2 n^2$$

$$= pn(1-p)$$

$$\equiv npq, \tag{4.67}$$

as desired. The standard deviation is then $\sqrt{npq}$.

SECOND SOLUTION: Instead of taking just one derivative, as we did in the second solution in Problem 4.4, we'll take two derivatives here. Starting with the binomial expansion,

$$(p+q)^n = \sum_{k=0}^{n} \binom{n}{k} p^k q^{n-k}, \tag{4.68}$$

we can take two derivatives with respect to p to obtain

$$n(n-1)(p+q)^{n-2} = \sum_{k=2}^{n} k(k-1) \binom{n}{k} p^{k-2} q^{n-k}. \tag{4.69}$$

If we now multiply both sides by p^2 and then set q to equal $1 - p$, we obtain

$$p^2 n(n-1)(1)^{n-1} = \sum_{k=2}^{n} k(k-1) \binom{n}{k} p^k (1-p)^{n-k}$$

$$\implies p^2 n(n-1) = \sum_{k=2}^{n} k(k-1) \cdot P(k). \tag{4.70}$$

The expectation value of $k(k-1)$ is therefore $p^2 n(n-1)$, in agreement with Eq. (4.65) in the first solution. The solution proceeds as above.

4.6. Hypergeometric distribution

(a) There are $\binom{N}{n}$ possible sets of n balls (drawn without replacement), and all of these sets are equally likely to be drawn. We therefore simply need to count the number of sets that have exactly k red balls. There are $\binom{K}{k}$ ways to choose k red balls from the K red balls in the box. And there are $\binom{N-K}{n-k}$ ways to choose the other $n-k$ balls (which we want to be blue) from the $N-K$ blue balls in the box. So the number of sets that have exactly k red balls is $\binom{K}{k}\binom{N-K}{n-k}$. The desired probability of obtaining exactly k balls when drawing n balls without replacement is therefore

$$P(k) = \frac{\binom{K}{k}\binom{N-K}{n-k}}{\binom{N}{n}} \qquad \text{(Hypergeometric distribution)} \qquad (4.71)$$

REMARK: Since the number of red balls, k, that you draw can't be larger than either K or n, we see that $P(k)$ is nonzero only if $k \leq \min(K,n)$. Likewise, the number of blue balls, $n-k$, that you draw can't be larger than either $N-K$ or n. So $P(k)$ is nonzero only if $n-k \leq \min(N-K,n) \implies n - \min(N-K,n) \leq k$. Putting these inequalities together, we see that $P(k)$ is nonzero only if

$$n - \min(N-K,n) \leq k \leq \min(K,n). \qquad (4.72)$$

If both K and $N-K$ are larger than n, then this reduces to the simple relation, $0 \leq k \leq n$. ♣

(b) If N and K are small, then the probabilities of drawing red/blue balls change after each draw. This is true because you aren't replacing the balls, so the ratio of red and blue balls changes after each draw.

However, if N and K are large, then the "without replacement" qualifier is inconsequential. The ratio of red and blue balls remains essentially unchanged after each draw. Removing one red ball from a set of a million red balls is hardly noticeable. The probability of drawing a red ball at each stage therefore remains essentially fixed at the value K/N. Likewise, the probability of drawing a blue ball at each stage remains essentially fixed at the value $(N-K)/N$. If we define the red-ball probability as $p \equiv K/N$, then the blue-ball probability is $1-p$. We therefore have exactly the setup that generates the binomial distribution, with red corresponding to success, and blue corresponding to failure. Hence we obtain the binomial distribution in Eq. (4.6).

Let's now show mathematically that the hypergeometric distribution in Eq. (4.71) reduces to the binomial distribution in Eq. (4.6). Expanding the binomial coefficients in Eq. (4.71) gives

$$P(k) = \frac{\dfrac{K!}{k!(K-k)!} \cdot \dfrac{(N-K)!}{(n-k)!((N-K)-(n-k))!}}{\dfrac{N!}{n!(N-n)!}}. \qquad (4.73)$$

If $K \gg k$, then we can say that

$$\frac{K!}{(K-k)!} = K(K-1)(K-2)\cdots(K-k+1) \approx K^k. \qquad (4.74)$$

This is true because all of the factors here are essentially equal to K, in a multiplicative sense. (The "$\gg$" sign in $K \gg k$ means "much greater than" in a multiplicative, not additive, sense.) We can make similar approximations to $(N - K)!/((N - K) - (n - k))!$ and $N!/(N - n)!$, so Eq. (4.73) becomes

$$P(k) \approx \frac{\dfrac{K^k}{k!} \cdot \dfrac{(N - K)^{n-k}}{(n - k)!}}{\dfrac{N^n}{n!}} = \frac{n!}{k!(n - k)!}\left(\frac{K}{N}\right)^k \left(\frac{N - K}{N}\right)^{n-k}$$

$$= \binom{n}{k} p^k (1 - p)^{n-k}, \tag{4.75}$$

where $p \equiv K/N$. This is the desired binomial distribution, which gives the probability of k successes in n trials, where the probability of success in each trial takes on the fixed value of p.

We made three approximations in the above calculation, and they relied on the three assumptions,

$$(1)\ K \gg k, \quad (2)\ N - K \gg n - k, \quad (3)\ N \gg n. \tag{4.76}$$

In words, these three assumptions are: (1) the number of red balls you draw is much smaller than the total number of red balls in the box, (2) the number of blue balls you draw is much smaller than the total number of blue balls in the box, and (3) the total number of balls you draw is much smaller than the total number of balls in the box. (The third assumption follows from the other two.) These three assumptions are what we mean by "N and K are very large."

4.7. **Expectation of the geometric distribution**

From Eq. (4.14), the probability that we need to wait just one iteration for the next success is p. For two iterations it is $(1 - p)p$, for three iterations it is $(1 - p)^2 p$, and so on. The expectation value of the number of iterations (that is, the waiting time) is therefore

$$1 \cdot p + 2 \cdot (1 - p)p + 3 \cdot (1 - p)^2 p + 4 \cdot (1 - p)^3 p + \cdots . \tag{4.77}$$

To calculate this sum, we'll use the trick we introduced in Problem 3.1 and write the sum as a geometric series starting with p, plus another geometric series starting with $(1-p)p$, and so on. And we'll use the fact that the sum of a geometric series with first term a and ratio r is $a/(1 - r)$. The expectation value in Eq. (4.77) then becomes

$$\begin{aligned}
p + (1 - p)p + (1 - p)^2 p + (1 - p)^3 p + \cdots \\
(1 - p)p + (1 - p)^2 p + (1 - p)^3 p + \cdots \\
(1 - p)^2 p + (1 - p)^3 p + \cdots \\
(1 - p)^3 p + \cdots
\end{aligned} \tag{4.78}$$

$$\vdots$$

This has the correct number of each type of term. For example, the $(1 - p)^2 p$ term appears three times. The first line above is a geometric series that sums to $a/(1 - r) = p/(1 - (1 - p)) = 1$. The second line is also a geometric series, and it sums to $(1-p)p/(1-(1-p)) = 1-p$. Likewise the third line sums to $(1-p)^2 p/(1-(1-p)) = (1 - p)^2$. And so on. The sum of the infinite number of lines in Eq. (4.79) therefore equals

$$1 + (1 - p) + (1 - p)^2 + (1 - p)^3 + \cdots . \tag{4.79}$$

But this itself is a geometric series, and it sums to $a/(1-r) = 1/(1-(1-p)) = 1/p$, as desired.

4.8. Properties of the exponential distribution

(a) The total probability equals the total area under the distribution curve. And this area is given by the integral of the distribution. The integral of $e^{-t/\tau}/\tau$ equals $-e^{-t/\tau}$, as you can verify by taking the derivative (and using the chain rule). The desired integral is therefore

$$\int_0^\infty \frac{e^{-t/\tau}}{\tau}\, dt = -e^{-t/\tau}\Big|_0^\infty = -e^{-\infty} - (-e^{-0}) = 1, \qquad (4.80)$$

as desired

(b) This is very similar to part (a), except that we now want the probability from 0 to t_{med} to equal $1/2$. That is,

$$\frac{1}{2} = \int_0^{t_{\text{med}}} \frac{e^{-t/\tau}}{\tau}\, dt = -e^{-t/\tau}\Big|_0^{t_{\text{med}}} = -e^{-t_{\text{med}}/\tau} - (-e^{-0}). \qquad (4.81)$$

This yields $e^{-t_{\text{med}}/\tau} = 1/2$. Taking the natural log of both sides then gives

$$-t_{\text{med}}/\tau = -\ln 2 \implies t_{\text{med}} = (\ln 2)\tau \approx (0.7)\tau. \qquad (4.82)$$

So the median value of t is $(0.7)\tau$. In other words, $(0.7)\tau$ is the value of t for which the two shaded areas in Fig. 4.28 are equal.

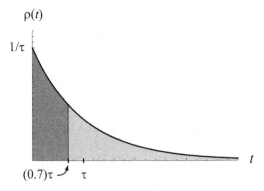

Figure 4.28: The areas on either side of the median are equal.

Note that the *median* value of t, namely $(0.7)\tau$, is *smaller* than the *mean* value (the expectation value) of t, namely τ. The reason for this is that the exponential distribution has a tail that extends to large values of t. These values of t drag the mean to the right, more so than the small values of t near zero drag it to the left (because the former are generally farther from t_{med} than the latter). Whenever you have an asymmetric distribution like this, the mean always lies on the "tail side" of the median.

(c) In the specific case of the exponential distribution, Eq. (4.55) in the solution to Problem 4.2 gives

$$\text{Expectation value} = \int_0^\infty t \cdot \frac{e^{-t/\tau}}{\tau}\, dt. \qquad (4.83)$$

You can evaluate this integral by performing "integration by parts," or you can just look it up. It turns out to be

$$\text{Expectation value} = -(t + \tau)e^{-t/\tau}\Big|_0^\infty$$
$$= -(\infty + \tau)e^{-\infty} + (0 + \tau)e^{-0}$$
$$= 0 + \tau, \tag{4.84}$$

as desired. In the first term here, we have used the fact that the smallness of $e^{-\infty}$ wins out over the largeness of ∞. You can check this on your calculator by replacing ∞ with, say, 100.

(d) Let's use T to denote the random variable whose value is t. Since the mean of the exponential distribution is τ, Eq. (3.34) tells us that the variance is $E(T^2) - \tau^2$. So we need to find $E(T^2)$. Eq. (4.55) gives

$$E(T^2) = \int_0^\infty t^2 \cdot \frac{e^{-t/\tau}}{\tau}\, dt. \tag{4.85}$$

Evaluating this by integration by parts is rather messy, so let's just look up the integral. It turns out to be

$$E(T^2) = -(t^2 + 2\tau t + 2\tau^2)e^{-t/\tau}\Big|_0^\infty$$
$$= -0 + (0 + 0 + 2\tau^2)e^{-0}$$
$$= 2\tau^2. \tag{4.86}$$

As in part (c), we have used the fact that the smallness of $e^{-\infty}$ makes the term associated with the upper limit of integration be zero. The variance is therefore

$$\text{Var}(T) = E(T^2) - \tau^2 = 2\tau^2 - \tau^2 = \tau^2. \tag{4.87}$$

The standard deviation is the square root of the variance, so it is simply τ, which interestingly is the same as the mean. As with all other quantities associated with the exponential distribution, the variance and standard deviation depend only on τ, because that is the only parameter that appears in the distribution.

4.9. Total probability

The sum over k ranges from 0 to ∞. The upper limit is ∞ because with continuous time (or space, or whatever), theoretically an arbitrarily large number of events can occur in a given time interval (although if k is much larger than a, then $P(k)$ is negligibly small). We have (invoking Eq. (7.7) from Appendix B to obtain the third line)

$$\sum_{k=0}^\infty P(k) = \sum_{k=0}^\infty \frac{a^k e^{-a}}{k!}$$
$$= e^{-a} \sum_{k=0}^\infty \frac{a^k}{k!}$$
$$= e^{-a} e^a$$
$$= 1, \tag{4.88}$$

as desired. You are encouraged to look at the derivation of Eq. (7.7) in Appendix B.

4.10. **Location of the maximum**

FIRST SOLUTION: In this solution we'll use the fact that the expression for $P(k)$ in Eq. (4.40) is actually valid for all positive values of k, not just integers (even though we're really only concerned with integers). This is due to the fact that it is possible to extend the meaning of $k!$ to non-integers. We can therefore treat $P(k)$ as a continuous distribution. The maximum value of this distribution might not occur at an integer value of k, but we'll be able to extract the appropriate value of k that yields the maximum when k is restricted to integers.

A convenient way to narrow down the location of the maximum of $P(k)$ is to set $P(k) = P(k+1)$. (In calculus, this is equivalent to finding the maximum by setting the first derivative equal to zero.) This will tell us roughly where the maximum is, because this relation can hold only if k and $k+1$ are on opposite sides of the maximum. This is true because the relation $P(k) = P(k+1)$ can't be valid on the right side of the curve's peak, because the curve is decreasing there, so all those points have $P(k) > P(k+1)$. Similarly, all the points on the left side of the peak have $P(k) < P(k+1)$. The only remaining possibility is that k is on the left side and $k+1$ is on the right side. That is, they are on opposite sides of the maximum.

Setting $P(k) = P(k+1)$ gives (after canceling many common factors to obtain the second line)

$$P(k) = P(k+1) \quad \Longrightarrow \quad \frac{a^k e^{-a}}{k!} = \frac{a^{k+1} e^{-a}}{(k+1)!}$$

$$\Longrightarrow \quad \frac{1}{1} = \frac{a}{k+1}$$

$$\Longrightarrow \quad k+1 = a$$

$$\Longrightarrow \quad k = a - 1. \tag{4.89}$$

The two relevant points on either side of the maximum, namely k and $k+1$, are therefore $a-1$ and a. So the maximum of the $P(k)$ plot (extended to non-integers) lies between $a-1$ and a. Since we're actually concerned only with integer values of k, the maximum is located at the integer that lies between $a-1$ and a (or at both of these values if a is an integer). In situations where a is large (which is often the case), the distinction between $a-1$ and a (or somewhere in between) isn't all that important, so we generally just say that the maximum of the probability distribution occurs roughly at a.

SECOND SOLUTION: We can avoid any issues about extending the Poisson distribution to non-integer values of k, by simply finding the integer value of k for which both $P(k) \geq P(k+1)$ and $P(k) \geq P(k-1)$ hold. $P(k)$ is then the maximum, because it is at least as large as the two adjacent $P(k \pm 1)$ values.

By changing the "=" sign in Eq. (4.89) to a "$\geq$" sign, we immediately see that $P(k) \geq P(k+1)$ implies $k \geq a-1$. And by slightly modifying Eq. (4.89), you can show that $P(k) \geq P(k-1)$ implies $a \geq k$. Combining these two results, we see that the integer value of k that yields the maximum $P(k)$ satisfies $a-1 \leq k \leq a$. The desired value of k is therefore the integer that lies between $a-1$ and a (or at both of these values if a is an integer), as we found in the first solution.

4.11. **Value of the maximum**

Since we know from the previous problem that the maximum of $P(k)$ occurs essentially at $k = a$, our goal is to find $P(a)$. Stirling's formula allows us to make a quick approximation to this value. Plugging $k = a$ into Eq. (4.40) and using Stirling's formula, $n! \approx n^n e^{-n} \sqrt{2\pi n}$, yields

$$P(a) = \frac{a^a e^{-a}}{a!} \approx \frac{a^a e^{-a}}{a^a e^{-a} \sqrt{2\pi a}} = \frac{1}{\sqrt{2\pi a}}. \qquad (4.90)$$

We see that the height is proportional to $1/\sqrt{a}$. So if a goes up by a factor of, say, 4, then the height goes down by a factor of 2.

It is easy to make quick estimates using this result. Consider the $a = 10$ plot in Fig. 4.21. The maximum is between 0.1 and 0.2, a little closer to 0.1. Let's say 0.13. And indeed, if $a = 10$ (for which Stirling's formula is quite accurate, from Table 2.6), Eq. (4.90) gives

$$P(10) \approx \frac{1}{\sqrt{2\pi(10)}} \approx 0.126. \qquad (4.91)$$

This is very close to the exact value of $P(10)$, which you can show is about 0.125. Since a is an integer here (namely, 10), Problem 4.10 tells us that $P(9)$ takes on this same value.

4.12. Expectation of the Poisson distribution

The expectation value is the sum of $k \cdot P(k)$, from $k = 0$ to $k = \infty$. However, the $k = 0$ term contributes nothing, so we can start the sum with the $k = 1$ term. Using Eq. (4.40), the expectation value is therefore

$$
\begin{aligned}
\sum_{k=1}^{\infty} k \cdot P(k) &= \sum_{k=1}^{\infty} k \cdot \frac{a^k e^{-a}}{k!} \\
&= \sum_{k=1}^{\infty} \frac{a^k e^{-a}}{(k-1)!} \qquad \text{(canceling the } k\text{)} \\
&= a \cdot \sum_{k=1}^{\infty} \frac{a^{k-1} e^{-a}}{(k-1)!} \qquad \text{(factoring out an } a\text{)} \\
&= a \cdot \sum_{j=0}^{\infty} \frac{a^j e^{-a}}{j!} \qquad \text{(defining } j \equiv k - 1\text{)} \\
&= a \cdot \sum_{j=0}^{\infty} P(j) \qquad \text{(using Eq. (4.40))} \\
&= a \cdot 1, \qquad \text{(total probability is 1)} \qquad (4.92)
\end{aligned}
$$

as desired. In the fourth line, we used the fact that since $j \equiv k - 1$, the sum over j starts with the $j = 0$ term, because the sum over k started with the $k = 1$ term. If you want to show explicitly that the total probability is 1, that was the task of Problem 4.9.

4.13. Variance of the Poisson distribution

As suggested in the statement of the problem, let's find the expectation value of $k(k-1)$. Since we've already done a calculation like this in Problem 4.12, we won't list out every step here as we did in Eq. (4.92). The $k = 0$ and $k - 1$ terms don't contribute anything to the expectation value of $k(k-1)$, so we can start the sum with the $k = 2$

term. We have (with $j \equiv k - 2$ in the 3rd line)

$$\sum_{k=2}^{\infty} k(k-1) \cdot P(k) = \sum_{k=2}^{\infty} k(k-1) \cdot \frac{a^k e^{-a}}{k!}$$

$$= a^2 \cdot \sum_{k=2}^{\infty} \frac{a^{k-2} e^{-a}}{(k-2)!}$$

$$= a^2 \cdot \sum_{j=0}^{\infty} \frac{a^j e^{-a}}{j!}$$

$$= a^2 \cdot \sum_{j=0}^{\infty} P(j)$$

$$= a^2 \cdot 1. \tag{4.93}$$

The expectation value of k^2 equals the expectation value of $k(k-1)$ plus the expectation value of k. The latter is just a, from Problem 4.12. So the expectation value of k^2 is $a^2 + a$. To obtain the variance, Eq. (3.34) tells us that we need to subtract off $\mu^2 = a^2$ from this result. The variance is therefore

$$(a^2 + a) - a^2 = a, \tag{4.94}$$

as desired. The standard deviation is then $\sqrt{a}$.

We will show in Section 5.3 that the standard deviation of the Poisson distribution equals $\sqrt{a}$ when a is large (when the Poisson looks like a Gaussian). But in this problem we demonstrated the stronger result that the standard deviation of the Poisson distribution equals $\sqrt{a}$ for *any* value of a, even a small one (when the Poisson *doesn't* look like a Gaussian).

REMARK: We saw in Problem 4.11 that for large a, the height of the bump in the Poisson $P(k)$ plot is $1/\sqrt{2\pi a}$, which is proportional to $1/\sqrt{a}$. The present $\sigma = \sqrt{a}$ result is consistent with this, because we know that the total probability must be 1. For large a, the $P(k)$ plot is essentially a continuous curve, so we need the total area under the curve to equal 1. A rough measure of the width of the bump is 2σ. The area under the curve equals (roughly) this width times the height. The product of 2σ and the height must therefore be of order 1. And this is indeed the case, because $\sigma = \sqrt{a}$ implies that $(2\sigma)(1/\sqrt{2\pi a}) = \sqrt{2/\pi}$, which is of order 1. This order-of-magnitude argument doesn't tell us anything about specific numerical factors, but it does tell us that the height and standard deviation must have inverse dependences on a. ♣

4.14. **Poisson accuracy**

Replacing a with pn in the Poisson distribution in Eq. (4.40), and setting $k = pn$ as instructed, gives

$$P_{\mathrm{P}}(pn) = \frac{(pn)^{pn} e^{-pn}}{(pn)!} . \tag{4.95}$$

Similarly, setting $k = pn$ in the exact binomial expression in Eq. (4.32) gives

$$P_{\mathrm{B}}(pn) = \binom{n}{pn} p^{pn} (1-p)^{n-pn}$$

$$= \frac{n!}{(pn)!(n-pn)!} p^{pn} (1-p)^{n-pn} . \tag{4.96}$$

The $(pn)!$ term here matches up with the $(pn)!$ term in $P_P(pn)$, so it will cancel in the ratio $P_P(pn)/P_B(pn)$. Let's apply Stirling's formula, $m! \approx m^m e^{-m} \sqrt{2\pi m}$, to the other two factorials in $P_B(pn)$. Since $n - pn = n(1 - p)$, we obtain (we'll do the simplification gradually here)

$$
\begin{aligned}
P_B(pn) &\approx \frac{n^n e^{-n} \sqrt{2\pi n}}{(pn)! \cdot (n(1-p))^{n(1-p)} e^{-n(1-p)} \sqrt{2\pi n(1-p)}} \cdot p^{pn} (1-p)^{n(1-p)} \\
&= \frac{n^n e^{-n}}{(pn)! \cdot n^{n(1-p)} e^{-n(1-p)} \sqrt{1-p}} \cdot p^{pn} \\
&= \frac{1}{(pn)! \cdot n^{-pn} e^{pn} \sqrt{1-p}} \cdot p^{pn} \\
&= \frac{1}{\sqrt{1-p}} \cdot \frac{(pn)^{pn} e^{-pn}}{(pn)!}.
\end{aligned}
\tag{4.97}
$$

This result fortuitously takes the same form as the $P_P(pn)$ expression in Eq. (4.95), except for the factor of $1/\sqrt{1-p}$ out front. The desired ratio is therefore simply

$$
\frac{P_P(pn)}{P_B(pn)} = \sqrt{1-p}.
\tag{4.98}
$$

This is the factor by which the peak of the Poisson plot is smaller than the peak of the (exact) binomial plot.

In the two plots in Fig. 4.20, the p values are $1/10$ and $1/100$, so the $\sqrt{1-p}$ ratios are $\sqrt{0.9} \approx 0.95$ and $\sqrt{0.99} \approx 0.995$. These correspond to percentage differences of 5% and 0.5%, or equivalently to fractional differences of $1/20$ and $1/200$. These are consistent with a visual inspection of the two plots; the 0.5% difference is too small to see in the second plot.

With the above $\sqrt{1-p}$ result, we can say that the Poisson approximation is a good one if $\sqrt{1-p}$ is close to 1, or equivalently if p is much smaller than 1. How much smaller? That depends on how good an approximation you want. If you want accuracy to 1%, then $p = 1/100$ works, but $p = 1/10$ doesn't.

REMARKS:

1. A helpful mathematical relation that is valid for small p is $\sqrt{1-p} \approx 1 - p/2$. (You can check this by plugging a small number like $p = 0.01$ into your calculator. Or you can square both sides to obtain $1 - p \approx 1 - p + p^2/4$, which is correct up to the quadratic $p^2/4$ difference, which is very small if p is small.) With this relation, our $\sqrt{1-p}$ result becomes $1 - p/2$. The difference between this result and 1 is therefore $p/2$. This makes it clear why we ended up with the above ratios of 0.95 and 0.995 for $p = 1/10$ and $p = 1/100$.

2. Note that our "goodness" condition for the Poisson approximation involves only p. That is, it is independent of n. This isn't terribly obvious. For a given value of p (say, $p = 1/100$), we will obtain the same accuracy whether n is, say, 10^3 or 10^5. Of course, the $a = pn$ expected values in these two cases are different (10 and 1000). But the ratio of $P_P(pn)$ to $P_B(pn)$ is the same (at least in the Stirling approximation).

3. In the language of balls and boxes, since $p = 1/b$, the $p \ll 1$ condition is equivalent to saying that the number of boxes satisfies $b \gg 1$. So the more boxes there are, the better the approximation. This condition is independent of the number n of balls (as long as n is large).

4. The result in Eq. (4.98) is valid even if the expected number of events pn is small, for example, 1 or 2. The is true because the $(pn)!$ terms cancel in the ratio of Eqs. (4.95) and (4.96), so we don't need to worry about applying Stirling's formula to a small number. The other two factorials, $n!$ and $(n - pn)!$, are large because we are assuming that n is large. ♣

4.15. **Bump or no bump**

We saw in Section 4.7.2 that the Poisson distribution is obtained by taking the $n \to \infty$ and $p \to 0$ limits of the binomial distribution (p took the form of $\lambda\epsilon$ in the derivation in Section 4.7.2). But in the $n \to \infty$ limit, the $p = 1/(n+1)$ condition for $P(0) = P(1)$ in the binomial case becomes $p \approx 1/n$. So $pn \approx 1$. But pn is just the average number of events a in the Poisson distribution. So $a \approx 1$ is the condition for $P(0) = P(1)$ in the Poisson case, as desired.

4.16. **Typos**

FIRST SOLUTION: Under the assumption that the typos occur randomly, the given setup calls for the Poisson distribution. If the expected number of typos in 50 pages is one, then the expected number of typos in a 350-page book is $a = 350/50 = 7$. So Eq. (4.40) gives the probability of zero typos in the book as

$$P(0) = \frac{a^0 e^{-a}}{0!} = e^{-a} = e^{-7} \approx 9 \cdot 10^{-4} \approx 0.1\%. \tag{4.99}$$

SECOND SOLUTION: (This is an approximate solution.) If there is one typo per 50 pages, then the expected number of typos per page is $1/50$. This implies that the probability that a given page has at least one typo is approximately 2%, which means that the probability that a given page has *zero* typos is approximately 98%. We are using the word "approximately" here, because the probability of zero typos on a given page must in fact be slightly larger than 98%. This is true because if it were exactly 98%, then in the 2% of the pages where a typo occurs, there might actually be two (or three, etc.) typos. Although these occurrences are rare in the present setup, they will nevertheless cause the expected number of typos per page to be (slightly) larger than $1/50$, in contradiction to the given assumption. The actual probability of having zero typos per page must therefore be slightly larger than 98%, so that slightly fewer than 2% of the pages have at least one typo.

However, if we work in the (reasonable) approximation where the probability of having zero typos per page equals 0.98, then the probability of having zero typos in 350 pages equals $(0.98)^{350} = 8.5 \cdot 10^{-4}$. This is close to the correct probability of $9 \cdot 10^{-4}$ in Eq. (4.99). Replacing 0.98 with a slightly larger number would yield the correct probability of $9 \cdot 10^{-4}$.

REMARKS: What should the probability of 0.98 (for zero typos on a given page) be increased to, if we want to obtain the correct probability of $9 \cdot 10^{-4}$ (for zero typos in the book)? Since the expected number of typos per page is $1/50$, we simply need to plug $a = 1/50$ into the Poisson expression for $P(0)$. This gives the true probability of having zero typos on a given page as

$$P(0) = \frac{a^0 e^{-a}}{0!} = e^{-a} = e^{-1/50} = 0.9802. \tag{4.100}$$

As expected, this is only a tiny bit larger than the approximate value of 0.98 that we used above. If we use the new (and correct) value of 0.9802, the result of our second solution is modified to $(0.9802)^{350} = 9 \cdot 10^{-4}$, which agrees with the correct answer in Eq. (4.99).

The relation between the (approximate) second solution and the (correct) first solution can be seen by writing our approximate answer of $(0.98)^{350}$ as

$$(0.98)^{350} = \left(1 - \frac{1}{50}\right)^{350} = \left(\left(1 - \frac{1}{50}\right)^{50}\right)^7 \approx (e^{-1})^7 = e^{-7}, \qquad (4.101)$$

which is the correct answer in Eq. (4.99). We have used the approximation in Eq. (7.4) to produce the e^{-1} term here. ♣

4.17. **Boxes with zero balls**

FIRST SOLUTION: The given information that 20 out of the 1000 boxes contain zero balls (on average) tells us that the probability that a given box contains zero balls is $P(0) = 20/1000 = 0.02$. The process at hand is approximately a Poisson process, just as the balls-in-boxes setup in the example on page 213 was. We therefore simply need to find the value of a in Eq. (4.40) that makes $P(0) = 0.02$. That is,

$$\frac{a^0 e^{-a}}{0!} = 0.02 \implies e^a = 50 \implies a = \ln 50 = 3.912. \qquad (4.102)$$

This a is the average number of balls in each of the 1000 boxes. The total number of balls in each trial is therefore $n = (1000)a = 3912$.

Note that once we know what a is, we can determine the number of boxes that contain other numbers of balls. For example $P(3) \approx (3.9)^3 e^{-3.9}/3! \approx 0.20$. So about 200 of the 1000 boxes end up with three balls, on average. $P(4)$ is about the same (a hair smaller). About 4.5 boxes (on average) end up with 10 balls, as you can show.

SECOND SOLUTION: We can solve the problem from scratch, without using the Poisson distribution. With $k = 0$, Eq. (4.33) tells us that the probability of obtaining zero balls in a given box is $P(0) = (1 - 1/1000)^n$. Setting this equal to 0.02 and using the approximation in Eq. (7.14) gives

$$(1 - 1/1000)^n = 0.02 \implies e^{-n/1000} = 0.02 \implies e^{n/1000} = 50$$
$$\implies n/1000 = \ln 50 \implies n = 3912. \qquad (4.103)$$

Alternatively, we can solve for n exactly, without using the approximation in Eq. (7.14). We want to find the n for which $(999/1000)^n = 0.02$. Taking the log of both sides gives

$$n \ln(0.999) = \ln(0.02) \implies n = \frac{-3.912}{-1.0005 \cdot 10^{-3}} = 3910. \qquad (4.104)$$

Our approximate answer of $n = 3912$ was therefore off by only 2, or equivalently 0.05%.

4.18. **Twice the events**

(a) This part of the problem is a repeat of Problem 4.11. The Poisson distribution is $P_a(k) = a^k e^{-a}/k!$, so the probability of obtaining a events is (using Stirling's formula for $a!$)

$$P_a(a) = \frac{a^a e^{-a}}{a!} \approx \frac{a^a e^{-a}}{a^a e^{-a} \sqrt{2\pi a}} = \frac{1}{\sqrt{2\pi a}}. \qquad (4.105)$$

(b) The average number of events during the time $2t$ is twice the average number during the time t. So we now have a Poisson process governed by an average of $2a$. The distribution is therefore $P_{2a}(k)$, and our goal is to calculate $P_{2a}(2a)$. In the same manner as in part (a), we find

$$P_{2a}(2a) = \frac{(2a)^{2a}e^{-2a}}{(2a)!} \approx \frac{(2a)^{2a}e^{-2a}}{(2a)^{2a}e^{-2a}\sqrt{2\pi(2a)}} = \frac{1}{\sqrt{4\pi a}}. \tag{4.106}$$

This is smaller than the result in part (a) by a factor of $1/\sqrt{2}$. In retrospect, we could have obtained the result of $1/\sqrt{4\pi a}$ by simply substituting $2a$ for a in the $1/\sqrt{2\pi a}$ result in part (a). The setup is the same here; we're still looking for the value of the distribution when k equals the average number of events. It's just that the average is now $2a$ instead of a.

(c) Since we're back to considering the original time t here, we're back to the Poisson distribution with an average of a. But since k is now $2a$, we want to calculate $P_a(2a)$. This equals

$$P_a(2a) = \frac{a^{2a}e^{-a}}{(2a)!} \approx \frac{a^{2a}e^{-a}}{(2a)^{2a}e^{-2a}\sqrt{2\pi(2a)}}$$

$$= \frac{1}{2^{2a}e^{-a}\sqrt{4\pi a}} = \left(\frac{e}{4}\right)^a \frac{1}{\sqrt{4\pi a}}. \tag{4.107}$$

This is smaller than the result in part (a) by a factor of $(e/4)^a/\sqrt{2}$. The $(e/4)^a$ part of this factor is approximately $(0.68)^a$, which is very small for large a. For example, if $a = 10$, then $(e/4)^a \approx 0.02$. And if $a = 100$, then $(e/4)^a \approx 1.7 \cdot 10^{-17}$.

For $a = 10$, the above three results are summarized in Fig. 4.29. The three dots indicate (from highest to lowest) the answers to parts (a), (b), and (c). This figure makes it clear why the answer to part (c) is much smaller than the other two answers; the $P_{10}(20)$ dot is on the tail of a curve, whereas the other two dots are near a peak. Although we have drawn the Poisson distributions as continuous curves, remember that the distribution applies only to integer values of k. The two highest dots aren't right at the peak of the curve, because the peak of the continuous curve is located at a value of k between $a - 1$ and a; see Problem 4.10.

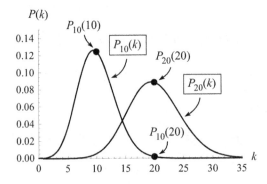

Figure 4.29: The Poisson curves for $a = 10$ and $a = 20$.

4.19. **P(0) the hard way**

The given interval (of time, space, or whatever) associated with the Poisson process has a expected events. Let's divide the interval into a very large number n of tiny intervals, each with a very small probability p of an event occurring. For simplicity, we are using p here instead of the $\lambda \epsilon$ we used at the beginning of Section 4.7.2. As in that section, we can ignore the distinction between the probability of an event in a tiny interval and the expected number of events in that interval, because we are assuming that $p \equiv \lambda \epsilon$ is very small; see Eq. (4.18).

The tasks of Problems 2.2 and 2.3 were to derive the "Or" rules for three and four events. Our goal here is basically to derive the "Or" rule for a large number n of independent events, each with a small probability p. These independent events are of course nonexclusive; we can certainly have more than one event occurring. Throughout this solution, you will want to have a picture like Fig. 2.17 in your mind. Although that picture applies to three events, it contains the idea for general n. Simple circles (each of which represents the probability that an event occurs in a given tiny interval) won't work for larger n, but it doesn't matter what the exact shapes are.

As in the solution to Problem 2.2(d), our goal is to determine the total area contained in the n partially overlapping regions (each with tiny area p) in the generalization of Fig. 2.17. The total area equals the probability of "Event 1 or Event 2 or ... Event n," which is the desired probability that at least one event occurs in the original interval. As in the solution to Problem 2.2(d), we can proceed as follows.

- If we add up the individual areas of all n tiny regions, we obtain np. (Each region represents the probability p that an event occurs in that particular tiny interval, with no regard for what happens with any of the other $n - 1$ tiny intervals.) But np equals the total expected number of events a in the original interval, because p is the expected number of events in each of the n tiny intervals. The sum of the individual areas of all n tiny regions therefore equals a. This a is the first term in the parentheses in Eq. (4.53).

- However, in adding up the individual areas of all n tiny regions, we have double counted each of the overlap regions where two events occur. The number of these regions is $\binom{n}{2} = n(n - 1)/2$, which essentially equals (in a multiplicative sense) $n^2/2$ for large n. The area (probability) of each double-overlap region is p^2, because that is the probability that two given events occur (with no regard for what else happens). The sum of the individual areas of the $n^2/2$ double-overlap regions is therefore $(n^2/2)p^2 = (np)^2/2 = a^2/2$. Since we have counted this area twice, and since we want to count it only once, we must correct for this by subtracting it off once. Hence the $-a^2/2!$ term in the parentheses in Eq. (4.53).

- We have now correctly determined the areas (probabilities) where *exactly one* or *exactly two* events occur. But what about the regions where three (or more) events occur? Each of these "triple" regions was counted $\binom{3}{1} = 3$ times when dealing with the "single" regions (because a triple region contains $\binom{3}{1}$ different single regions), but then uncounted $\binom{3}{2} = 3$ times when dealing with the "double" regions (because a triple region contains $\binom{3}{2}$ different double regions). We have therefore counted each triple region $\binom{3}{1} - \binom{3}{2} = 0$ times. There are $\binom{n}{3} = n(n - 1)(n - 2)/3! \approx n^3/3!$ of these regions. The area of each region is p^3, because that is the probability that three given events occur (with no regard for what else happens). The sum of the individual areas of the $n^3/3!$ triple regions is therefore $(n^3/3!)p^3 = (np)^3/3! = a^3/3!$. Since we have not counted this area at all, and since we want to count it once, we must correct for this by adding it on once. Hence the $+a^3/3!$ term in the parentheses in Eq. (4.53).

- One more iteration for good measure: We have now correctly determined the areas (probabilities) where *exactly one* or *exactly two* or *exactly three* events occur. But what about the regions where four (or more) events occur? Each of these "quadruple" regions was counted $\binom{4}{1} = 4$ times when dealing with the single regions, then uncounted $\binom{4}{2} = 6$ times when dealing with the double regions, then counted $\binom{4}{3} = 4$ times when dealing with the triple regions. We have therefore counted each quadruple region $\binom{4}{1} - \binom{4}{2} + \binom{4}{3} = 2$ times. There are $\binom{n}{4} = n(n-1)(n-2)(n-3)/4! \approx n^4/4!$ of these regions. The area of each region is p^4, because that is the probability that four given events occur (with no regard for what else happens). The sum of the individual areas of the $n^4/4!$ quadruple regions is therefore $(n^4/4!)p^4 = (np)^4/4! = a^4/4!$. Since we have counted this area twice, and since we want to count it only once, we must correct for this by subtracting it off once. Hence the $-a^4/4!$ term in the parentheses in Eq. (4.53).

Continuing in this manner gives the entire area in Fig. 2.17, or rather, the entire area in the analogous figure for the case of n events instead of three. In the $n \to \infty$ limit, we will obtain an infinite number of terms inside the parentheses in Eq. (4.53). All of the multiple counting is removed, so each region is counted exactly once. The total area represents the probability that at least one event occurs. Subtracting this from 1 gives the probability $P(0)$ in Eq. (4.53) that zero events occur.

As mentioned in the remark in the solution to Problem Eq. (2.3), we have either overcounted or undercounted each region *once* at every stage. This is the *inclusion–exclusion principle,* and it follows from the binomial expansion of $0 = (1-1)^m$. Using the expansion in Eq. (1.21) with $a = 1$ and $b = -1$, we have

$$(1-1)^m = \binom{m}{0} - \binom{m}{1} + \binom{m}{2} - \binom{m}{3} + \cdots + \binom{m}{m-1}(-1)^{m-1} + \binom{m}{m}(-1)^m. \quad (4.108)$$

The lefthand side equals zero, and the $\binom{m}{0}$ and $\binom{m}{m}$ terms equal 1, so we obtain

$$\binom{m}{1} - \binom{m}{2} + \binom{m}{3} - \binom{m}{m-1}(-1)^{m-1} = 1 + (-1)^m. \quad (4.109)$$

From the pattern of reasoning in the above bullet points, the lefthand side here is the number of times we have already counted each m-tuple region, in our handling of all of the 'lesser" regions – the single regions up through the $(m-1)$-tuple regions. (We are assuming inductively that we have overcounted or undercounted by 1 at each earlier stage.) The righthand side is either 2 or 0, depending on whether m is even or odd. We have therefore either overcounted or undercounted each m-tuple region by 1, which is consistent with the above results for $m = 2, 3$, and 4. There are $\binom{n}{m}$ of the m-tuple regions, each of which has an area of p^m. So at each stage, we need to either subtract or add an area (probability) of $\binom{n}{m}p^m \approx (n^m/m!)p^m = (np)^m/m! = a^m/m!$. These are the terms in parentheses in Eq. (4.53).

REMARK: In the end, the solution to this problem consists of the reasoning in the remark in the solution to Problem Eq. (2.3), combined with the fact that if n is large, we can say that $\binom{n}{m}p^m \approx (n^m/m!)p^m$, which equals $a^m/m!$. Now, taking into account all of the above double (and triple, etc.) counting is of course a much more laborious way to find $P(0)$ than simply using Eq. (4.40). Equivalently, the double-counting solution is more laborious than using Eq. (4.32) with $k = 0$, which quickly gives $P(0) = (1 - p)^n \approx e^{-pn} = e^{-a}$, using Eq. (7.14). (Eq. (4.32) is equivalent to Eq. (4.34), which led to Eq. (4.40).) The reasoning behind Eq. (4.32) involved directly finding the probability that *zero* events occur, by multiplying together all of the probabilities $(1 - p)$ that each event doesn't occur. This is clearly a much quicker method than the double-counting method of finding the probability that *at least one* event occurs, and then subtracting that from 1. This double-counting method is exactly the *opposite* of the helpful "art of not" strategy we discussed in Section 2.3.1! ♣

4.20. **Probability of at least 1**

(a) In finding the probability that *at least one* ball ends up in the given box, our strategy (in both parts of this problem) will be to find the probability that *zero* balls end up in the given box, and then subtract this probability from 1. The process at hand is approximately a Poisson process, just as the balls-in-boxes setup in the example on page 213 was. So from the Poisson distribution in Eq. (4.40), the probability that zero balls end up in the given box is $P(0) = a^0 e^{-a}/0! = e^{-a}$. The probability that at least one ball ends up in the given box is then $1 - e^{-a}$. This is an approximate result, because the process isn't exactly Poisson.

In the given setup, we have $n = 10^6$ balls and $b = 10^9$ boxes. So the average number of balls in a given box is $a = n/b = 1/1000$. Since this number is small, we can use the approximation in Eq. (7.9) (with $x \equiv -a$) to write $e^{-a} \approx 1 - a$. The desired probability that at least one ball ends up in the given box is therefore

$$1 - e^{-a} \approx 1 - (1 - a) = a = \frac{1}{1000}. \tag{4.110}$$

This makes sense. The expected number, a, of balls is small, which means that double (or triple, etc.) events are rare. The probability that at least one ball ends up in the given box is therefore essentially equal to $P(1)$. Additionally, since double (or triple, etc.) events are rare, we have $P(1) \approx a$, because the expected number of balls can be written as $a = P(1) \cdot 1 + P(2) \cdot 2 + \cdots \implies P(1) \approx a$. The two preceding sentences tell us that the probability that at least one ball ends up in the given box is approximately equal to a, as desired.

(b) The probability that a particular ball ends up in the given box is $1/b$, where $b = 10^9$ is the number of boxes. So the probability that the particular ball *doesn't* end up in the given box is $1 - 1/b$. This holds for all $n = 10^6$ of the balls, so the probability that *zero* balls end up in the given box is $(1 - 1/b)^n$. (This is just Eq. (4.33) with $k = 0$.) The probability that *at least one* ball ends up in the given box is therefore $1 - (1 - 1/b)^n$. This is the exact answer.

We can now use the $(1 + \alpha)^n \approx e^{n\alpha}$ approximation in Eq. (7.14) to simplify the answer. (We're using α in place of the a in Eq. (7.14), because we've already reserved the letter a for the average number of balls, n/b, here.) With $\alpha \equiv -1/b$, Eq. (7.14) turns the $1 - (1 - 1/b)^n$ probability into

$$1 - (1 - 1/b)^n \approx 1 - e^{-n/b} = 1 - e^{-a}. \tag{4.111}$$

The $e^{-a} \approx 1 - a$ approximation then turns this into a, as in part (a).

REMARK: We have shown that for small $a = n/b$, the probability that at least one ball ends up in the given box is approximately a. This result of course doesn't hold for non-small a because, for example, if we consider the $a = 1$ case, there certainly isn't a probability of 1 that at least one ball ends up in the given box. And we would obtain a nonsensical probability larger than 1 if $a > 1$. From either Eq. (4.110) or Eq. (4.111), the correct probability (in the Poisson approximation) that at least one ball ends up in the given box is $1 - e^{-a}$. For non-small a, we can't use the $e^{-a} \approx 1 - a$ approximation to turn $1 - e^{-a}$ into a. ♣

4.21. Comparing probabilities

(a) The three events are independent. So with $p = 1/1000$, the desired probability is simply p^3, which equals 10^{-9}.

(b) The three trials of the process are independent, so the desired probability is again p^3, where $p \approx 1/1000$ is the probability that exactly one ball lands in the given box in a given trial of the process. So we again obtain an answer of 10^{-9}. This setup is basically the same as the setup in part (a).

(c) If we perform a single trial of throwing a million balls into a billion boxes, the probability that three *specific* balls end up in the given box is $(1/b)^3$ (where $b = 10^9$), because each ball has a $1/b$ chance of landing in the box.[6] There are $\binom{n}{3}$ ways to pick the three specific balls from the $n = 10^6$ balls, so the probability that exactly three balls end up in the box is $\binom{n}{3}/b^3$. We can simplify this result by making an approximation to the binomial coefficient. Using the fact that $n - 1$ and $n - 2$ are both essentially equal (multiplicatively) to n if n is large, we have

$$\binom{n}{3}\frac{1}{b^3} = \frac{n(n-1)(n-2)}{3!}\frac{1}{b^3} \approx \frac{n^3}{3!}\frac{1}{b^3}$$

$$= \frac{1}{3!}\left(\frac{n}{b}\right)^3 = \frac{(10^{-3})^3}{3!} = \frac{10^{-9}}{3!} . \tag{4.112}$$

(d) The process in part (c) is approximately a Poisson process with $a = n/b = 1/000$. The probability that exactly three balls end up in the given box is therefore given by Eq. (4.40) as

$$P(3) = \frac{a^3 e^{-a}}{3!} . \tag{4.113}$$

Since $a = 1/000$ is small, the e^{-a} factor is essentially equal to 1, so we can ignore it. We therefore end up with

$$P(3) \approx \frac{a^3}{3!} = \frac{(10^{-3})^3}{3!} = \frac{10^{-9}}{3!} , \tag{4.114}$$

in agreement with the result in part (c).

In all of the parts to this problem, there is of course nothing special about the number 3 in the statement of the problem. If 3 is replaced by a general number k, then the results in parts (c) and (d) simply involve $k!$ instead of $3!$. (Well, technically k needs to be small compared with n, but that isn't much of a restriction in the present setup with $n = 10^6$.)

---

[6]There is technically a nonzero probability that other balls also land in the box. But this probability is negligible, so we don't have to worry about subtracting it off, even though we want *exactly* three balls in the box. Equivalently, the binomial distribution also involves a factor of $(1 - 1/b)^{n-3}$ (which ensures that the other $n - 3$ balls don't land in the box), but this factor is essentially equal to 1 in the present setup.

(e) The result in part (c) is smaller than the result in part (b) by a factor of $1/3! = 1/6$. Let's explain intuitively why this is the case.

In comparing the setups in parts (b) and (c), let's compare the respective probabilities (labeled $p_3^{(b)}$ and $p_3^{(c)}$) that three *specific* balls (labeled A, B, and C) end up in the given box. Although we solved part (b) in a quicker manner (by simply cubing p), we'll need to solve it here in the same way that we solved part (c), in order to compare the two setups. Note that in comparing the setups, it suffices to compare the probabilities for three specific balls, because both setups involve the same number of groups of three specific balls, namely $\binom{n}{3}$. So the total probabilities in each case are $\binom{n}{3}p_3^{(b)}$ and $\binom{n}{3}p_3^{(c)}$, with the $\binom{n}{3}$ factor being common to both.

Consider first the setup in part (c), with the single trial. There is only one way that all three of A, B, and C can end up in the box: If you successively throw down the n balls, then when you get to ball A, it must end up in the box (which happens with probability $1/b$); and then when you get to ball B, it must also end up in the box (which again happens with probability $1/b$); and finally when you get to ball C, it must also end up in the box (which again happens with probability $1/b$). The probability that all three balls end up in the box is therefore $p_3^{(c)} = (1/b)^3$. (This is just a repeat of the reasoning we used in part (c).)

Now consider the setup in part (b), with the three trials. There are now *six* ways that the three balls can end up in the box, because there are $3!$ permutations of the three balls. Ball A can end up in the box in the first of the three trials of n balls (which happens with probability $1/b$), and then B can end up in the box in the second trial (which again happens with probability $1/b$), and then C can end up in the box in the third trial (which again happens with probability $1/b$). We'll label this scenario as ABC. But the order in which the balls go into the boxes in the three successive trials can take five other permutations too, namely ACB, BAC, BCA, CAB, CBA. Each of the six possible permutations occurs with probability $(1/b)^3$, so the probability that all three balls (A, B, and C) end up in the box equals $p_3^{(b)} = 6(1/b)^3$. This explains why the answer to part (b) is six times the answer to part (b).

As mentioned above, if we want to determine the total probabilities in each setup, we just need to multiply each of $p_3^{(b)}$ and $p_3^{(c)}$ by the number $\binom{n}{3} \approx n^3/3!$ of groups of three balls. This was our strategy in part (c), and the result was $(n/b)^3/3!$. In part (b) this gives $(n^3/3!)(6/b^3) = (n/b)^3 = p^3$, in agreement with our original (quicker) solution. Note that it isn't an extra factor of $3!$ in the denominator that makes the answer to part (c) be smaller; parts (b) and (c) both have the $3!$ arising from the $\binom{n}{3}$ binomial coefficient. Rather, the answer to part (c) is smaller because it *doesn't* have the extra $3!$ in the numerator arising from the different permutations.

REMARK: Alternatively, you can think in terms of probabilities instead of permutations. In part (c) the probability (as we noted above) that three specific balls end up in the box is $(1/b)(1/b)(1/b)$, because each of the three balls must end up in the box when you throw it down. In contrast, in part (b) the probability that three specific balls end up in the box is $(3/b)(2/b)(1/b)$, because in the first trial of n balls, any of the three specific balls can end up in the box. And then in the second trial, one of the two other balls must end up in the box. And finally in the third trial, the remaining one of the three balls must end up in the box. The probability in part (b) is therefore larger by a factor of $3! = 6$.

Intuitively, it makes sense that the probability in part (b) is larger, because in part (c) if ball A doesn't end up in the box when you throw it down, you are guaranteed failure (for the three specific balls A, B, and C). But in part (b) if ball A doesn't end up in the box in the first of the three trials of n balls, you still have two more chances (with balls B and C) in that trial to get a ball in the box. So you have three chances to put one of the balls in the box in the first trial. And likewise you have two chances in the second trial. ♣

4.22. **Area under a Gaussian curve**

Let I be the desired integral. Then following the hint, we have

$$I^2 = \left(\sqrt{\frac{b}{\pi}} \int_{-\infty}^{\infty} e^{-bx^2} dx\right)\left(\sqrt{\frac{b}{\pi}} \int_{-\infty}^{\infty} e^{-by^2} dy\right)$$

$$= \frac{b}{\pi} \int_{-\infty}^{\infty} \int_{-\infty}^{\infty} e^{-b(x^2+y^2)} \, dx \, dy. \qquad (4.115)$$

If we convert from Cartesian to polar coordinates, then $x^2 + y^2$ becomes r^2 (by the Pythagorean theorem), and the area element $dx \, dy$ in the plane becomes $r \, dr \, d\theta$. This expression follows from the fact that we can imagine covering the plane with infinitesimal rectangles with sides of length dr in the radial direction and $r \, d\theta$ (the general form of an arclength) in the tangential direction.

The original double Cartesian integral runs over the entire x-y plane, so the new double polar integral must also run over the entire plane. The polar limits of integration are therefore 0 to ∞ for r, and 0 to 2π for θ. The above integral then becomes

$$I^2 = \frac{b}{\pi} \int_0^{2\pi} \int_0^{\infty} e^{-br^2} r \, dr \, d\theta. \qquad (4.116)$$

The θ integral simply gives 2π. The indefinite r integral is $-e^{-br^2}/2b$, as you can verify by differentiating this. The factor of r in the area element is what makes this integral doable, unlike the original Cartesian integral. We therefore have

$$I^2 = \frac{b}{\pi} \cdot 2\pi \cdot \left(-\frac{e^{-br^2}}{2b}\right)\Big|_0^{\infty}$$

$$= \frac{b}{\pi} \cdot 2\pi \cdot \frac{-1}{2b} \cdot (0 - 1)$$

$$= 1. \qquad (4.117)$$

So $I = \sqrt{1} = 1$, as desired. Note that if we didn't have the factor of $\sqrt{b/\pi}$ in the distribution, we would have ended up with

$$\int_{-\infty}^{\infty} e^{-bx^2} dx = \sqrt{\frac{\pi}{b}}. \qquad (4.118)$$

This is a useful general result.

The above change-of-coordinates trick works if we're integrating over a circular region centered at the origin. (An infinitely large circle covering the entire plane falls into this category.) If we want to calculate the area under a Gaussian curve with the limits of the x integral being arbitrary finite numbers a and b, then our only option is to evaluate the integral numerically. (The change-of-coordinates trick doesn't help with the rectangular region that arises in this case.) For example, if we want the limits to be $\pm\sigma = \pm 1/\sqrt{2b}$, then we must resort to numerics to show that the area is approximately 68% of the total area.

4.23. **Variance of the Gaussian distribution**

FIRST SOLUTION: With $\mu = 0$, the variance of the second expression for $f(x)$ in Eq. (4.42) is

$$E(X^2) = \int x^2 f(x)\, dx = \sqrt{\frac{1}{2\pi\sigma^2}} \int_{-\infty}^{\infty} x^2 e^{-x^2/2\sigma^2}\, dx. \qquad (4.119)$$

We can evaluate this integral by using integration by parts. That is, $\int fg' = fg - \int f'g$. If we write the x^2 factor as $x \cdot x$, then with $f \equiv x$ and $g' \equiv xe^{-x^2/2\sigma^2}$, we can integrate g' to obtain $g = -\sigma^2 e^{-x^2/2\sigma^2}$. So we have

$$\int_{-\infty}^{\infty} x \cdot xe^{-x^2/2\sigma^2}\, dx = x \cdot \left(-\sigma^2 e^{-x^2/2\sigma^2}\right)\Big|_{-\infty}^{\infty} - \int_{-\infty}^{\infty} 1 \cdot \left(-\sigma^2 e^{-x^2/2\sigma^2}\right) dx$$

$$= 0 + \sigma^2 \int_{-\infty}^{\infty} e^{-x^2/2\sigma^2}\, dx. \qquad (4.120)$$

The 0 comes from the fact that the smallness of $e^{-\infty^2}$ wins out over the largeness of the factor of ∞ out front. The remaining integral can be evaluated by invoking the general result in Eq. (4.118). With $b \equiv 1/2\sigma^2$ the integral is $\sqrt{2\pi\sigma^2}$. So Eq. (4.120) gives

$$\int_{-\infty}^{\infty} x^2 e^{-x^2/2\sigma^2}\, dx = \sigma^2 \sqrt{2\pi\sigma^2}. \qquad (4.121)$$

Plugging this into Eq. (4.119) then gives

$$E(X^2) = \sqrt{\frac{1}{2\pi\sigma^2}} \cdot \sigma^2 \sqrt{2\pi\sigma^2} = \sigma^2, \qquad (4.122)$$

as desired.

SECOND SOLUTION: This solution involves a handy trick for calculating integrals of the form $\int_{-\infty}^{\infty} x^{2n} e^{-bx^2}\, dx$. Using the $\int_{-\infty}^{\infty} e^{-bx^2}\, dx = \sqrt{\pi} b^{-1/2}$ result from Eq. (4.118) and successively differentiating both sides with respect to b, we obtain

$$\int_{-\infty}^{\infty} e^{-bx^2}\, dx = \sqrt{\pi} b^{-1/2},$$

$$\int_{-\infty}^{\infty} x^2 e^{-bx^2}\, dx = \frac{1}{2} \sqrt{\pi} b^{-3/2},$$

$$\int_{-\infty}^{\infty} x^4 e^{-bx^2}\, dx = \frac{3}{4} \sqrt{\pi} b^{-5/2}, \qquad (4.123)$$

and so on. On the lefthand side, it is indeed legal to differentiate the integrand (the expression inside the integral) with respect to b. If you have your doubts about this, you can imagine writing the integral as a sum over, say, a million terms. It is then certainly legal to differentiate each of the million terms with respect to b. In short, the derivative of the sum is the sum of the derivatives.

The second line in Eq. (4.123) is exactly the integral we need when calculating the variance. With $b \equiv 1/2\sigma^2$, the second line gives

$$\int_{-\infty}^{\infty} x^2 e^{-x^2/2\sigma^2}\, dx = \frac{1}{2} \sqrt{\pi} \left(\frac{1}{2\sigma^2}\right)^{-3/2} = \sqrt{2\pi}\, \sigma^3, \qquad (4.124)$$

in agreement with Eq. (4.121).

Chapter 5

Gaussian approximations

In this chapter we will concentrate on three of the distributions we studied in Chapter 4, namely the binomial, Poisson, and Gaussian distributions. In Section 5.1 we show how a binomial distribution reduces to a Gaussian distribution when the numbers involved are large. Section 5.2 covers the *law of large numbers*, which says that in a very large number of trials, the observed fraction of events will be very close to the theoretical probability. In Section 5.3 we show how a Poisson distribution reduces to a Gaussian distribution when the numbers involved are large. In Section 5.4 we tie everything together. This leads us in Section 5.5 to the *central limit theorem*, which is the statement that no matter what distribution you start with, the sum (or average) of the outcomes of many trials will be approximately Gaussian. As in Chapter 4, parts of this chapter are a bit mathematical, but there's no way around this if we want to do things properly. We will invoke some results from Appendix C.

5.1   Binomial and Gaussian

In Section 4.5 we discussed the binomial distribution, in particular the binomial distribution that arises from a series of coin flips. The probability distribution for the total number of Heads in, say, 30 flips takes the form of the left plot in Fig. 4.10. The shape of this plot looks suspiciously similar to the shape of the Gaussian plot in Fig. 4.25, so you might wonder if the binomial distribution is actually a Gaussian distribution (or more precisely, if the discrete binomial points lie on a continuous Gaussian curve). It turns out that for small numbers of coin flips, this isn't quite true. But for large numbers of flips, a binomial distribution takes essentially the form of a Gaussian distribution. The larger the number of flips, the closer it comes to a Gaussian.

For three different numbers of coin flips (2, 6, and 20), Fig. 5.1 shows the comparison between the exact binomial distribution (the dots) and the Gaussian approximation (the curve), which we'll derive below in Eq. (5.13). The coordinate on the x axis is the number of Heads relative to the expected value (which is half the number of flips). So for n flips, the possible x values range from $-n/2$ to $n/2$. The Gaussian

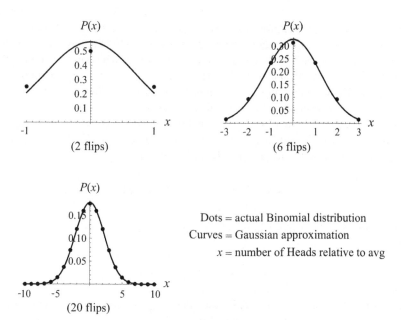

Figure 5.1: Comparison of the binomial distribution and the Gaussian approximation, for various numbers of coin flips. x is the number of Heads relative to the expected number.

approximation is clearly very good for 20 flips. And it gets even better for larger numbers of flips.

We will now demonstrate why the binomial distribution takes essentially the form of a Gaussian distribution when the number of flips is large. For convenience, we'll let the number of flips be $2n$, just to keep some factors of $1/2$ from cluttering things up. We will assume that n is large.

We'll need two bits of mathematical machinery for this derivation. The first is Stirling's formula, which we introduced in Section 2.6. It says that if n is large, then $n!$ is approximately given by

$$n! \approx n^n e^{-n} \sqrt{2\pi n}. \tag{5.1}$$

It's a good idea at this point to go back and review Section 2.6. The second thing we'll need is the approximation in Eq. (7.15) in Appendix C:

$$(1 + a)^m \approx e^{ma} e^{-ma^2/2}. \tag{5.2}$$

We're using m instead of n here, because we've already reserved n for half the number of flips. You are encouraged to read Appendix C at this point (after reading Appendix B), to see where this approximation comes from. However, feel free to just accept it for now if you want. But in that case, you should at least verify with a calculator that it works fairly well for, say, $a = 0.1$ and $m = 30$.

The following derivation is a bit mathematical, but the result (that a binomial distribution can be approximated by a Gaussian distribution) is well worth it. We'll demonstrate this result just for coin flips (that is, a binomial distribution with $p = 1/2$), but it actually holds for any p; see the discussion following the remarks below.

We'll start with the binomial distribution in Eq. (4.8), which gives the probability of obtaining k Heads in n coin flips. However, since we're letting the number of coin flips be $2n$ here, the n in Eq. (4.8) gets replaced by $2n$. Also, let's replace k by $n + x$, which just means that we're defining x to be the number of Heads relative to the expected number (which is n). Writing the number of Heads as $n + x$ will make our calculations *much* simpler than if we had stuck with k. With these adjustments, Eq. (4.8) becomes (with the subscript B for binomial)

$$P_B(x) = \frac{1}{2^{2n}} \binom{2n}{n+x} \qquad \text{(for $2n$ coin flips)} \qquad (5.3)$$

We will now show that if n is large, $P_B(x)$ takes the approximate form,

$$P_B(x) \approx \frac{e^{-x^2/n}}{\sqrt{\pi n}}, \qquad (5.4)$$

which is the desired Gaussian. This takes the same form as the first Gaussian expression in Eq. (4.42), with $b = 1/n$ and $\mu = 0$.

So here we go – get ready for some math! But it's nice math, in the sense that a huge messy equation will undergo massive cancelations and yield a nice simple result. The first step is to use Stirling's approximation to rewrite each of the three factorials in the binomial coefficient in Eq. (5.3). This gives

$$\binom{2n}{n+x} = \frac{(2n)!}{(n+x)!(n-x)!} \qquad (5.5)$$

$$\approx \frac{(2n)^{2n} e^{-2n} \sqrt{2\pi(2n)}}{[(n+x)^{n+x} e^{-(n+x)} \sqrt{2\pi(n+x)}] \cdot [(n-x)^{n-x} e^{-(n-x)} \sqrt{2\pi(n-x)}]} .$$

Canceling all the e's and a few other factors gives

$$\binom{2n}{n+x} \approx \frac{(2n)^{2n} \sqrt{n}}{(n+x)^{n+x} (n-x)^{n-x} \sqrt{\pi} \sqrt{n^2 - x^2}} . \qquad (5.6)$$

Let's now divide both the numerator and denominator by n^{2n}. In the denominator, we'll do this by dividing the first and second factors by n^{n+x} and n^{n-x}, respectively. The result is

$$\binom{2n}{n+x} \approx \frac{2^{2n} \sqrt{n}}{\left(1 + \frac{x}{n}\right)^{n+x} \left(1 - \frac{x}{n}\right)^{n-x} \sqrt{\pi} \sqrt{n^2 - x^2}} . \qquad (5.7)$$

It's now time to apply the approximation in Eq. (5.2). With the a and m in that relation defined to be $a \equiv x/n$ and $m \equiv n + x$, we have (using the notation $\exp(y)$ for e^y, to avoid writing lengthy exponents)

$$\left(1 + \frac{x}{n}\right)^{n+x} \approx \exp\left((n+x)\left(\frac{x}{n}\right) - \frac{1}{2}(n+x)\left(\frac{x}{n}\right)^2\right). \qquad (5.8)$$

When we multiply things out here, we find that there is a $-x^3/2n^2$ term. However, we'll see below that the x's we'll be dealing with are much smaller than n, which means that the $-x^3/2n^2$ term is much smaller than the other terms. So we'll ignore it. We are then left with

$$\left(1 + \frac{x}{n}\right)^{n+x} \approx \exp\left(x + \frac{x^2}{2n}\right). \tag{5.9}$$

Although the $x^2/2n$ term here is much smaller than the x term (assuming $x \ll n$), we will in fact need to keep it, because the x term will cancel in Eq. (5.11) below. (The $-x^3/2n^2$ term would actually cancel too, for the same reason.) In a similar manner, we obtain

$$\left(1 - \frac{x}{n}\right)^{n-x} \approx \exp\left(-x + \frac{x^2}{2n}\right). \tag{5.10}$$

Using these results in Eq. (5.7), we find

$$\binom{2n}{n+x} \approx \frac{2^{2n}\sqrt{n}}{\exp\left(x + \frac{x^2}{2n}\right)\exp\left(-x + \frac{x^2}{2n}\right)\sqrt{\pi}\sqrt{n^2 - x^2}}. \tag{5.11}$$

When combining (adding) the exponents, the x and $-x$ cancel. Also, under the assumption that $x \ll n$, we can say that $\sqrt{n^2 - x^2} \approx \sqrt{n^2 - 0} = n$. (As with any approximation claim, if you don't trust this, you can simply plug in some numbers and see how well it works. For example, you can let $n = 10{,}000$ and $x = 100$, which satisfy the $x \ll n$ relation.) Eq. (5.11) then becomes

$$\binom{2n}{n+x} \approx \frac{2^{2n}\sqrt{n}}{e^{x^2/n}\sqrt{\pi}\,n}. \tag{5.12}$$

Finally, if we substitute Eq. (5.12) into Eq. (5.3), the 2^{2n} factors cancel, and we are left with the desired result (with the subscript G for Gaussian),

$$\boxed{P_B(x) \approx \frac{e^{-x^2/n}}{\sqrt{\pi n}} \equiv P_G(x)} \qquad \text{(for } 2n \text{ coin flips)} \tag{5.13}$$

This is the probability of obtaining $n+x$ Heads in $2n$ coin flips. If we want to switch back to having the number of flips be n instead of $2n$, then we just need to replace n with $n/2$ in Eq. (5.13). The result is (with x now being the deviation from $n/2$ Heads)

$$\boxed{P_B(x) \approx \frac{e^{-2x^2/n}}{\sqrt{\pi n/2}} \equiv P_G(x)} \qquad \text{(for } n \text{ coin flips)} \tag{5.14}$$

Whether you use Eq. (5.13) or Eq. (5.14), the coefficient of π and the inverse of the coefficient of x^2 are both equal to half the number of flips.

If you want to write the above results in terms of the actual number k of Heads, instead of the number x of Heads relative to the expected number, you can just replace x with either $k - n$ in Eq. (5.13), or $k - n/2$ in Eq. (5.14).

The most important part of the above results is the n in the denominator of the exponent, because this determines the width of the distribution. We'll talk about this in Section 5.2, but first some remarks.

REMARKS:

1. In the above derivation, we claimed that if n is large (as we are assuming), then any values of x that we are concerned with are much smaller than n. This allowed us to simplify various expressions by ignoring certain terms. Let's be explicit about how the logic of the $x \ll n$ assumption proceeds.

   What we showed above (assuming n is large) is that *if* the $x \ll n$ condition is satisfied, *then* Eq. (5.13) is valid. And the fact of the matter is that if n is large, we'll never be interested in values of x that don't satisfy $x \ll n$ (and hence for which Eq. (5.13) might not be valid), because the associated probabilities are negligible. This is true because if, for example, $x = 10\sqrt{n}$ (which certainly satisfies $x \ll n$ if n is large, which means that Eq. (5.13) is indeed valid), then the $e^{-x^2/n}$ exponential factor in Eq. (5.13) equals $e^{-10^2} = e^{-100} \approx 4 \cdot 10^{-44}$, which is completely negligible. (Even if x is only $2\sqrt{n}$, the $e^{-x^2/n}$ factor equals $e^{-2^2} = e^{-4} \approx 0.02$.) Larger values of x will yield even smaller probabilities, because we know that the binomial coefficient in Eq. (5.3) decreases as x gets farther from zero; recall Pascal's triangle in Section 1.8.1. These probabilities might not satisfy Eq. (5.13), but we don't care, because they're so small.

2. In the terminology of Eq. (5.14) where the number of coin flips is n, the plots in Fig. 5.1 correspond to n equalling 2, 6, and 20. So in the third plot, for example, the continuous curve is a plot of $P_G(x) = e^{-x^2/10}/\sqrt{10\pi}$.

3. $P_G(x)$ is an even function of x. That is, x and $-x$ yield the same value of the function; it is symmetric around $x = 0$. This is true because x appears only through its square. This evenness makes intuitive sense, because we're just as likely to get, say, four Heads above the average as four Heads below the average.

4. We saw in Eq. (2.66) in Section 2.6 that the probability that exactly half (that is, n) of $2n$ coin flips come up Heads equals $1/\sqrt{\pi n}$. This result is a special case of the $P_G(x)$ result in Eq. (5.13), because if we plug $x = 0$ (which corresponds to n Heads) into Eq. (5.13), we obtain $P_G(x) = e^{-0}/\sqrt{\pi n} = 1/\sqrt{\pi n}$.

5. Note that we really did need the $e^{-ma^2/2}$ factor in the approximation in Eq. (5.2). If we had used the less accurate version, $(1+a)^m \approx e^{ma}$ from Eq. (7.14) in Appendix C, we would have had incorrect x^2/n terms in Eqs. (5.9) and (5.10), instead of the correct $x^2/2n$ terms.

6. If we compare the Gaussian result in Eq. (5.14) with the second of the Gaussian expressions in Eq. (4.42), we see that they agree if $\sigma = \sqrt{n/4}$. This correspondence makes *both* the prefactor and the coefficient of x^2 in the exponent agree. The standard deviation of our Gaussian approximation in Eq. (5.14) (for the binomial distribution for n coin flips) is therefore $\sigma = \sqrt{n/4}$. This agrees (as it must) with the exact binomial standard deviation we obtained in Eq. (3.48).

   Before going through the above derivation, it certainly wasn't obvious that a binomial should reduce to a Gaussian when n is large. However, the previous paragraph shows that *if* it reduces to a Gaussian, *then* the n's must appear exactly as they do in Eq. (5.14), because we know that the standard deviation (which is the σ in Eq. (4.42)) must agree with the $\sqrt{n/4}$ value that we already found in Eq. (3.48).

7. Since the area (probability) under the Gaussian distribution in Eq. (4.42) is 1 (see Problem 4.22), and since Eq. (5.14) takes the same form as Eq. (4.42), the area under the distribution in Eq. (5.14) must likewise be 1. Of course, we already knew

this, because Eq. (5.14) is an approximation to the binomial distribution, whose total probability is 1. ♣

If the two probabilities involved in a binomial distribution are p and $1-p$ instead of the two $1/2$'s in the case of a coin toss, then the probability of k successes in n trials is given in Eq. (4.6) as $P(k) = \binom{n}{k}p^k(1-p)^{n-k}$. (We've gone back to using n to represent the total number of trials, instead of the $2n$ we used in Eq. (5.3).) For example, if we're concerned with the number of 5's we obtain in n rolls of a die, then $p = 1/6$.

It turns out that for large n, the binomial distribution $P(k)$ is essentially a Gaussian distribution for *any* value of p, not just the $p = 1/2$ value we discussed above. The Gaussian is centered around the expected value of k (namely pn), as you would expect. The derivation of this Gaussian form follows the same steps as above. But it gets rather messy, so we'll just state the result: For large n, the probability of obtaining $k = pn + x$ successes in n trials is approximately equal to

$$P_G(x) \approx \frac{e^{-x^2/[2np(1-p)]}}{\sqrt{2\pi np(1-p)}} \qquad \text{(for } n \text{ biased coin flips)} \qquad (5.15)$$

If $p = 1/2$, this reduces to the result in Eq. (5.14), as it should.

Eq. (5.15) implies that the bump in the plot of $P_G(x)$ is symmetric around $x = 0$ (or equivalently, around $k = pn$) for *any* p, not just $p = 1/2$. This isn't so obvious, because for $p \neq 1/2$, the bump isn't centered around $n/2$. That is, the *location* of the bump is lopsided with respect to $n/2$. So you might think that the *shape* of the bump should be lopsided too. But it isn't. (Well, the tail extends farther to one side, but $P_G(x)$ is essentially zero in the tails.) Fig. 5.2 shows a plot of Eq. (5.15) for $p = 1/6$ and $n = 60$, which corresponds to rolling a die 60 times and seeing how many, say, 5's you get. The $x = 0$ point corresponds to having $pn = (1/6)(60) = 10$ rolls of a 5. The bump is quite symmetric (although technically not exactly). This is consistent with what we noted about the binomial distribution in the remark at the end of the example in Section 3.4.

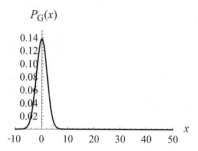

Figure 5.2: The probability distribution for the number of 5's in 60 dice rolls. x is the deviation from the expected number (which is 10). The bump in the distribution is essentially symmetric.

As in the sixth remark above, if we compare the Gaussian result in Eq. (5.15) with the second of the Gaussian expressions in Eq. (4.42), we see that they agree if $\sigma = \sqrt{np(1-p)} \equiv \sqrt{npq}$. Again, this correspondence makes *both* the prefactor and the coefficient of x^2 in the exponent agree. The standard deviation of our Gaussian approximation in Eq. (5.15) (for the binomial distribution for n biased coin flips) is therefore $\sigma = \sqrt{npq}$. This agrees (as it must) with the exact binomial standard deviation we obtained in Eq. (3.47).

As we also noted in the sixth remark, if someone claims (correctly) that a general binomial distribution involving probability p reduces to a Gaussian, then they must also claim that the $np(1-p)$ factors appear exactly as they do in Eq. (5.15), because we know that the standard deviation (which is the σ in Eq. (4.42)) must agree with the $\sqrt{np(1-p)} \equiv \sqrt{npq}$ value that we already found in Eq. (3.47).

5.2   The law of large numbers

The *law of large numbers* is, in a sense, the law that makes the subject of probability a useful one, in that it allows us to make meaningful predictive statements about future outcomes. The law can be stated in various ways, but we'll go with:

- Law of large numbers:

  If you repeat a random process a very large number of times, then the observed fraction of times that a certain event occurs will be very close to the theoretical probability.

More precisely, consider the probability, p_d (with the "d" for "differ"), that the observed fraction differs from the theoretical probability by more than a specified small number, say $\delta = 0.01$ or 0.001. Then the law of large numbers says that p_d goes to zero as the number of trials becomes large. Said in a more down-to-earth way, if you perform enough trials, the observed fraction will be pretty much what it "should" be.

REMARK: The probability p_d in the preceding paragraph deals with the results of a large number (call it n_1) of trials of a given random process (such as a coin flip). If you want to experimentally measure p_d, then you need to perform a large number (call it n_2) of sets, *each of which* consists of a large number n_1 of coin flips (or whatever). For example, we might be concerned with the fraction of Heads that show up in $n_1 = 10{,}000$ coin flips. If we ask for the probability p_d that this fraction differs from 50% by more than 1%, then we could do, say, $n_2 = 100{,}000$ sets of $n_1 = 10{,}000$ flips (which means a billion flips in all!) and then make a list or a histogram of the resulting n_2 observed fractions. The fraction of these fractions that are smaller than 49% or larger than 51% is our desired probability p_d. The larger n_2 is, the closer our result for p_d will be to its true value (which happens to be 5%; see Problem 5.3). This is how you experimentally determine p_d. The law of large numbers says that if you make n_1 larger and larger (which means that you need to make n_2 larger too), then p_d approaches zero. ♣

The clause "a very large number of times" is critical in the law. If you flip a coin only, say, 10 times, then you of course cannot be nearly certain that you will obtain Heads half (or very close to half) of the time. In fact, the probability of obtaining exactly five Heads is only $\binom{10}{5}/2^{10} \approx 25\%$.

You will note that the above statement of the law of large numbers is essentially the same as the definition of probability presented at the beginning of Section 2.1.

Things therefore seem a bit circular. Is the law of large numbers a theorem or a definition? This problem can be (somewhat) remedied by stating the law as, "If you repeat a random process a very large number of times, then the observed fraction of times that a certain event occurs will approach a definite value." Then, given that a definite value is approached, we can define that value to be the probability. The law of large numbers is therefore what allows probability to be well defined. (If this procedure doesn't allay your concerns about circularity, rest assured, it shouldn't. See the "On average" subsection in Appendix A for some discussion of this.)

We won't give a formal proof of the law, but we'll look at a coin-flipping setup in detail. This should convince you of the truth of the law. We'll basically do the same type of analysis here that we did in Section 3.4, where we discussed the standard deviation of the mean. But now we'll work with Gaussian distributions, in particular the one in Eq. (5.13), where the number of flips is $2n$. Comparing the Gaussian expression $P_G(x)$ in Eq. (5.13) with the second of the Gaussian expressions in Eq. (4.42), we see that the standard deviation when the number of flips is $2n$ equals $\sigma = \sqrt{n/2}$. This is consistent with the fact that the standard deviation when the number of flips is n equals $\sigma = \sqrt{n/4}$.

Fig. 5.3 shows plots of $P_G(x)$ for $n = 10$, 100, and 1000. So the numbers of coin flips are 20, 200, and 2000. As n gets larger, the curve's height shrinks, because Eq. (5.13) says that the height is proportional to $1/\sqrt{n}$. And the width expands, because σ is proportional to $\sqrt{n}$. Because these two factors are reciprocals of each other, this combination of shrinking and expanding doesn't change the area under the curve. This is consistent with the fact that the area is always equal to 1.

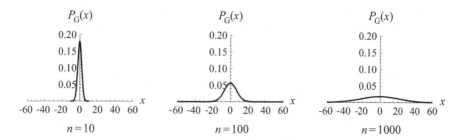

Figure 5.3: Illustration of how the Gaussian distribution in Eq. (5.13) depends on n. The number of coin flips is $2n$, and x is the deviation from the expected number of Heads (which is n).

The critical fact about the $\sqrt{n}$ expansion factor in the width is that although it increases as n increases, *it doesn't increase as fast as n does*. In fact, compared with n, it actually *decreases* by a factor of $1/\sqrt{n}$. This means that if we plot $P_G(x)$ with the horizontal axis running from $-n$ to n (instead of it being fixed as in Fig. 5.3), then the width of the curve actually *shrinks* by a factor of $1/\sqrt{n}$ (relative to n). Fig. 5.4 shows this effect. In this figure, both the width (relative to n) and the height of the curves are proportional to $1/\sqrt{n}$ (the height behaves the same as in Fig. 5.3), so all of the curves have the same shape. They just have different sizes; they differ successively by a factor of $1/\sqrt{10}$. The area under each curve is still equal to 1, though, because of the different scales on the x axis.

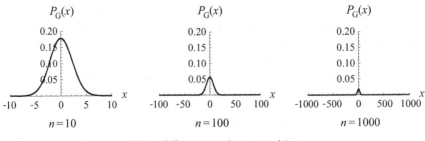

(Note differerent scales on x axis)

Figure 5.4: Repeat of the curves in Fig. 5.3, but now with the full range of possible values on the x axis.

A slightly more informative curve to plot is the ratio of $P_G(x)$ to its maximum height at $x = 0$. This modified plot makes it easier to see what's happening with the width. Since the maximum value of the Gaussian distribution in Eq. (5.13) is $1/\sqrt{\pi n}$, we're now just plotting $e^{-x^2/n}$. So all of the curves have the same value of 1 at $x = 0$. If we let the horizontal axis run from $-n$ to n as in Fig. 5.4, we obtain the plots shown in Fig. 5.5. These are simply the plots in Fig. 5.4, except that they're stretched in the vertical direction so that they all have the same height. We see that the bump gets thinner and thinner (on the scale of n) as n increases. (Each successive bump is thinner by a factor $1/\sqrt{10}$.) This implies that the *percentage* deviation from the average of n Heads gets smaller and smaller as n increases.

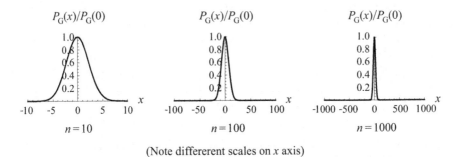

(Note differerent scales on x axis)

Figure 5.5: Repeat of the curves in Fig. 5.4, measured relative to the maximum height.

We can now understand why the law of large numbers holds. Equivalently (for the case of coin flips), we can now understand the reason behind the claim we made at the end of Section 2.1, when we said that the observed fraction of Heads gets closer and closer to the actual probability of 1/2, as the number of trials gets larger and larger. We stated that if you flip a coin 100 times (which corresponds to $n = 50$ here), the probability of obtaining 49, 50, or 51 Heads is only about 24%. This is roughly 3 times the 8% result in Eq. (2.67), because the probabilities for 49 and 51 are roughly the same as for 50.

This probability of 24% is consistent with the first plot in Fig. 5.6, where we've indicated the 49% and 51% tick marks (which correspond to $x = \pm 1$) on the x axis.

If we make a histogram of the probabilities (in order to interpret the probability as an area), then the natural thing to do is to have the "bin" for 49 go from 48.5 to 49.5, etc. So if we're looking at 49, 50, or 51 Heads, we're actually concerned with (approximately) the area between 48.5 and 51.5. This is the shaded area shown, with a width of 3. (This shaded area might not look like it is 24% of the total area, but it really is!) The distinction between the tick marks and the shaded area (which extends 0.5 beyond the tick marks) matters in the present $n = 50$ case, but it is inconsequential when n is large, because the distribution is effectively continuous.

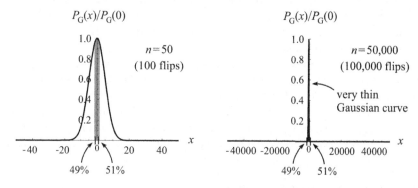

(Note differerent scales on x axis)

Figure 5.6: Illustration of the law of large numbers. If the number of coin flips is very large (as it is in the second plot), then the percentage of Heads is nearly certain to be very close to 50%.

At the end of Section 2.1, we also stated that if you flip a coin 100,000 times (which corresponds to $n = 50,000$), the probability of obtaining Heads between 49% and 51% of the time is 99.999999975%. This is consistent with the second plot in Fig. 5.6, because essentially all of the area under the curve lies between the 49% and 51% marks (which correspond to $x = \pm 1000$). The standard deviation for $n = 50,000$ is $\sqrt{n/2} = \sqrt{25,000} = 158$. So the 51% mark corresponds to about six standard deviations from the mean. There is virtually no chance of obtaining a result more than 6σ from the mean. In contrast, in the case with $n = 50$, the standard deviation is $\sqrt{n/2} = \sqrt{25} = 5$. Most of the area under the curve lies outside the 49% and 51% marks (where $x = \pm 1$), or rather, the 48.5% and 51.5% marks.

The law of large numbers states that if p_d is the probability that the observed fraction differs from the theoretical probability by more than a specified small number δ, then p_d goes to zero as the number of trials becomes large. The right plot in Fig. 5.6 demonstrates this for $\delta = 1\% = 0.01$ and 100,000 flips. From the 99.999999975% probability mentioned above, there is only a $p_d = 0.000000025\%$ probability of ending up outside the 49%–51% range. Although we've demonstrated the law of large numbers only in the case of coin flips (a binomial process with $p = 1/2$), it holds for any random process that is performed a large number of times.

The law of large numbers is an extremely important result, and it all comes down to the fact that although the standard deviation of our Gaussian coin-flip distribution grows with n (it is $\sigma = \sqrt{n/2}$ for $2n$ flips), it grows only like the *square root* of n, so it *shrinks* in comparison with the full spread of outcomes (which is $2n$). Said a different way, although the width of the distribution grows in an *additive* sense (this is sometimes called an "absolute" sense), it decreases in a *multiplicative* sense (compared with n). It is the latter of these effects that is relevant when calculating percentages.

This is exactly the same observation that we made back in Section 3.4 when we discussed the standard deviation of the mean. This makes sense, of course, because the percentage of Heads we've been talking about here is exactly the same thing as the average number of Heads per flip that we talked about in Section 3.4. So technically everything in this section is just a repeat of what we did Section 3.4. But it never hurts to see something twice!

The law of large numbers is what makes polls more accurate if more people are interviewed, and why casinos nearly always come out ahead. It is what makes it prohibitively unlikely for all of the air molecules in a room to end up on one side of the room, and why a piece of paper on your desk doesn't spontaneously combust. The list of applications is essentially endless, and it would be an understatement to say that the world would be a very different place without the law of large numbers.

5.3   Poisson and Gaussian

We showed in Section 5.1 that the binomial distribution in Eq. (5.3) becomes the Gaussian distribution in Eq. (5.13) in the limit where the number of trials is large. We will now show that the Poisson distribution in Eq. (4.40) becomes a Gaussian distribution in the limit of large a, where a is the expected number of successes in a given interval (of time, space, or whatever).

Note that it wouldn't make sense to take the limit of a large number of trials here, as we did in the binomial case, because the number of trials isn't specified in the Poisson distribution. The only parameter that appears is the expected number of successes, a. However, in the binomial case, a large number n of trials implies a large expected number of successes (because the expected number pn grows with n). So the large-a limit in the Poisson case is analogous to the large-n limit in the binomial case.

As in the binomial case, we will need to use the two approximations in Eqs. (5.1) and (5.2). Applying Stirling's formula to the $k!$ in Eq. (4.40) gives (with the subscript P for Poisson)

$$P_P(k) = \frac{a^k e^{-a}}{k!}$$

$$\approx \frac{a^k e^{-a}}{k^k e^{-k} \sqrt{2\pi k}}. \tag{5.16}$$

The result in Problem 4.10 is that the maximum of $P_P(k)$ occurs at a (or technically between $a - 1$ and a, but for large a this distinction is inconsequential). So let's see how $P_P(k)$ behaves near $k = a$. To this end, we'll define x by $k \equiv a + x$. So x is the number of successes relative to the average, a. This is analogous to the

$k \equiv n + x$ definition we used in Section 5.1. As it did there, working with x here will make our calculations much simpler. In terms of x, Eq. (5.16) becomes

$$P_P(x) \approx \frac{a^{a+x} e^{-a}}{(a+x)^{a+x} e^{-a-x} \sqrt{2\pi(a+x)}}. \tag{5.17}$$

We can cancel a factor of e^{-a}. And we can divide both the numerator and denominator by a^{a+x}. Furthermore, we can ignore the x in the square root, because we'll find below that the x's we're concerned with are small compared with a. The result is

$$P_P(x) \approx \frac{1}{\left(1 + \frac{x}{a}\right)^{a+x} e^{-x} \sqrt{2\pi a}}. \tag{5.18}$$

It's now time to use the approximation in Eq. (5.2). With the a in Eq. (5.2) defined to be x/a here, and with the m defined to be $a + x$, Eq. (5.2) gives

$$\left(1 + \frac{x}{a}\right)^{a+x} \approx \exp\left((a+x)\left(\frac{x}{a}\right) - \frac{1}{2}(a+x)\left(\frac{x}{a}\right)^2\right). \tag{5.19}$$

Multiplying this out and ignoring the small $-x^3/2a^2$ term (because we'll find below that $x \ll a$), we obtain

$$\left(1 + \frac{x}{a}\right)^{a+x} \approx \exp\left(x + \frac{x^2}{2a}\right). \tag{5.20}$$

This is just Eq. (5.9) with $n \to a$. Substituting Eq. (5.20) into Eq. (5.18) gives

$$P_P(x) \approx \frac{1}{e^x e^{x^2/2a} e^{-x} \sqrt{2\pi a}}, \tag{5.21}$$

which simplifies to

$$\boxed{P_P(x) \approx \frac{e^{-x^2/2a}}{\sqrt{2\pi a}} \equiv P_G(x)} \tag{5.22}$$

This is the desired Gaussian. If you want to write this result in terms of the actual number k of successes, instead of the number x of successes relative to the average, then the definition $k \equiv a + x$ gives $x = k - a$, so we have

$$\boxed{P_P(k) \approx \frac{e^{-(k-a)^2/2a}}{\sqrt{2\pi a}} \equiv P_G(k)} \tag{5.23}$$

As we noted in the last remark in Section 4.8, the Poisson distribution (and hence the Gaussian approximation to it) depends on only one parameter, a. And as with the Gaussian approximation to the binomial distribution, the Gaussian approximation to the Poisson distribution is symmetric around $x = 0$ (equivalently, $k = a$).

Fig. 5.7 shows a comparison between the exact $P_P(k)$ function in the first line of Eq. (5.16), and the approximate $P_G(k)$ function in Eq. (5.23). The approximation works quite well for $a = 20$ and extremely well for $a = 100$; the curve is barely noticeable behind the dots.

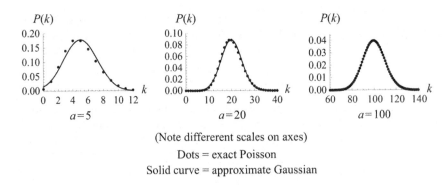

(Note differerent scales on axes)

Dots = exact Poisson

Solid curve = approximate Gaussian

Figure 5.7: Comparison of the Poisson distribution in the first line of Eq. (5.16), and the Gaussian approximation in Eq. (5.23), for different values of a.

If we compare the Gaussian distribution in Eq. (5.23) with the second expression in Eq. (4.42), we see that the Gaussian is centered at $\mu = a$ (of course) and that the standard deviation is $\sigma = \sqrt{a}$. Again, since the Poisson distribution depends on only the one parameter a, we already knew that the standard deviation has to be a function of a. But it takes some work to show that it equals $\sqrt{a}$. Of course, as in the sixth remark in Section 5.1, we know that *if* the Poisson distribution reduces to a Gaussian, *then* the a's must appear exactly as they do in Eq. (5.22), because we know that the standard deviation (which is the σ in Eq. (4.42)) must agree with the $\sqrt{a}$ value that we already found in Problem 4.13.

Note that although $\sqrt{a}$ grows with a, it doesn't grow as fast as a itself. So as a grows, the width of the bump in a Poisson distribution becomes thinner compared with the distance a from the origin to the center of the bump. This is indicated in Fig. 5.8, where we show the Poisson distributions for $a = 100$ and $a = 1000$. Note the different scales on the axes.

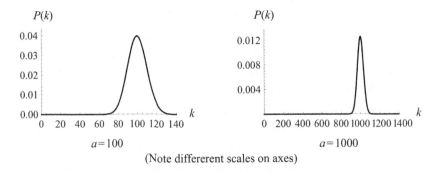

(Note differerent scales on axes)

Figure 5.8: As a grows, the Poisson bump's width (which is proportional to $\sqrt{a}$) becomes thinner compared with the distance a from the origin to the center of the bump.

We claimed at a few points in the above derivation that if a is large (as we are assuming), then any values of x that we are concerned with are much smaller than a. The logic behind this statement is exactly the same as the logic in the first remark in Section 5.1, because a appears in Eq. (5.22) in basically the same way that n

appears in Eq. (5.13). In a nutshell, for large a, the only values of x for which the $P_G(x)$ in Eq. (5.22) is nonnegligible are ones that are much smaller than a. Values of x that are larger than this might lead to probabilities that don't satisfy Eq. (5.22), but we don't care, because these probabilities are so small.

5.4   Binomial, Poisson, and Gaussian

We have seen how the binomial, Poisson, and Gaussian distributions are related to each other in various limits. In Section 4.7.2 we showed how the binomial leads to the Poisson in the small-p and large-n limit. In Section 5.1 we showed how the binomial reduces to the Gaussian in the large-n limit. And in Section 5.3 we showed how the Poisson reduces to the Gaussian in the large-a limit. The summary of these relations is shown in Fig. 5.9.

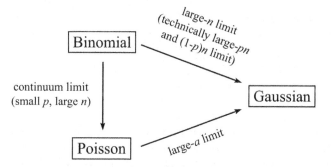

Figure 5.9: How the binomial, Poisson, and Gaussian distributions relate in various limits.

The detailed descriptions of the three relations are the following.

- (Section 4.7.2) The vertical arrow on the left side of Fig. 5.9 indicates that the Poisson distribution is obtained from the binomial distribution by taking the continuum limit. By this we mean the following. Consider a given time (or space, etc.) interval t. Imagine that instead of having trials take place at a rate of n_0 per time t (each with probability p_0 of success), we have them take place at a rate of $10n_0$ per time t (each with probability $p_0/10$ of success), or at a rate of $100n_0$ per time t (each with probability $p_0/100$ of success). And so on, with larger rates n and smaller probabilities p, with the product pn held fixed at p_0n_0. All of these scenarios have the same average of $a = p_0n_0$ successes occurring per time t. And all of them are governed by the binomial distribution. But the more that time is subdivided (that is, the more continuously that the trials take place), the closer the probability distribution (for the number of successes per time t) comes to the Poisson distribution given in Eq. (4.40), with $a = p_0n_0$. We can imagine taking the $n \to \infty$ and $p \to 0$ limits, with the product pn held fixed at a.

- (Section 5.1) The upper-right diagonal arrow in Fig. 5.9 indicates that the Gaussian distribution is obtained from the binomial distribution by taking the large-n limit, where n is the number of trials performed. In Eq. (5.14) we derived this result for $p = 1/2$, and then in Eq. (5.15) we stated the result for a general value of p.

However, for general p, the condition under which the binomial reduces to the Gaussian turns out to be not just that n is large, but rather that both pn and $(1 - p)n$ are large. The need for this stricter condition becomes apparent if you work through the (rather messy) derivation of the more general result in Eq. (5.15).

Note that since p is at most 1, large pn (or $(1 - p)n$) necessarily implies large n. But the converse isn't true. That is, large n doesn't necessarily imply large pn and $(1 - p)n$. If p is an "everyday" number (such as $p = 1/2$ for a coin flip), then large n does in fact imply large pn and $(1 - p)n$. But if p is very small, then n needs to be extremely large ("doubly" large, in a sense), in order to make pn large. For example, if $p = 10^{-3}$, then $n = 10^3$ doesn't make pn large. We need n to be much larger, say, 10^5 or 10^6. A similar statement holds with p replaced with $1 - p$.

Since pn and $(1 - p)n$ are the expected numbers of success and failures in the binomial process involving n Bernoulli trials, we see that the condition under which the binomial reduces to the Gaussian is that both of these expected values are large. If neither p nor $1 - p$ is exceedingly small, then this condition reduces to the condition that n is large.

- (Section 5.3) The lower-right diagonal arrow in Fig. 5.9 indicates that the Gaussian distribution is obtained from the Poisson distribution by taking the large-a limit, where a is the expected number of events that happen during the particular interval (of time, space, etc.) that you are considering. We derived this result in Eq. (5.22). The large-a limit in the Poisson-to-Gaussian case is consistent with the large-pn (and $(1 - p)n$) limit in the binomial-to-Gaussian case, because both a and pn are the expected number of events/successes.

5.5   The central limit theorem

There are two paths in Fig. 5.9 that go from the binomial distribution to the Gaussian distribution. One goes directly by taking the large-pn and $(1 - p)n$ limits (which are simply the large-n limit if p isn't extremely close to 0 or 1). The other goes via the Poisson distribution by first taking the continuum limit, and then taking the large-a limit. The fact that all of the arrows in Fig. 5.9 eventually end up at the Gaussian (equivalently, that no arrows point away from the Gaussian) is consistent with the *central limit theorem*. There are different forms of this theorem, but in the most common form, it says that under a reasonable set of assumptions:

- Central limit theorem:

  If you perform a large number of trials of a random process, then the probability distribution for the sum (or average) of the outcomes is approximately a Gaussian (or "normal") distribution. The greater the number of trials, the better the Gaussian approximation.

The formal proof of this theorem involves some heavy-duty math, so we won't give it here. We'll instead just look at some examples that hopefully will convince you of the theorem's validity.

Let's start with the coin-flipping scenarios in Fig. 5.1. The central limit theorem requires that the trials have *numerical* values. So technically the outcomes of Heads and Tails aren't applicable. But if we assign the value 1 to Heads and 0 to Tails, then we have a Bernoulli process with proper numerical values. The sum of the outcomes of many of these Bernoulli trials is then simply the number of Heads, which is just what appears on the x axis (relative to the expected number) in Fig. 5.1. For two trials (flips), the probability distribution doesn't match up too well with a Gaussian. But for six flips, it matches up reasonably well. And for 20 flips, it matches up extremely well.

So far, there is nothing new here. Coin flips are governed by the binomial distribution (which arises from the sum of n Bernoulli trials), and we already know that a binomial distribution reduces to a Gaussian distribution when n is large. The power of the central limit theorem comes from the fact that we can start with *any* arbitrary distribution (not just a Bernoulli one), and if we perform a large number of trials, the sum will be approximately Gaussian distributed.

For example, imagine rolling a large number of dice and looking at the probability distribution for their sum.[1] The probability distribution for a single die consists of six points on a horizontal line, because all six numbers have equal probabilities of $1/6$. But the central limit theorem says that if we roll 100 dice, the distribution for the sum will be (essentially) a Gaussian centered around 350, since the average for each roll is 3.5. We can therefore start with a flat-line distribution, and then if we perform enough trials, we get a Gaussian distribution for the sum. If you want to experimentally verify this, you will need to consider a large number of sets of trials, with each set consisting of 100 trials (rolls). This is a task best left for a computer and a random number generator!

Note that (as stated in the theorem) we need the number of trials (die rolls, coin flips, etc.) to be large. If you roll only one die, then the plot of the probability distribution for the sum (which is just the single number showing) simply consists of six points on a horizontal line. This row of six points certainly does *not* look like a Gaussian curve. If you instead roll two dice, then as an exercise you can show that Table 1.5 implies that the distribution for the sum takes the shape of a triangle that is peaked at $2 \cdot 3.5 = 7$. This triangle isn't a Gaussian either. But it's closer to a Gaussian than a flat line. If you roll three dice, the distribution for the sum (which is peaked at $3 \cdot 3.5 = 10.5$) takes a curved shape that starts to look like a Gaussian; see Fig. 5.10.[2] With 10 dice, the distribution takes a Gaussian shape, for all practical purposes. The meaning of the word "large" in the first line of the statement of the central limit theorem depends on the process at hand. But in most cases, 10 or 20

---

[1] We're now doing something new. With the exception of a brief mention of the sum of two dice on page 11, all of our previous encounters with dice in this book have involved the number of times a particular face comes up. We generally haven't dealt with the *sum* of the dice.

[2] These histograms were generated numerically. Each bin is associated with the value at its lower end. Technically these histograms aren't probability distributions, because we're plotting the actual number of times each sum occurs, instead of the probability that it occurs. But the probability that each sum occurs is obtained by just dividing the number of times it occurs by the 10^6 sets of rolls.

trials of the process (rolls here) are plenty sufficient to yield an essentially Gaussian distribution.

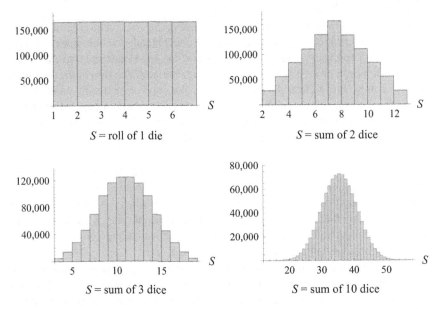

Figure 5.10: Illustration of the central limit theorem. These histograms were generated numerically with $n_s = 10^6$ sets of n_t dice rolls, for $n_t = 1, 2, 3, 10$. Each bin in the histograms is associated with the value at its lower end. If the number n_t of trials in each set is reasonably large, then the probability distribution is essentially Gaussian, as illustrated by the last histogram.

The above examples dealt with the *sum* of the values of the random variable. But the central limit theorem holds for the *average* of the values too, of course, because the average is obtained by dividing the sum by a particular number, namely the number n_t of trials (dice rolls, etc.). So if the histogram of the sum takes a Gaussian form, then so does the histogram of the average. The numbers on the x axis are simply reduced by a factor of n_t. If we work with averages, the histograms in Fig. 5.10 will all be centered around 3.5.

The numbers n_t and n_s

We should clarify the distinction between the two types of large numbers that arise when talking about the central limit theorem:

- The first is the number n_t of trials that generate each data point. (Each data point is the sum or average of the results of the n_t trials.) For example, n_t might be 10 dice rolls, or 50 coin flips. The distribution for the sum of the random variables associated with these n_t trials has an (approximately) Gaussian shape if n_t is large. Usually $n_t \approx 20$ is sufficiently large.

- The second is the number n_s of sets, each consisting of n_t trials, that you must consider if you want to experimentally measure the distribution of the data points. Each set generates a data point (which is the sum of the results of the n_t trials in that particular set). If n_s isn't large, then you won't get good statistics; the measured distribution will be choppy. For all of the numerically generated histograms in Fig. 5.10, we used $n_s = 10^6$. This number was large enough so that all of the histograms have essentially the same shape as the actual theoretical probability distribution.

An important difference between n_t and n_s is the following. The true (theoretical) probability distribution for the sum of n_t trials depends on n_t, of course (along with the specifics of what each trial involves – coins, dice, or whatever). However, the true distribution has nothing to do with n_s. This number is simply the number of sets, each consisting of n_t trials, that you are considering if you are trying to experimentally determine the true distribution for the sum of the n_t trials. But the true distribution (which depends on n_t) exists whether or not you try to determine it by considering an arbitrary number n_s of sets.

As an example of why n_s must be large (if you want to accurately determine the true distribution), consider $n_t = 10$ dice rolls. The probability distribution for the sum of the 10 dice is essentially a Gaussian (even though 10 isn't a terribly large number) that is centered at $10 \cdot 3.5 = 35$, as we saw in the fourth plot in Fig. 5.10. If you want to experimentally verify that this is indeed the distribution, it won't do much good to consider only $n_s = 100$ sets of $n_t = 10$ rolls. The distribution of the 100 observed data points (sums of 10 dice) might look like the first histogram in Fig. 5.11. (As in Fig. 5.10, each bin in these histograms is associated with the value at its lower end.) This isn't much of a Gaussian. But if we increase the number of sets to $n_s = 1000, 10,000,$ or $100,000$, we obtain the three other histograms shown, which progressively look more and more like a Gaussian. We see that a nice Gaussian is obtained with $n_t = 10$ (which isn't that large) and $n_s = 100,000$ (which is quite large). So perhaps the numbers n_t and n_s can be better described with, respectively, the words "at least medium-ish" and "large." Note that since the $n_s = 10^5$ plot in Fig. 5.11 is already quite smooth, nothing much was gained by increasing n_s to 10^6 in the fourth plot (with $n_t = 10$) in Fig. 5.10. These two plots are essentially the same (up to a factor of 10 on the vertical axis). See Problem 5.6 for the exact shape.

One more important point: Figs. 5.10 and 5.11 both show a progression of histograms that become more and more Gaussian, so we should reiterate exactly what each figure illustrates. In Fig. 5.10, the progression of histograms is the statement of the central limit theorem: the probability distribution approaches a Gaussian as the number of trials n_t (whose sum or average we are taking) grows. Because the $n_s = 10^6$ value we used is so large, all of the histograms have essentially the same shape as the actual probability distributions. In contrast, in Fig. 5.11 the progression of histograms is simply the statement that we need to consider a large number n_s of data points if we want to produce a good (not noisy) approximation to the actual probability distribution, which in the present case happens to be essentially Gaussian, due to (1) the central limit theorem and (2) the reasonably large number $n_t = 10$.

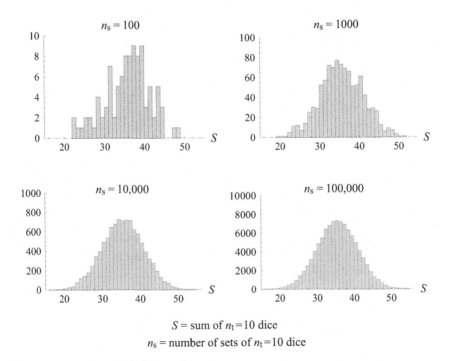

S = sum of n_t=10 dice

n_s = number of sets of n_t=10 dice

Figure 5.11: Illustration of why n_s needs to be large. If n_s isn't large, then the observed distribution doesn't look like the actual probability distribution. This figure is *not* an illustration of the central limit theorem. It is an illustration of the fact that the data is "noisy" when n_s is small. Each bin in the histograms is associated with the value at its lower end.

Two more examples

The central limit theorem holds for any underlying probability distribution (subject to some reasonable assumptions). We know from Fig. 5.10 that the theorem holds for the sum (or average) of many dice rolls, where the underlying distribution is a flat line of six points. And we also know from our binomial-to-Gaussian derivation in Section 5.1 that the theorem holds for the sum (or average) of the number of Heads that appear in many coin flips, where the underlying distribution is a Bernoulli one. But the theorem also holds for other underlying probability distributions that don't look as nice. For example, consider the discrete distribution shown in Fig. 5.12. The probabilities for the three possible outcomes are $p(2) = 0.6$, $p(3.2) = 0.1$, and $p(7) = 0.3$.

You can quickly show that the expectation value of this distribution is 3.62. The central limit theorem says that the probability distribution for the average of, say, 100 numbers chosen from the distribution is a Gaussian centered at 3.62. And indeed, Fig. 5.13 shows a Gaussian histogram of $n_s = 100,000$ numerically generated data points, each of which is the average of $n_t = 100$ numbers chosen from the distribution. The histogram is centered at about 3.6.

All of the examples so far in this section have involved discrete distributions. But the central limit theorem holds for continuous distributions too. Fig. 5.14 shows

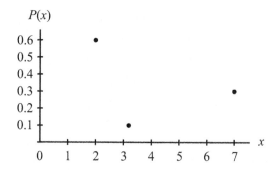

Figure 5.12: An arbitrary probability distribution with three possible outcomes.

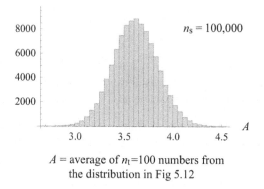

A = average of n_t=100 numbers from
the distribution in Fig 5.12

Figure 5.13: A histogram of n_s = 100,000 averages of n_t = 100 numbers chosen from the distribution in Fig. 5.12.

a Gaussian histogram of n_s = 100,000 numerically generated data points, each of which is the average of n_t = 50 numbers taken from a uniform distribution ranging from 0 to 1. The average of this distribution is simply 0.5, which is correctly where the Gaussian is centered. The task of Problem 5.7 is to verify that the histograms in Figs. 5.13 and 5.14 have the correct standard deviations.

We mentioned above right after the statement of the central limit theorem that due to the math involved, we haven't included a proof. But hopefully the above examples have convinced you of the theorem's validity.

5.6 Summary

- For a large number of trials, n, a binomial distribution reduces to a Gaussian distribution. We showed this for coin flips, but it also holds for a binomial distribution governed by a general probability p. The standard deviation of the Gaussian is $\sqrt{np(1-p)}$.

- The *law of large numbers* states that the measured probability over a large number of trials will be essentially equal to the theoretical probability. This

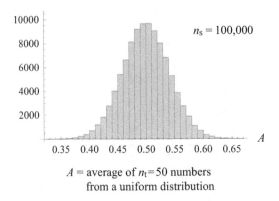

A = average of n_t = 50 numbers
from a uniform distribution

Figure 5.14: A histogram of n_s = 100,000 averages of n_t = 50 numbers chosen from a uniform distribution (from 0 to 1).

law is a consequence of the fact that a Gaussian distribution has the property that the larger the number of trials, the thinner the distribution's bump, relative to the whole span of possible outcomes.

- In the limit of a large expected number of events, a, a Poisson distribution reduces to a Gaussian distribution. The standard deviation of the Gaussian is $\sqrt{a}$.

- The *central limit theorem* says (in its most common form) that if you perform a large number of trials of a random process, the probability distribution for the sum (or average) of the outcomes is approximately a Gaussian distribution.

5.7 Exercises

See **www.people.fas.harvard.edu/~djmorin/book.html** for a supply of problems without included solutions.

5.8 Problems

Section 5.1: Binomial and Gaussian

5.1. **Equal percentages** ∗∗

In the last paragraph of Section 2.1, the same percentage 99.999999975%, appeared twice. Explain why you know that these two percentages must be the same, even if you don't know what the common value is.

5.2. **Rolling sixes** ∗∗

In the solution to Problem 2.13 (known as the Newton-Pepys problem), we noted that the answer to the question, "If $6n$ dice are rolled, what is the probability of obtaining at least n 6's?," approaches $1/2$ in the $n \to \infty$ limit. Explain why this is the case.

5.3. Coin flips ⁎⁎

If you flip 10^4 coins, how surprised would you be if the observed percentage of Heads differs from the expected value of 50% by more than 1%? Answer the same question for 10^6 coins. (These numbers are large enough so that the binomial distribution can be approximated by a Gaussian.)

5.4. Identical distributions ⁎⁎

A thousand dice are rolled. Fig. 5.15 shows the probability distribution (given by Eq. (5.15)) for the number of 6's that appear, relative to the expected number (which is 167). How many *coins* should you flip if you want the probability distribution for the number of Heads that appear (relative to the expected number) to look exactly like the distribution in Fig. 5.15 (at least in the Gaussian approximation)?

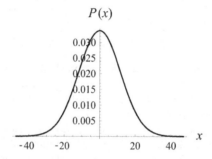

Figure 5.15: The probability distribution for the number of 6's in 1000 dice rolls, relative to the expected number, 167.

Section 5.2: The law of large numbers

5.5. Gambler's fallacy ⁎

Assume that after 20 coin flips, you have obtained only five Heads. The probability of this happening is small (about 1.5%, since $\binom{20}{5}/2^{20} = 0.0148$), but not negligible. Since the law of large numbers says that the fraction of Heads approaches 50% as the number of flips gets large, should you expect to see more Heads than Tails in future flips?

Section 5.5: The central limit theorem

5.6. Finding the Gaussian ⁎⁎

What is the explicit form of the Gaussian function $f(x)$ that matches up with the fourth histogram in Fig. 5.11? Assume that $n_t = 10$ is large enough so that the Gaussian approximation does indeed hold.

5.7. **Standard deviations** ∗∗

Calculate the theoretically predicted standard deviations of the histograms in Figs. 5.13 and 5.14, and check that your results are consistent with a visual inspection of the histograms. You will need the result from Problem 4.3 for Fig. 5.14.

5.9 Solutions

5.1. **Equal percentages**

Looking back at the given paragraph, our goal is to show that the probability of obtaining Heads between 49% and 51% of the time in 10^5 coin flips equals the probability of obtaining Heads between 49.99% and 50.01% of the time in 10^9 coin flips. In the first case, the 1% deviation from average corresponds to $10^5/10^2 = 10^3$ Heads. In the second case, the 0.01% deviation from average corresponds to $10^9/10^4 = 10^5$ Heads. Eq. (3.48) gives the standard deviation of the number of Heads that show up in n tosses of a fair coin as $\sigma = \sqrt{n/4}$. In the present two cases, this yields standard deviations of $\sqrt{10^5/4} = 158$ and $\sqrt{10^9/4} = 15{,}800$. The numbers of standard deviations corresponding to the above two spreads of 10^3 and 10^5 Heads are therefore

$$\frac{10^3}{158} = 6.3 \quad \text{and} \quad \frac{10^5}{15{,}800} = 6.3. \tag{5.24}$$

Since these numbers are equal, the probabilities of lying within the two specified ranges must be equal. We have used the fact that the Gaussian approximation is valid in both scenarios, which implies that the distribution relative to the mean is completely determined by the standard deviation.

If you want to show that the common probability equals 99.999999975%, you can numerically either add up the exact binomial probabilities in the given ranges, or integrate the Gaussian approximations over the given ranges. A computer will be necessary for either option, of course. (Performing the sums or integrals over the complementary regions outside the given ranges works just as well, or even better.)

5.2. **Rolling sixes**

From Eq. (5.15) and the surrounding discussion, we know that for a large number of rolls, the binomial distribution for the number of 6's that appear is essentially a Gaussian distribution centered at the mean (which is n, if there are $6n$ rolls). Since the Gaussian distribution is symmetric around the mean, we conclude that there is a $1/2$ chance that the number of 6's is greater than or equal to the mean value, n, as desired.

Technically, the probability is slightly larger than $1/2$. This is true because if we split the Gaussian distribution exactly down the middle, then we're including only *half* of the probability of obtaining *exactly* n 6's. We need to include *all* of this probability, because we're concerned with the probability of *at least* n 6's. The probability of obtaining at least n 6's in $6n$ rolls therefore equals $1/2$ plus half of the probability of obtaining exactly n 6's. But if n is large, the probability of obtaining exactly n 6's is small, so it doesn't matter much if we ignore half of it.

If n is small, the above logic doesn't hold. This is consistent with the fact that the probability of obtaining at least n 6's can be appreciably more than $1/2$, as we saw in Problem 2.13. For small n, the above logic breaks down partly because the probability of obtaining *exactly* n 6's is appreciable, and partly because the Gaussian approximation doesn't hold for small n.

5.3. **Coin flips**

Eq. (3.48) gives the standard deviation of the number of Heads that show up in n tosses of a fair coin as $\sigma = \sqrt{n/4}$. For $n = 10^4$ and 10^6 this yields $\sigma = 50$ and 500. And since 1% of 10^4 and 10^6 equals 10^2 and 10^4, these ±1% spreads are equal to $10^2/50 = 2$ and $10^4/500 = 20$ standard deviations.

As noted in Fig. 4.26, the probability of being within 2σ of the mean is 95%. So in the case of 10^4 coins, there is a 5% probability that the percentage of Heads differs from the expected value of 50% by more than 1%. 5% is small but not negligible, so if you observe a deviation larger than 1%, you will probably be mildly surprised.

In contrast, the probability of being within 20σ of the mean is *exactly* 1, for all practical purposes. We mentioned near the end of Section 4.8 that the probability of being within five standard deviations of the mean is about 0.9999994. Tables of these probabilities don't even bother going anywhere near 20σ, because the probability is so close to 1. So in the case of 10^6 coins, there is a 0% probability that the percentage of Heads differs from the expected value of 50% by more than 1%. Therefore, even if you said that you would be "extremely, outrageously, massively surprised" if the deviation from the mean exceeded 1%, that still doesn't do justice to the unlikelyhood. You are simply not going to end up 20σ from the mean, period; see the remark below. The law of large numbers is a powerful thing!

What about 10^5 coins, which is between the above two cases? From Problem 5.1, we know that the probability that the percentage of Heads differs from 50% by more than 1% equals 0.000000025%. So if we increase the number of coins from 10^4 to 10^5, the probability of being outside the ±1% marks drops from a reasonable 5% to essentially zero. And then for 10^6 coins, the probability is exactly zero for all practical purposes.

REMARK: Let's produce a (very rough) upper bound on the probability of being outside 20σ. If $x = 20\sigma$, then the second expression in Eq. (4.42) gives a probability density of

$$f(20\sigma) = \frac{e^{-(20\sigma)^2/2\sigma^2}}{\sqrt{2\pi\sigma^2}} = \frac{e^{-20^2/2}}{\sigma\sqrt{2\pi}} \approx \frac{10^{-87}}{\sigma\sqrt{2\pi}}. \tag{5.25}$$

If $x = 21\sigma$, you can show that the above 10^{-87} factor becomes 10^{-96}. So $f(21\sigma)$ is completely negligible compared with $f(20\sigma)$. We can therefore assume that $f(21\sigma)$ is exactly zero. To obtain an upper bound on the area of the distribution that lies outside 20σ, we can assume that $f(x)$ takes on the constant value of $f(20\sigma)$ between $x = 20\sigma$ and $x = 21\sigma$, and then suddenly drops to zero. Of course, it *doesn't* take on this constant value; it decreases fairly quickly to nearly zero. But all we care about here is obtaining an upper bound on the area; a significant overestimate is fine for our purposes. So assuming a constant value of $f(20\sigma)$ between $x = 20\sigma$ and $x = 21\sigma$, the area in this span of one standard deviation is $\sigma \cdot f(20\sigma)$, which from Eq. (5.25) equals $10^{-87}/\sqrt{2\pi}$. Doubling this (to account for the span between -20σ and -21σ) gives $\sqrt{2/\pi} \cdot 10^{-87}$ as an upper bound on the area. We can therefore say that a (generous) upper bound on the probability of being outside 20σ is 10^{-87}. The actual probability obtained numerically from the exact binomial distribution is $5.5 \cdot 10^{-89}$, which is about 20 times smaller than 10^{-87}.

To get an idea of how ridiculously small this probability is, imagine (quite hypothetically) gathering together as many people as there are protons and neutrons in the earth (roughly $4 \cdot 10^{51}$), and imagine each person running the given experiment (flipping 10^6 coins) once a second for the entire age of the universe (roughly $4 \cdot 10^{17}$ seconds). And then repeat this whole process a quintillion (10^{18}) times. This will yield $1.6 \cdot 10^{87}$ runs of the experiment, in which case (working with our high 10^{-87} estimate) you might expect one or two runs to have percentages of Heads that differ from 50% by more than 1%. But again, this is a high estimate, given the actual probability of $5.5 \cdot 10^{-89}$. Although most people might think that there is a nonnegligible probability of obtaining more than $510,000$ or fewer than $490,000$ Heads in 10^6 coin flips, the probability is in fact zero, for all practical purposes. (As another example, you can show that the same 20σ result applies when flipping a trillion coins and ending up outside the 49.999% to 50.001% range.) The above probability of 10^{-87} isn't just small; it is *ridiculously* small. The moral of all this is that unless you think in terms of the standard deviation (which is proportional to $\sqrt{n}$), it's hard to get any intuition for these types of setups. People have a tendency to think linearly, that is, to assume that a reasonable deviation from the mean might be, say, $n/10$ or $n/100$, independent of the size of n. This linear thinking will lead you astray. ♣

5.4. Identical distributions

Fig. 5.15 is a plot of the $P(x)$ in Eq. (5.15), with $n_d = 1000$ and $p_d = 1/6$. (The "d" here is for dice.) $P(x)$ is completely determined by the product $np(1-p)$, because this product appears in both the exponent and the denominator in Eq. (5.15). We therefore want to find the value of n_c (with "c" for coin) such that

$$n_c p_c (1 - p_c) = n_d p_d (1 - p_d). \tag{5.26}$$

Since $p_c = 1/2$, this gives

$$n_c \cdot \frac{1}{2} \cdot \frac{1}{2} = n_d \cdot \frac{1}{6} \cdot \frac{5}{6} \implies n_c = \frac{5}{9} n_d. \tag{5.27}$$

In the given case with $n_d = 1000$, this yields $n_c = 556$. The exact binomial distributions for the two processes aren't exactly identical, of course, but they are both extremely close to the common Gaussian approximation in Eq. (5.15).

The common value of $np(1-p)$ is 139. The standard deviation is the square root of this (by comparing Eq. (5.15) with Eq. (4.42)), so $\sigma \approx 12$. This is consistent with a visual inspection of Fig. 5.15. Note that the expected number of Heads is $556/2 = 278$, but this number is irrelevant here, because we're concerned only with the distribution relative to the average. The means $p_d n_d = 167$ and $p_c n_c = 278$ are necessarily different, because there is no way to simultaneously make the $np(1-p)$ values equal *and* the pn values equal, since these quantities differ by the factor $1 - p$.

REMARK: At first glance, it might not be obvious that an n_c should exist that yields the same distribution relative to the mean. But it is clear once you realize that both distributions are Gaussians, and that (ignoring the μ in Eq. (4.42) since we're looking at the distributions relative to the mean) the Gaussians depend on only *one* parameter, σ. So if we can generate the same σ, then we can generate the same distribution. And we can indeed generate the same σ, because $\sigma_{coin} = \sqrt{n/4}$ from Eq. (3.48), so we just need to pick the appropriate n. ♣

5.5. Gambler's fallacy

No. Each coin flip is independent of the flips that have already taken place. Therefore, there is no reason to expect more Heads than Tails in future flips. Past flips are irrelevant.

This incorrect interpretation (that there will be more Heads than Tails in future flips) of the law of large numbers arises from the confusion between additive and multiplicative differences. If you obtain five Heads in 20 flips, then you are five Heads below average. If you keep flipping more coins, then on average you will always be five Heads below average. However, if you are worried about remaining below average, these five Heads are the least of your worries if you end up flipping a million coins. The standard deviation for $n = 10^6$ coin flips is $\sqrt{n/4} = 500$, so you are nearly certain to end up much farther than five Heads from average (above or below).

The deficiency of five Heads will always be there (on average, if you have many clones of this scenario), but it represents a smaller and smaller fraction of the number of Heads, as the number of flips gets larger and larger. The *numbers* of Heads and Tails do *not* somehow conspire to equalize in the long run. On the contrary, the numbers tend to diverge, with their difference generally being on the order of the standard deviation (proportional to $\sqrt{n}$). However, this difference becomes a smaller and smaller fraction of the number of flips (namely n), which means that the *fractions* of Heads and Tails both approach $1/2$. But there is certainly no conspiring going on. All future outcomes are independent of all past outcomes.

The incorrect interpretation (that there will be more Heads than Tails in future flips) of the law of large numbers is known as the *gambler's fallacy*. Alternatively, it can simply be called wishful thinking.

5.6. **Finding the Gaussian**

From the first example in Section 3.2, we know that the variance of a single die roll is 2.92. The standard deviation is therefore $\sqrt{2.92} = 1.71$. Eq. (3.45) then tells us that the standard deviation of the sum of the rolls of $n_t = 10$ dice is $\sigma = \sqrt{10}(1.71) = 5.4$. This is the σ that appears in the second Gaussian expression in Eq. (4.42). And the mean μ of the sum of 10 dice rolls is $10 \cdot 3.5 = 35$. So the desired Gaussian function is

$$f(x) = (100,000) \sqrt{\frac{1}{2\pi(5.4)^2}} \, e^{-(x-35)^2/2(5.4)^2}. \tag{5.28}$$

The factor of 100,000 out front arises because the histograms in Fig. 5.11 deal with the actual *number* of outcomes. So we need to multiply the *probability* distribution in Eq. (4.42) by $n_s = 100,000$ to obtain the histogram.

However, if we want to be picky, we must remember that each histogram bin in Fig. 5.11 is associated with the value at its lower end. And since each bin has width 1, the histogram is shifted by 0.5 to the right from where it would be if each bin were centered on the associated value of x. So we actually want the μ in Eq. (4.42) to be 35.5. (This correction has nothing to do with the probability concepts we're covering here. It's just a figment of the way we plotted the histograms. The true mean is simply $\mu = 35$.) The function that matches up with the histogram is therefore

$$f(x) = (100,000) \sqrt{\frac{1}{2\pi(5.4)^2}} \, e^{-(x-35.5)^2/2(5.4)^2}. \tag{5.29}$$

Fig. 5.16 shows a plot of this function superimposed on the histogram. The other three histograms in Fig. 5.11 come from the same underlying probability distribution (because they all involve 10 dice). But they're less smooth because the smaller n_s values allow a few random fluctuations to strongly influence the histograms.

5.7. **Standard deviations**

Consider Fig. 5.13 first. The underlying distribution has probabilities $p(2) = 0.6$, $p(3.2) = 0.1$, and $p(7) = 0.3$. The mean is therefore

$$(0.6)(2) + (0.1)(3.2) + (0.3)(7) = 3.62. \tag{5.30}$$

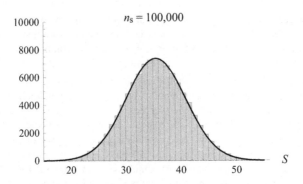

Figure 5.16: The Gaussian approximation to the fourth histogram in Fig. 5.11.

The standard deviation is then

$$\sqrt{(0.6)(2-3.62)^2 + (0.1)(3.2-3.62)^2 + (0.3)(7-3.62)^2} = 2.24. \qquad (5.31)$$

From Eq. (3.53) the standard deviation of the average of 100 numbers taken from this distribution is $\sigma_{avg} = (2.24)/\sqrt{100} = 0.224$. This is consistent with Fig. 5.13 (the spacing between tick marks on the x axis is 0.05). Remember from Fig. 4.25 that at one standard deviation from the mean, the distribution is $1/\sqrt{e} = 0.61$ as tall as the peak.

Now consider Fig. 5.14. From Problem 4.3, the standard deviation of the uniform distribution from 0 to 1 is $\sigma = 1/\sqrt{12} = 0.29$. And then from Eq. (3.53) the standard deviation of the average of 50 numbers taken from this distribution is $\sigma_{avg} = (0.29)/\sqrt{50} = 0.041$. This is consistent with Fig. 5.14 (the spacing between tick marks on the x axis is 0.01).

Chapter 6

Correlation and regression

In this chapter, we will consider how two different random variables may be related, or *correlated*. In Section 6.1 we give some examples of what it means for two variables to be correlated or uncorrelated. In Section 6.2 we present a model for how two variables can be correlated, based on the given underlying probability distributions. We then get quantitative in Section 6.3 and derive expressions for the *correlation coefficient, r.* One of these expressions involves the *covariance* of the two variables. We show in Section 6.4 how we can take advantage of a correlation to make an improved prediction for the Y value associated with a given X value. In Section 6.5 we calculate the joint probability density $\rho(x, y)$ in terms of σ_x, σ_y, and r, in the case where the underlying distributions are Gaussian. We find that the curves of constant $\rho(x, y)$ are ellipses. We analyze these ellipses in Section 6.6.

In Section 6.7 we discuss the all-important *regression lines*, which give the expected value of Y, given X (or the expected value of X, given Y). We then present in Section 6.8 two examples on the use of regression lines. A ubiquitous effect here is *regression toward the mean*. Finally, in Section 6.9 we analyze the best-fit (or *least-squares*) line. We find that this line is none other than the regression line. Indeed, the regression line is often *defined* as the least-squares line. We have chosen to take a different route in this chapter and introduce the regression line by considering the underlying probability distributions that produce the random variable Y. This route makes it easier to see what's going on "under the hood." But it's good to see that we end up with the same regression line, independent of what route we take.

6.1 The concept of correlation

Consider a pair of random variables X and Y. For example, X might be an object's mass measured in kilograms, and Y might be its mass measured in grams. Or X might be a person's height, and Y might be his/her shoe size. Or X might be the alphabetical placement of the second letter in a person's last name ($A = 1$, $B = 2$, etc.), and Y might be his/her cholesterol level.

One of the main issues we will address in this chapter is the degree to which knowledge of X helps predict Y (or vice versa). Equivalently, we will address the degree to which two variables are *correlated*. The larger the correlation, the more that one variable helps predict the other. We'll be precise about this in Section 6.3 when we define the correlation coefficient, usually denoted by the letter r. To get a qualitative feel for what a correlation means, let's consider the three examples mentioned above, which range from perfect correlation to no correlation at all.

- PERFECT CORRELATION: An example of perfect correlation is the mass of an object expressed in kilograms or grams. If we know the mass X in kilograms, then we also know the mass Y in grams. We simply need to multiply by 1000. That is, $Y = 1000X$. One kilogram equals 1000 grams, 2.73 kilograms equals 2730 grams, etc. Knowledge of the mass in kilograms allows us to state exactly what the mass is in grams. (The converse is also true, of course. Knowledge of the mass in grams allows us to state exactly what the mass is in kilograms. Just divide by 1000.) If we take a group of objects and determine their masses in kilograms and grams, and then plot the results, we will obtain something like the plot shown in Fig. 6.1. (We'll assume for the present purpose that any measurement errors are negligible.) All of the points lie on a straight line. This is a consequence of the perfect correlation.

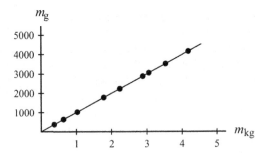

Figure 6.1: The mass in grams is perfectly correlated with the mass in kilograms.

- SOME CORRELATION: An example of nonzero but imperfect correlation is the second example mentioned above, involving height and shoe size. (Men's and women's shoe sizes use different scales, so let's just look at men's sizes here. Also, some manufacturer sizes run large or small, but we'll ignore that issue.) We certainly don't expect perfect correlation between height and shoe size, because that would mean we would be able to exactly predict a person's shoe size based on height (or vice versa). This isn't possible, of course, because all people who are six feet tall certainly don't have the same shoe size. Additionally, there can't be perfect correlation because shoe sizes use a discrete scale, whereas heights are continuous.

But is there at least some correlation? That is, does knowledge of a person's height allow us to make a better guess of his shoe size, compared with our guess if we had no knowledge of the height? Well, 6-footers certainly have a larger shoe size than 5-footers *on average*, so the answer should be yes. Of course, we might well find a 5-footer whose feet are larger than a 6-footer's. But on average, a person's shoe size increases with height. A scatter plot of some data is shown in Fig. 6.2. (I asked a sampling of students for their height and shoe size. Height is measured to the nearest inch. Since men's and women's sizes use different scales, I used only the data for 26 male students.) From the data, the average shoe size of all 26 people is 10.4, whereas the average shoe size of a 6-footer is 11.4. So if you want to make a guess for the shoe size of a 6-footer, you'll do better by guessing 11.4 than 10.4. After we introduce the correlation coefficient in Section 6.3, we'll be able to be quantitative in Section 6.4 about how much better the guess is (at least for a large number of data points).

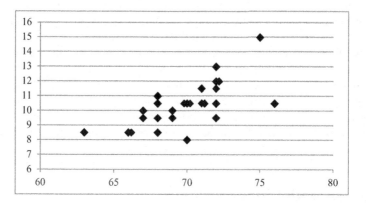

Figure 6.2: A scatter plot of shoe size versus height (in inches).

- ZERO CORRELATION: An example of zero correlation is the third example mentioned above, involving the alphabetical placement ($A = 1$, $B = 2$, etc.) of the second letter in the last name, along with cholesterol level. It is highly doubtful that there is much of a correlation here. Would knowing that the second letter of a last name is "i" help you in predicting the cholesterol level? Negligibly, at best. Of course, certain names (Murphy, Smith, Li, etc.) are common in certain ethnicities, and it is undoubtedly the case that different ethnicities have slightly different cholesterol levels (on average) due to differing genes and diet. But let's assume that this effect is small and is washed out by other effects. So for the sake of argument, we'll assume that there is no correlation here. However, this example should convince you that small (or perhaps even large) correlations might pop up in situations where at first glance it's hard to imagine any correlation!

The first two of the above examples involve a *positive* correlation; an increase in X corresponds to an *increase* in Y (on average). The line (or general blob) of points in the scatter plot has an upward slope. It is also possible to have a *negative* correlation, where an increase in X corresponds to a *decrease* in Y (on average). The line (or general blob) of points in the scatter plot will then have a downward slope. An example of negative correlation is vitamin C intake and the incidence of scurvy. The more vitamin C you take, the less likely you are to have scurvy – at least on the low end of the intake scale; on the upper end it doesn't matter how much you take.

Note that *correlation* does not necessarily imply *causation*. In the case of vitamin C and scurvy, there does happen to be causation; taking more vitamin C helps prevent you from getting scurvy. But in the case of height and shoe size, it isn't that being tall *causes* your feet to be larger, any more than having large feet causes you to be taller. (The situation is symmetrical, so if you want to argue causation, you'll be hard pressed to say which is causing which.) Instead, what's going on is that there is a *third* thing, namely genetics (and diet too), that causes *both* your height and foot size to be larger or smaller (on average). Another example along these lines consists of the number of times that people in a given town on any given day put on sunglasses, along with the number of times they apply sunscreen. There is a positive correlation between these two things, but neither one causes the other. Instead, they are both caused by a third thing – sunshine!

We'll deal only with linear correlation in this chapter, although there are certainly examples of nonlinear correlation. A simple example is the relation between the area of a square and its side length: area = (side length)^2. This relation is quadratic, not linear. Another example is the relation between job income and a person's age. Three-year-olds don't earn much from working a job, and neither do 100-year-olds (usually). So the plot of average income vs. age must start at zero, then increase to some maximum, and then decrease back to zero.

6.2   A model for correlation

Let's now try to understand the general way in which two random variables can be correlated. This understanding will lead us to the correlation coefficient r in Section 6.3. For the purpose of making some pretty plots, we'll assume in the present discussion that our two random variables each have Gaussian (normal) distributions. This assumption isn't necessary; our mathematical results will hold for any distributions. Indeed, when dealing with actual real-world data, it is often the case that one or both of the variables are not normally distributed. The correlation coefficient is still defined perfectly well by Eq. (6.6) or Eq. (6.9) below. However, due to the central limit theorem (see Section 5.5), many real-life random variables are approximately normally distributed.

Consider a random variable X that is normally distributed with mean zero and standard deviation σ_x:

$$X: \quad \mu = 0, \quad \sigma = \sigma_x. \tag{6.1}$$

We have chosen the mean to be zero just to make our calculations and figures cleaner. All of the results below hold more generally for any mean.

Consider another random variable Y that is correlated (to some extent) with X. By this we mean that Y is partially determined (in a linear manner) by X and partially determined by another random variable Z (assumed to be normally distributed) that is independent of X. Z can in turn be the sum of many other random variables, all independent of X. We're lumping the effect of all these variables into one variable Z. We can be quantitative about the dependence of Y on X and Z by writing Y as

$$\boxed{Y = mX + Z} \tag{6.2}$$

where m is a numerical factor. To keep things simple, we will assume that the mean of Z is also zero. So if the standard deviation of Z is σ_z, we have

$$Z: \quad \mu = 0, \quad \sigma = \sigma_z. \tag{6.3}$$

Note that if we take the mean (expectation value) of Eq. (6.2), we see that the various means are related by

$$\mu_y = m\mu_x + \mu_z, \tag{6.4}$$

where we have used the fact that the expectation value of the sum equals the sum of the expectation values; see Eq. (3.7). Since we are assuming $\mu_x = \mu_z = 0$ here, Eq. (6.4) implies that μ_y is also equal to zero.

In Eq. (6.2), we are producing Y from two known (and independent) distributions X and Z. To be explicit, the meaning of Eq. (6.2) is the following. Pick an x value of the random variable X and multiply the result by m to obtain mx. Then pick a z value of the random variable Z and add it to mx to obtain $y = mx + z$. This is the desired value of y. We can label this ordered pair of (X, Y) values as (x_1, y_1). We then repeat the process with new values of X and Z to obtain a second (X, Y) pair (x_2, y_2). And so on, for as many pairs as we like.

As an example, Y could be the measured weight of an object, X could be the true weight, and Z could be the error introduced by the measurement process (reading the scale, behavior of the scale depending on a slightly lopsided placement of the object on it, etc.). These variables might not have Gaussian distributions, but again, that assumption isn't critical in our discussion. In this example, $m = 1$.

We should mention that although Eq. (6.2) is the starting point for deriving most of the correlation results in this chapter, rarely is it the starting point in practice. That is, rarely are you given the underlying X and Z distributions. Instead, you are invariably given some data, and you need to calculate the correlation coefficient r via Eq. (6.9) below. But the key to deriving Eq. (6.9) is realizing that we can write Y as $mX + Z$ (at least in the case of linear correlation), even if we don't know exactly what X and Z are.

To see what sort of correlation Eq. (6.2) produces between X and Y, let's consider two special cases, in order to get a general idea of the effects of m and Z.

Perfect correlation ($\sigma_z = 0$)

If the standard deviation of Z is $\sigma_z = 0$, then Z always just takes on the value $z = 0$, because we're assuming that the mean of Z is zero. (More generally, Z takes on a constant value z_0.) So Eq. (6.2) reduces to $Y = mX$. That is, Y is a fixed number m times X; all values of x and y are related by $y = mx$. This means that all of the (x, y) points in the scatter plot lie on the straight line $y = mx$, as shown in Fig. 6.3 for 100 random points generated numerically from a Gaussian distribution X. We have arbitrarily chosen $m = 0.5$ and $\sigma_x = 1$. In the present case of a straight line, we say that X and Y are perfectly (or completely) correlated. The value of Y is completely determined by the value of X. There is no additional random variable Z to mess up this complete determination.

In the case where σ_z is small but nonzero, we obtain a strong but not perfect correlation. Fig. 6.4 shows a plot of 200 points in the case where σ_z equals $(0.1)\sigma_x$.

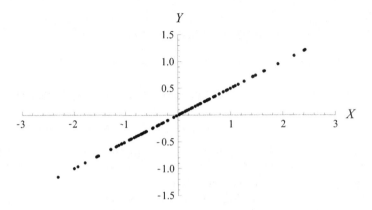

Figure 6.3: Perfect correlation.

We have again chosen $m = 0.5$ and $\sigma_x = 1$ (and hence $\sigma_z = 0.1$). We have generated the points by picking 200 random values from each of the Gaussian distributions X and Z, and then forming $Y = mX + Z$. In the present case of small σ_z, knowledge of X is very helpful in predicting Y, although it doesn't predict Y exactly.

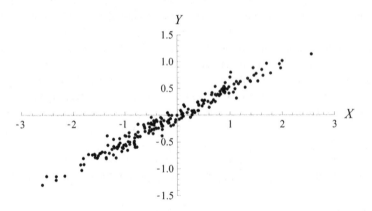

Figure 6.4: Strong correlation.

Zero correlation ($m = 0$)

If $m = 0$, then Eq. (6.2) reduces to $Y = Z$. And since Z is independent of X, this means that Y is also independent of X. Fig. 6.5 shows a plot of 2000 points in the case where $m = 0$. We have arbitrarily chosen $\sigma_x = 2$ and $\sigma_z = 1$. We have generated the points by picking 2000 random values from each of the Gaussian distributions X and Z, and then setting Y equal to Z.

It is clear from Fig. 6.5 that X and Y are completely uncorrelated. The distribution for Y is independent of the value of X. That is, for any given value of X, the Y values are normally distributed around $Y = 0$, with the same standard deviation

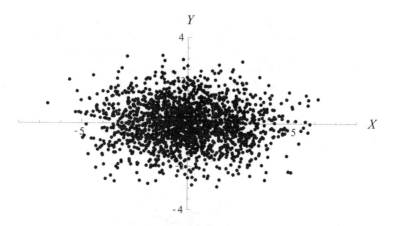

Figure 6.5: Zero correlation.

(which equals σ_z). In other words, the probability (or rather, the probability density) of obtaining a certain value of Y, *given* a particular value of X, is independent of the X value. This probability is given by the Gaussian distribution for Z, since $Y = Z$ in the present case where $m = 0$.

If we imagine drawing vertical shaded strips at two different values of X, as shown in Fig. 6.6 (which is the same as Fig. 6.5, except with 10,000 points), then the distributions of Y values in these two strips are the same, except for an overall scaling factor. This scaling factor is simply the probability (or rather, the probability density) of obtaining each of the given values of X. Larger values of $|X|$ are less likely, due to the $e^{-x^2/2\sigma_x^2}$ factor in the Gaussian distribution. So there are fewer dots in the right strip. But *given* a value of X, the probability distribution for Y (in this $m = 0$ case) is simply the probability distribution for Z, which is independent of X.

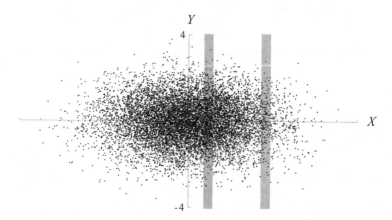

Figure 6.6: If $m = 0$, the distribution of Y values within a vertical strip is independent (aside from an overall scaling factor) of the location of the strip.

In the case were m is small but nonzero, we obtain a weak correlation. Fig. 6.7 shows a plot of 2000 points in the case where $m = 0.2$, again with $\sigma_x = 2$ and $\sigma_z = 1$. In this case, knowledge of X helps a little bit in predicting the Y value. It doesn't help much in the region near the origin; the plot doesn't display much of a tilt there (it looks basically the same as Fig. 6.5 near the origin). But for larger values of X, there is a clear bias in the values of Y. More points lie above the X axis on the right side of the plot, and more points lie below the X axis on the left side.

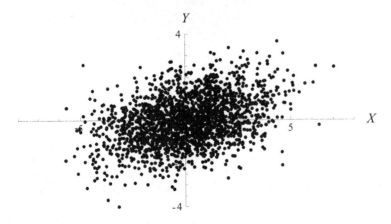

Figure 6.7: Weak correlation.

REMARKS:

1. All of the above scatter plots are centered at the origin because we assumed that the means of X and Z are zero, which implies that the mean of Y is zero, from Eq. (6.4). If Z instead had a nonzero mean μ_z, then the blob of points would be shifted upward by μ_z. If X had a nonzero mean μ_x, then the blob would be shifted rightward by μ_x and also upward $m\mu_x$.

2. In the above discussions, we treated X as the independent variable and Y as the dependent variable, and we looked at the extent to which X determined Y. However, if someone gives you one of the above scatter plots, you could quite reasonably tilt your head sideways and consider X to be a "function" of Y, and then look at the extent to which Y determines X. We will discuss this alternative way of relating the variables in Sections 6.5 and 6.7.

3. We noted in Fig. 6.6 that the relative distribution of Y values within a vertical strip is independent of the location of the strip. This fact holds not only in Fig. 6.6 where there is zero correlation, but also (in a slightly modified sense) in the case of nonzero correlation, even when there is strong correlation as in Fig. 6.4. Although it might seem like the spread (the standard deviation) of Y values gets smaller out in the tails of the plot in Fig. 6.4, the spread is in fact the same for all values of X. The $Y = mX + Z$ expression tells us that for any given value of X, the Y values are centered at mX (instead of zero; this is the aforementioned slight modification) and have the *same* standard deviation of σ_z around this value. The spread *seems* to be larger in the middle of the plot, but only because there are more points there. ♣

6.3 The correlation coefficient, r

We will now show how to produce the correlation coefficient r from quantities associated with the righthand side of Eq. (6.2), namely m, σ_x, and σ_z. To do this, we will need to determine the standard deviation of $Y = mX + Z$. We know that mX has a standard deviation $m\sigma_x$ and Z has a standard deviation of σ_z. And we know from Eq. (3.42) that the standard deviation of the sum of two independent variables (as X and Z are) is obtained by adding the two standard deviations in quadrature. (The variables need not be Gaussian for this to be true.) Therefore, Y is described by:

$$Y: \quad \mu = 0, \quad \sigma_y = \sqrt{m^2\sigma_x^2 + \sigma_z^2}. \tag{6.5}$$

The $\mu = 0$ value follows from Eq. (6.4), since we are assuming $\mu_x = \mu_z = 0$.

Let's check some limiting cases of Eq. (6.5). In one extreme where $\sigma_z = 0$ (complete correlation between X and Y), we have $\sigma_y = m\sigma_x$. All of the standard deviation of Y comes from X; none of it comes from Z. In the other extreme where $m = 0$ (no correlation between X and Y), we have $\sigma_y = \sigma_z$. All of the standard deviation of Y comes from Z; none of it comes from X.

For general values of m and σ_z, we *define* the correlation coefficient r to be the fraction of σ_y that can be attributed to X (assuming a linear model). Since the part of σ_y that can be attributed to X is $m\sigma_x$, this fraction is

$$\boxed{r \equiv \frac{m\sigma_x}{\sigma_y} = \frac{m\sigma_x}{\sqrt{m^2\sigma_x^2 + \sigma_z^2}}} \quad \text{(correlation coefficient)} \tag{6.6}$$

Equivalently, r^2 equals the fraction of the variance of Y that can be attributed to X.

The use of the expression for r in Eq. (6.6) requires knowledge of m, along with σ_x and either σ_y or σ_z. If we are given m and the underlying X and Z distributions that make up Y, then we can use Eq. (6.6) to find r. But as mentioned earlier, we are usually just given a collection of data points in the x-y plane, without being given m or the exact X and Z distributions. How do we find r in that case?

Covariance

To find r if we are given a collection of data points, we need to define the *covariance* of two random variables. The covariance of X and Y is denoted by $\text{Cov}(X,Y)$ and is defined to be

$$\boxed{\text{Cov}(X,Y) \equiv E[(X - \mu_x)(Y - \mu_y)]} \tag{6.7}$$

Note that if we set Y equal to X, then the covariance of X and Y ($= X$) simplifies to $E[(X - \mu_x)^2]$, which from Eq. (3.19) is simply the variance of X. The covariance can therefore be thought of as a generalization of the variance. Like the correlation, the covariance gives a measure of how much two variables linearly depend on each other. In one extreme where $Y = X$, we have $\text{Cov}(X,Y) = \sigma_x^2$. In the other extreme where X and Y are independent variables, we have $\text{Cov}(X,Y) = 0$. This is true because for independent variables, Eq. (3.16) tells us that the expectation value in Eq. (6.7) equals the product of the expectation values $E(X - \mu_x)$ and $E(Y - \mu_y)$. And these are equal to $E(X) - \mu_x$ and $E(Y) - \mu_y$, which are both zero by the definition of the μ's. There is actually yet a further extreme, namely $Y = -X$. In this case we have $\text{Cov}(X,Y) = -\sigma_x^2$.

In the situations we'll be dealing with, we'll usually take the means μ_x and μ_y to be zero, in which case Eq. (6.7) reduces to

$$\text{Cov}(X,Y) \equiv E(XY) \qquad (\text{if } \mu_x = \mu_y = 0). \tag{6.8}$$

Having defined the covariance, we now claim that the definition of r in Eq. (6.6) is equivalent to

$$\boxed{r \equiv \frac{\text{Cov}(X,Y)}{\sigma_x \sigma_y}} \tag{6.9}$$

To demonstrate this equivalence, we just need to replace Y with $mX + Z$ in Eq. (6.9). We'll assume that μ_x and μ_y (and hence μ_z, from Eq. (6.4)) are zero; the general case proceeds similarly. We obtain

$$r = \frac{\text{Cov}(X,Y)}{\sigma_x \sigma_y} = \frac{\text{Cov}(X, mX + Z)}{\sigma_x \sigma_y} = \frac{E[X(mX + Z)]}{\sigma_x \sigma_y}$$

$$= \frac{mE(X^2) + E(XZ)}{\sigma_x \sigma_y} = \frac{m\sigma_x^2 + 0}{\sigma_x \sigma_y} = \frac{m\sigma_x}{\sigma_y}, \tag{6.10}$$

which is the expression for r in Eq. (6.6). We have used the fact that since X and Z are independent, Eq. (3.16) allows us to write $E(XZ) = E(X)E(Z) = 0 \cdot 0 = 0$. We have also used Eq. (3.50) to say that $E(X^2) = \sigma_x^2$, since $\mu_x = 0$. The upshot here is that Eq. (6.9) reduces to Eq. (6.6) because $\text{Cov}(X,Y)$ picks out the part of Y that comes from X and gets rid of the part that comes from Z. This leaves us with $m\sigma_x^2$, so dividing by $\sigma_x \sigma_y$ gives the desired ratio of standard deviations in Eq. (6.6). Note that neither Eq. (6.6) nor Eq. (6.9) requires that the underlying distributions be Gaussian.

Compared with Eq. (6.6), the advantage of Eq. (6.9) is that it doesn't involve m. Eq. (6.9) is therefore the one you want to use if you are simply given a set of data points (x_i, y_i) instead of the underlying distributions in Eq. (6.2). Although we defined $\text{Cov}(X,Y)$ in Eqs. (6.7) and (6.8) for known distributions, $\text{Cov}(x,y)$ can also be defined for a set of data points. It's just that instead of talking about the *expectation value* of XY (assuming that the means are zero), we talk about the *average* value of the $x_i y_i$ products, where the average is taken over all of the given (x_i, y_i) data points. If we have n points (x_i, y_i), then the covariance in the general case of nonzero means is

$$\boxed{\text{Cov}(x,y) \equiv \frac{1}{n} \sum (x_i - \overline{x})(y_i - \overline{y})} \qquad (\text{for data points}) \tag{6.11}$$

If the averages $\overline{x}$ and $\overline{y}$ are zero, then the covariance is just the average of the products $x_i y_i$, that is, $\text{Cov}(x,y) = (1/n) \sum x_i y_i$.

In defining r for a set of data points, the σ_x and σ_y standard deviations in Eq. (6.9) are replaced with the $\tilde{s}_x$ and $\tilde{s}_y$ standard deviations from Eq. (3.60), calculated for the specific sets of points, x_i and y_i. So the correlation coefficient is given by

$$r \equiv \frac{\text{Cov}(x,y)}{\tilde{s}_x \tilde{s}_y} = \frac{\sum (x_i - \bar{x})(y_i - \bar{y})}{\sqrt{\sum (x_i - \bar{x})^2} \sqrt{\sum (y_i - \bar{y})^2}} \qquad \text{(for data points)} \qquad (6.12)$$

Note that no factors of n remain in this expression, because the factor of n in $\text{Cov}(x,y)$ (see Eq. (6.11)) cancels with the factors of $\sqrt{n}$ in each of $\tilde{s}_x$ and $\tilde{s}_y$ (see Eq. (3.60)).

If you are considering the n data points to be a subset of a larger population, then it is more appropriate to use the *sample standard deviations* s_x and s_y instead of $\tilde{s}_x$ and $\tilde{s}_y$. The sample standard deviations are defined via the sample variance s^2 in Eq. (3.73) with an $n - 1$ in the denominator. Likewise, it is more appropriate to use the *sample covariance*, defined analogously with an $n - 1$ instead of an n in the denominator of Eq. (6.11). However, using these "sample" quantities (with $n - 1$ instead of n) doesn't affect the final result in Eq. (6.12), because the $n - 1$ factors cancel, just as the n factors did. The expression for r on the righthand side of Eq. (6.12) is therefore valid in any case. We'll do an example involving r and $\text{Cov}(x,y)$ below on page 290.

REMARKS:

1. We chose to initially define r by Eq. (6.6) instead of by Eq. (6.9) (which is more common in a practice), because Eq. (6.6) makes it clear what the meaning of r is. It is the fraction of σ_y that can be attributed to X. If most of σ_y comes from X and not Z, then X and Y have a high correlation. If most of σ_y comes from Z and not X, then X and Y have a low correlation.

2. The correlation coefficient r is independent of the means of X, Y, and Z. This follows from the fact that none of the quantities in Eq. (6.6) or Eq. (6.9) (m, σ_x, σ_y, σ_z, or $\text{Cov}(X,Y)$) depend on the means. Changing the means simply shifts the whole blob of points around in the X-Y plane.

3. The correlation coefficient r doesn't depend on a uniform scaling of X or Y. That is, r doesn't depend on a uniform stretching of the X or Y axes. This is true because if we define new variables $X' \equiv aX$ and $Y' \equiv bY$ (which imply $\mu_{x'} = a\mu_x$ and $\mu_{y'} = b\mu_y$), then you can quickly use Eq. (6.7) to show that $\text{Cov}(X',Y')$ is larger than $\text{Cov}(X,Y)$ by the factor ab. Likewise, $\sigma_{x'}\sigma_{y'}$ is larger than $\sigma_x\sigma_y$ by the same factor ab, from two applications of Eq. (3.41). The r in Eq. (6.9) therefore doesn't change. Basically, stretching each of the axes in a scatter plot by arbitrary amounts doesn't change how well the value of X helps predict the value of Y.

4. Eq. (6.9) is symmetric in X and Y. This means that if we switch the independent and dependent variables in a scatter plot and imagine X being partially dependent on Y (instead of Y being partially dependent on X), then the correlation coefficient is the same. This isn't terribly obvious, given the lack of symmetry in the relation in Eq. (6.2), where Z is independent of X, not Y. We'll have more to say about this symmetry in Sections 6.5 and 6.7 below.

5. From Eq. (6.6) we see that m can be written as $m = r\sigma_y/\sigma_x$. In terms of the covariance, m is therefore

$$m = r\frac{\sigma_y}{\sigma_x} = \frac{\text{Cov}(X,Y)}{\sigma_x\sigma_y} \cdot \frac{\sigma_y}{\sigma_x} = \frac{\text{Cov}(X,Y)}{\sigma_x^2}. \qquad (6.13)$$

6. An alternative expression for the covariance in Eq. (6.7) can be derived by expanding the product $(X - \mu_x)(Y - \mu_y)$. Using $E(X) \equiv \mu_x$ and $E(Y) \equiv \mu_y$, we have

$$\begin{aligned}
\text{Cov}(X,Y) &\equiv E[(X - \mu_x)(Y - \mu_y)] \\
&= E[XY - \mu_y X - \mu_x Y + \mu_x \mu_y] \\
&= E(XY) - \mu_y E(X) - \mu_x E(Y) + \mu_x \mu_y \\
&= E(XY) - \mu_y \mu_x - \mu_x \mu_y + \mu_x \mu_y \\
&= E(XY) - \mu_x \mu_y.
\end{aligned} \tag{6.14}$$

This reduces to Eq. (3.34) when $X = Y$. ♣

Examples with various *r* values

Fig. 6.8 shows examples of scatter plots for six different values of r. All of the (numerically generated) plots have $\sigma_x = 2$ and $\sigma_y = 1$, and there are 1000 points in each. Note that it takes a sizeable r to obtain a scatter plot that looks significantly different from the $r = 0$ case; the $r = 0.3$ plot looks roughly the same. The plots in this figure give you a visual sense of what a particular r means, so you should keep them in mind whenever you're given an r value. If someone says, "The r value is 0.7, and that seems pretty high, so I can be fairly certain of what Y will be, given X," then you will know that this person is mistaken. When $r = 0.7$, there is still a sizeable spread in the Y values for a given X.

What is considered to be a "good" or "high" value of r? Well, that depends on what data you're dealing with. If you're a social scientist and you find an $r = 0.7$ correlation between a certain characteristic and say, the number of months that a person has been unemployed, then that is a very significant result. You have just found a characteristic that helps substantially in predicting the length of unemployment. (But keep in mind that correlation does not necessarily imply causation. Although you have found something that helps in predicting, it might not help in explaining.) However, if you're a physicist and you find a $r = 0.7$ correlation between the distance d an object falls (in vacuum, dropped from rest) and the square of the falling time t, then that is a terrible result. Something has gone severely wrong, because the data points should (at least up to small experimental errors) lie on the straight line given by $d = (g/2)t^2$, where g is the acceleration due to gravity.

All of the plots in Fig. 6.8 have positive values of r. The plots for negative values look the same except that the blobs of points have downward slopes. For example, a scatter plot with $r = -0.7$ is shown in Fig. 6.9. Since r is negative, Eq. (6.6) implies that m is also, so Eq. (6.2) tells us that an increase in X yields a decrease in Y (on average). Hence the negative slope.

In Figs. 6.3 though 6.7, the three specified parameters that were used to numerically generate the plots were

$$\sigma_x, \sigma_z, m, \tag{6.15}$$

whereas in Figs. 6.8 and 6.9 the three specified parameters were

$$\sigma_x, \sigma_y, r. \tag{6.16}$$

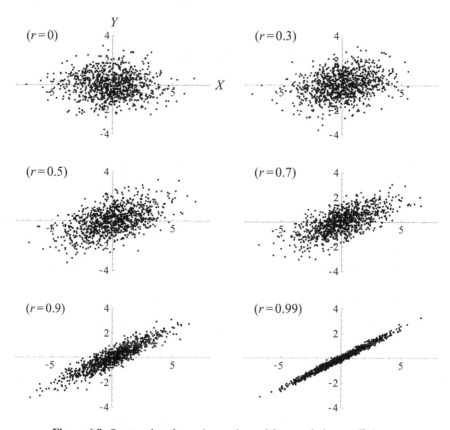

Figure 6.8: Scatter plots for various values of the correlation coefficient r.

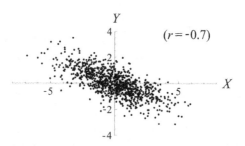

Figure 6.9: A scatter plot with negative correlation.

Both sets of parameters contain the same information, expressed in different ways (although both sets contain σ_x). It is easy to go from one set to the other. Given the set in Eq. (6.15), the σ_y and r values in Eq. (6.16) can be found via Eq. (6.6):

$$\sigma_y = \sqrt{m^2\sigma_x^2 + \sigma_z^2} \qquad \text{and} \qquad r = \frac{m\sigma_x}{\sqrt{m^2\sigma_x^2 + \sigma_z^2}} \qquad (6.17)$$

For example, Fig. 6.7 was generated from the parameters $m = 0.2$, $\sigma_x = 2$, and $\sigma_z = 1$. So you can quickly show that Eq. (6.17) gives $\sigma_y = 1.08$ and $r = 0.37$.

To go the other way, the above expression for σ_y can be rewritten as $\sigma_z^2 = \sigma_y^2 - m^2\sigma_x^2$. But Eq. (6.6) tells us that $m^2\sigma_x^2 = r^2\sigma_y^2$, so we obtain $\sigma_z^2 = (1 - r^2)\sigma_y^2$. Therefore, given the set of parameters in Eq. (6.16), the σ_z and m values in Eq. (6.15) can be found via

$$\boxed{\sigma_z = \sigma_y\sqrt{1 - r^2}} \qquad \text{and} \qquad \boxed{m = \frac{r\sigma_y}{\sigma_x}} \qquad (6.18)$$

For example, the $r = 0.3$ plot in Fig. 6.8 was generated from the parameters $r = 0.3$, $\sigma_x = 2$ and $\sigma_y = 1$. So Eq. (6.18) gives $\sigma_z = 0.95$ and $m = 0.15$.

From here on, we will usually describe scatter plots in terms of r (and σ_x and σ_y) instead of m (and σ_x and σ_z). But you can always switch back and forth between r and m by using Eqs. (6.17) and (6.18). However, we are by no means finished with m. This quantity is extremely important, in that it is the slope of the *regression line*, which is the topic of Section 6.7.

As mentioned above, it is more common to be given a scatter plot, or equivalently a list of (x_i, y_i) pairs, than it is to be given Eq. (6.2) along with the underlying distributions X and Z. So let's explicitly list out the procedure for finding all of the parameters you might want to know, given a scatter plot of points. We'll be general here and not assume that the means of X and Y are zero. Here are the steps:

1. Calculate the means $\bar{x}$ and $\bar{y}$ of the x_i and y_i data points.

2. Calculate the standard deviations $\tilde{s}_x$ and $\tilde{s}_y$ via Eq. (3.60).

3. Calculate the covariance via Eq. (6.11).

4. Calculate r via Eq. (6.12).

5. Calculate m from Eq. (6.18), with the σ's replaced with $\tilde{s}$'s.

Example: Consider the 20 points (X, Y) listed in Table 6.1 and plotted in Fig. 6.10. (These points don't have any significance; I just made them up.) What is the correlation coefficient between X and Y?

X	12	7	10	3	18	13	17	6	9	12
Y	10	13	6	4	25	14	20	7	14	15

X	13	14	5	7	16	11	8	13	15	9
Y	18	9	7	15	26	16	12	12	17	10

Table 6.1: 20 points (X, Y).

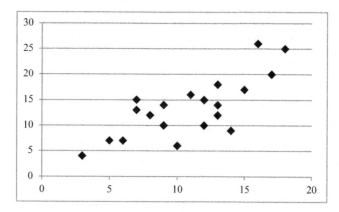

Figure 6.10: The scatter plot of the points in Table 6.1.

Solution: The quickest way to analyze the data is to use Excel or something similar, by making a column for the X values and another column for the Y values. Now, technically if we're concerned only with the correlation coefficient r, then one Excel function, CORREL, gets the job done. But let's pretend we don't know that. Then to calculate r, we need to find the standard deviations $\tilde{s}_x$ and $\tilde{s}_y$ by using the STDEV.P function (they turn out to be 4.02 and 5.72), and also the covariance by using the COVARIANCE.P function (it turns out to be 17.45). The correlation coefficient r is then found via Eq. (6.12) to be $r = 0.76$. Eq. (6.18) then gives $m = 1.08$. The ".P" in these functions stands for "population."

Alternatively, you can use the STDEV.S and COVARIANCE.S functions, which have factors of $n - 1$ instead of n in the denominator. (The ".S" stands for "sample.") You will obtain the same result for r, because all the $(n - 1)$'s cancel out in Eq. (6.12), just as the n's do when using the ".P" functions.

If you have access to only pencil/paper or a basic calculator, then the process will of course take longer. You will need to work through the whole list of steps preceding this example. The means $\overline{x}$ and $\overline{y}$ happen to be 10.9 and 13.5.

6.4 improving the prediction for Y

As we have seen in a number of plots, the larger the correlation coefficient r is, the more the knowledge of a particular X value helps in predicting the associated Y value. In this section we will be quantitative about this. We will determine exactly how much the prediction is improved, given r. In the following discussion, we will assume that the X, Y, and Z distributions in the $Y = mX + Z$ relation are all known.

If we want to predict the value of Y *without* taking into account the associated value of X, then intuitively the most reasonable prediction is the mean value μ_y of the entire Y distribution. We'll justify this choice below in Eq. (6.22). However, if we *do* take into account the associated value of X, and if there is a nonzero correlation between X and Y, then we can make a prediction for Y that is better than the above μ_y prediction. What is the value of this better prediction, and how much better is it than μ_y? In answering this, we'll need to define what we mean by how "good" a prediction is. We'll do this by instead defining how "bad" a prediction is.

Considering the entire Y distribution

Consider first the case where we are looking at the entire Y distribution. That is, we are *not* looking at a specific value of X. Imagine that we state our prediction for Y (call it y_p) and then pick a large number n of actual Y values y_i. We then calculate the variance of these values, *measured relative to our prediction*. That is, we calculate

$$\tilde{s}_p^2 \equiv \frac{1}{n}\sum_{i=1}^{n}(y_i - y_p)^2. \tag{6.19}$$

In the limit where we pick an infinite number of y_i values, the preceding expression becomes an expectation value,

$$\boxed{\sigma_p^2 \equiv E\left[(Y - y_p)^2\right]} \qquad \text{(badness of prediction)} \tag{6.20}$$

We will take this variance as a measure of how bad our prediction is. The larger the variance, the worse the prediction. If y_p isn't anywhere near the various y_i values, then our prediction is clearly a poor one. And consistent with this, the variance σ_p^2 is large. We could, of course, choose a different definition of badness, but the one involving the variance in Eq. (6.20) is the standard definition. See the remark in the solution to Problem 6.10 for some discussion of this.

Given the above definition of badness, the best y_p prediction is the one that minimizes σ_p^2. To find this y_p value, we'll need to do a little calculus and take the derivative of σ_p^2 with respect to y_p, and then set the result equal to zero. If we expand the square in Eq. (6.20), we obtain

$$\sigma_p^2 = E[Y^2] - 2E[Y]y_p + y_p^2. \tag{6.21}$$

Setting the derivative with respect to y_p equal to zero then gives

$$-2E[Y] + 2y_p = 0 \implies y_p = E[Y] \equiv \mu_y. \tag{6.22}$$

We see that σ_p^2 is minimized when y_p equals the expectation value $E[Y]$, that is, the mean μ_y. We therefore want to choose the mean μ_y as our prediction. As mentioned above, this is probably what your intuition would have told you to do anyway!

$$\boxed{\text{Best prediction} = \text{Mean}} \tag{6.23}$$

In the case of the $y_p = \mu_y$ best prediction, the variance σ_p^2 in Eq. (6.20) is simply the actual variance σ_y^2 of the Y distribution. So σ_y^2 is a measure of how "bad" our best prediction is if we *don't* take into account any information about the X value:

$$\sigma_y^2 = \text{badness of best guess } \mu_y, \text{ with no knowledge of } X. \tag{6.24}$$

Considering a specific value of X

Now consider the case where we *do* take into account a specific value of X; call it x_0. Given x_0, the only possible Y values y_i that we can possibly pick when forming the variance in Eq. (6.19) are the ones in the shaded strip in Fig. 6.11. (In principle, the strip should be very thin.) In drawing a scatter plot that is centered at the origin like this, we are tacitly assuming that the means of X and Y are zero. This can always be accomplished by measuring X and Y relative to their means, if they happen to be nonzero. Eq. (6.4) then implies that Z also has a mean of zero.

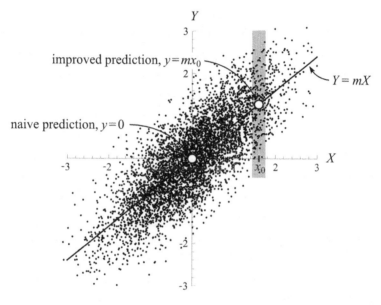

Figure 6.11: Given that X equals x_0, the best prediction for Y is mx_0 (the upper-right white dot). This is a better prediction than the naive $Y = 0$ prediction (the lower-left white dot) relevant to the entire distribution.

The mean of the Y values in the shaded strip is mx_0, because $Y = mX + Z$, and because $\mu_z = 0$. The best y_p prediction is therefore mx_0. This true because the general result in Eq. (6.23) still applies. The logic leading up to that equation remains valid; it's just that we're now taking the expectation value of *only the part of the Y distribution that lies in the shaded strip*, instead of the complete Y distribution (the entire blob of points). In the present case where we are incorporating our knowledge of x_0, the Y in Eq. (6.20) should technically be replaced with Y_{x_0}, or some similar notation, to indicate that we are concerned only with Y values that are associated with (or nearly with) x_0.

In the shaded strip in Fig. 6.11, the variance σ_p^2 in Eq. (6.20) (measured relative to our y_p prediction of mx_0) equals $E[(Y - mx_0)^2]$. But $Y - mx_0$ equals Z in the shaded strip, because $Y = mX + Z$. The variance σ_p^2 is therefore simply $E[Z^2] \equiv \sigma_z^2$. So σ_z^2 is a measure of how "bad" our best prediction is if we *do* take into account the particular value of X:

$$\sigma_z^2 = \text{badness of best guess } mx_0, \text{ knowing that } X = x_0. \qquad (6.25)$$

mx_0 is the Y value associated with the upper-right white dot in Fig. 6.11. Our earlier prediction of 0 (or more generally μ_y) is the Y value associated with the lower-left white dot.

Given our definition of badness in terms of the variance σ_p^2 in Eq. (6.20), the ratio of the variances associated with the above two predictions in Eqs. (6.24) and (6.25) is a measure of how much better one prediction is than another. That is, the ratio

$$\frac{\text{Var}(Z)}{\text{Var}(Y)} = \frac{\sigma_z^2}{\sigma_y^2} \qquad (6.26)$$

is the desired measure of how much better our prediction is if we take into account the particular value of X. From Eq. (6.18) we know that $\sigma_z = \sigma_y \sqrt{1 - r^2}$, so the ratio in Eq. (6.26) equals

$$\boxed{\frac{\sigma_z^2}{\sigma_y^2} = 1 - r^2} \qquad \text{(improvement of prediction)} \qquad (6.27)$$

This ratio is the factor by which the variance of a large number of data points (measured relative to our prediction) is reduced if we use our knowledge of X. For example, if $r = 1$, then the factor is 0. This makes sense. With perfect correlation, our prediction is perfect, so the variance is reduced to nothing. In the other extreme where $r = 0$, the factor is 1. This also makes sense. With no correlation, knowledge of the X value doesn't help in predicting the Y value, so the variance isn't reduced at all. If, say, $r = 0.5$, then $1 - r^2 = 0.75$, which means that our prediction is only slightly improved (that is, the variance is only slightly reduced) if we use our knowledge of X.

Note that since Eq. (6.27) involves the square of r, the sign of r doesn't matter. The improvement factor $1 - r^2$ is the same for, say, $r = -0.5$ and $r = 0.5$. This is clear from looking at a scatter plot. The only difference is that a positive-r blob of points tilts upward while a negative-r blob tilts downward.

6.5   Calculating $\rho(x, y)$

The scatter plots in Fig. 6.8, along with most of the other scatter plots in this chapter, were generated numerically by using Gaussian distributions for X and Z. (From Problem 6.4, it follows that Y is also Gaussian.) A quick glance at the plots in Fig. 6.8 indicates that all of the blobs of points have ellipse-like shapes. And indeed, if X and Z are Gaussians, then the probability distributions in the plane are in fact exactly elliptical. By this we mean that if we look at all points in the x-y plane that have the same probability density $\rho(x, y)$, then these points all lie on an ellipse. Since we're dealing with a 2-D plane, $\rho(x, y)$ is a probability density per unit *area*. That is, if we multiply $\rho(x, y)$ by a small area in the plane, we obtain the probability of lying in that small region.

Let's rigorously demonstrate the above ellipse claim. In this section, unlike in previous sections, the Gaussian assumption for X and Z will be necessary. We'll start with the $Y = mX + Z$ relation. Imagine picking a random value from the X distribution, along with a random value from the Z distribution. If we end up with a particular point (x, y) in the plane, then we must have picked an X value of x and a Z value of $y - mx$. Since X and Z are independent variables, the joint probability of these two outcomes is simply the product of the individual probabilities. Of course, the probability of obtaining *exactly* a specific value of X or Z is zero, because we're dealing with continuous distributions. We should therefore really be talking about probability densities and tiny areas in the plane. So to be formal, we can say that the probability of obtaining X and Y values that lie in a tiny area $dx\, dy$ around the point (x, y) is

$$\rho(x, y)\, dx\, dy = \big(\rho(X = x)\, dx\big) \cdot \big(\rho(Z = y - mx)\, dz\big), \tag{6.28}$$

where dz is the interval of z that corresponds to the dy interval of y. But since the coefficients of Y and Z in the relation $Y = mX + Z$ are equal, dz is simply equal to dy (for a given x). Using the second Gaussian expression in Eq. (4.42) for $\rho(x)$ and $\rho(z)$, and assuming $\mu_x = \mu_y = 0$, Eq. (6.28) becomes

$$\rho(x, y)\, dx\, dy = \frac{1}{\sqrt{2\pi}\sigma_x} e^{-x^2/2\sigma_x^2}\, dx \cdot \frac{1}{\sqrt{2\pi}\sigma_z} e^{-(y-mx)^2/2\sigma_z^2}\, dy. \tag{6.29}$$

Our goal is to produce an expression for $\rho(x, y)$ that involves only x and y, without any reference to z. So we must get rid of the two σ_z's in Eq. (6.29). Additionally, let's get rid of m in favor of the correlation coefficient r. We can rewrite Eq. (6.29) as

$$\rho(x, y) = \frac{1}{2\pi\sigma_x\sigma_z} \exp\left(-\frac{x^2}{2\sigma_x^2} - \frac{(y - mx)^2}{2\sigma_z^2}\right). \tag{6.30}$$

From Eq. (6.18) we know that $\sigma_z = \sigma_y\sqrt{1 - r^2}$ and $m = r\sigma_y/\sigma_x$. So

$$\rho(x, y) = \frac{1}{2\pi\sigma_x\sigma_y\sqrt{1 - r^2}} \exp\left(-\frac{x^2}{2\sigma_x^2} - \frac{(y - (r\sigma_y/\sigma_x)x)^2}{2(1 - r^2)\sigma_y^2}\right). \tag{6.31}$$

Let's simplify the exponent here. We can rewrite it as

$$-\frac{1}{2(1 - r^2)}\left(\frac{(1 - r^2)x^2}{\sigma_x^2} + \frac{y^2 - 2rxy\sigma_y/\sigma_x + r^2x^2\sigma_y^2/\sigma_x^2}{\sigma_y^2}\right). \tag{6.32}$$

The r^2x^2/σ_x^2 terms cancel, yielding

$$-\frac{1}{2(1 - r^2)}\left(\frac{x^2}{\sigma_x^2} + \frac{y^2}{\sigma_y^2} - \frac{2rxy}{\sigma_x\sigma_y}\right). \tag{6.33}$$

Our final result for the joint probability density $\rho(x, y)$ is therefore

$$\boxed{\rho(x, y) = \frac{1}{2\pi\sigma_x\sigma_y\sqrt{1 - r^2}} \exp\left(-\frac{1}{2(1 - r^2)}\left(\frac{x^2}{\sigma_x^2} + \frac{y^2}{\sigma_y^2} - \frac{2rxy}{\sigma_x\sigma_y}\right)\right)} \tag{6.34}$$

If we multiply this $\rho(x,y)$ by a small area in the plane, we obtain the probability that a randomly chosen (X,Y) point lies in that area. The double integral of $\rho(x,y)$ over the entire x-y plane must be 1, because the total probability is 1. If you want to explicitly verify this, you can "complete the square" of the quadratic function of y (or x) in the exponent of $\rho(x,y)$. Of course, when you do this, you'll just be working backward through the above steps, which means that you'll end up with the $\rho(x,y)$ in Eq. (6.29). But the double integral of that $\rho(x,y)$ over the entire x-y plane is certainly 1, because it involves a Gaussian dx integral and a Gaussian dy integral, each of which we know is 1. (With the form in Eq. (6.29), the dy integral should be done first.)

The result for $\rho(x,y)$ in Eq. (6.34) contains the complete information of our setup. Everything we might want to figure out can be determined from $\rho(x,y)$. It contains exactly the same (complete) information as our original description of the setup, namely that Y is given by $mX + Z$, where X and Z are Gaussian distributions with means of zero and standard deviations of σ_x and σ_z.

Eq. (6.34) tells us that the curves of constant probability density are ellipses. This is true because the exponent in $\rho(x,y)$ (which contains all of the x and y dependence; there is none in the prefactor) takes the form of $Ax^2 + By^2 + Cxy$. And we'll just accept here the well-known fact that a curve described by the equation $Ax^2 + By^2 + Cxy = D$ is an ellipse. If $C = 0$, then the axes of the ellipse are parallel to the coordinate axes. But if C is nonzero, then the ellipse is tilted. Since $C \propto r$, we see that the ellipse is tilted whenever there is a nonzero correlation between X and Y.

If the distributions for X and Z aren't Gaussian, then the constant-$\rho(x,y)$ curves aren't ellipses. So whenever we talk about ellipses in the following sections, we are assuming that the underlying distributions are Gaussian.

We now come to a very important point, which is so important that we'll put it in a box:

> The probability density $\rho(x,y)$ in Eq. (6.34) is *symmetric* in x and y.

More precisely, if x and σ_x are switched with y and σ_y, then $\rho(x,y)$ is unchanged. (We have used the fact that the expression for r in terms of the covariance, given in Eq. (6.9), is symmetric in x and y.) This symmetry of $\rho(x,y)$ is by no means obvious from looking at our original $Y = mX + Z$ expression, because Z is independent of X and not Y, which makes things appear asymmetric.

But given that we now know that $\rho(x,y)$ is symmetric, let's switch x and y in the $Y = mX + Z$ relation and see what we get. The point here is that whatever relation we get, it must have the same probability distribution $\rho(x,y)$ (that is, the same shape of the blob of points in the x-y plane), because $\rho(x,y)$ is symmetric in x and y. To switch x and y in the relation $Y = mX + Z$, we must first make the x and y dependences explicit by writing m as $r\sigma_y/\sigma_x$, and σ_z as $\sigma_y\sqrt{1-r^2}$, from Eq. (6.18). The relation $Y = mX + Z$ can then be written in the more explicit form,

$$Y = \left(\frac{r\sigma_y}{\sigma_x}\right)X + Z_{X\text{-ind}}^{\sigma_y\sqrt{1-r^2}} \tag{6.35}$$

where we have indicated the standard deviation of the (X-independent) distribution Z. Switching the x's and y's gives (again using the fact that r is symmetric in x and y)

$$X = \left(\frac{r\sigma_x}{\sigma_y}\right)Y + Z_{Y-\text{ind}}^{\sigma_x\sqrt{1-r^2}} \qquad (6.36)$$

The (new and different) Z here is independent of Y and has a standard deviation of $\sigma_x\sqrt{1-r^2}$. The above Z notation might seem a bit awkward, but it is important to indicate the two ways in which the two Z's differ (standard deviation, and which other variable they are independent of).

So what did we just show? All three of the equations Eq. (6.34), Eq. (6.35), and Eq. (6.36) have equivalent information. Eq. (6.34) puts X and Y on equal footing, whereas Eq. (6.35) treats Y as being dependent on X and $Z_{X-\text{ind}}^{\sigma_y\sqrt{1-r^2}}$, and Eq. (6.36) treats X as being dependent on Y and $Z_{Y-\text{ind}}^{\sigma_x\sqrt{1-r^2}}$. But they all say the same thing, and they all produce the same probability density $\rho(x,y)$ and hence the same shape of the blob of points in the x-y plane.

If you don't trust the above symmetry reasoning, you can show that Eq. (6.34) follows from Eq. (6.36) by starting with Eq. (6.36) and then working through the same steps as in Eq. (6.29) through Eq. (6.34). Of course, you will quickly discover that redoing the algebra is unnecessary, because all you're doing is switching x and y. Since the final result for $\rho(x,y)$ is symmetric in x and y, the switch doesn't affect $\rho(x,y)$. The expressions in Eqs. (6.35) and (6.36) will be critical when we discuss the regression lines in Section 6.7.

6.6 The standard-deviation box

Assuming that all of our variables are Gaussians with means of zero, consider the constant-$\rho(x,y)$ ellipse shown in Fig. 6.12. If we are given nothing but the tilted ellipse in the figure, can we determine the value of m in the $Y = mX + Z$ relation that produces this ellipse? Indeed we can, in the following manner.

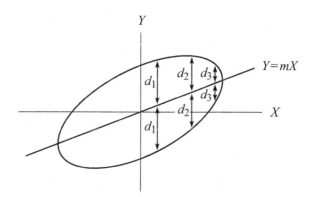

Figure 6.12: For a given value of X, the Y values on a constant-ρ ellipse are symmetrically located above and below the $Y = mX$ point.

Recall that for a given value of X, the Y values are symmetrically distributed above and below the $Y = mX$ point, because $Y = mX + Z$, and because Z is a Gaussian with mean zero. Since this holds for all values of X, the $Y = mX$ line must be the line that bisects the vertical span between any two vertically-aligned points on the ellipse, as indicated in Fig. 6.12.[1] The $Y = mX$ line must therefore intersect the ellipse at the ellipse's leftmost and rightmost points, where the slope is vertical. This is true because if the slope weren't vertical at an intersection point, then the distance d either above or below the intersection point would be zero, while the other distance below or above the point would be nonzero.

We can check that the numbers work out in Fig. 6.12. We generated this ellipse by arbitrarily choosing $\sigma_x = (1.5)\sigma_y$ and $r = 0.56$. If you plug these values into Eq. (6.34) and then set the exponent in $\rho(x, y)$ equal to an arbitrary (negative) constant, you will produce an ellipse with the shape shown. (The common value of $\rho(x, y)$ associated with the ellipse doesn't matter, because that just determines the overall size of the ellipse, and not the shape.) From Eq. (6.18) we know that $m = r\sigma_y/\sigma_x$, which gives $m = 0.37$ here. And this is indeed the slope of the tilted line in Fig. 6.12.

Let's now consider an ellipse with a particular value of $\rho(x, y)$, namely $\rho(x, y) = e^{-1/2}\rho(0,0)$. We are now interested in the actual size of the ellipse. From Eq. (6.34) we know that $\rho(0,0) = 1/(2\pi\sigma_x\sigma_y\sqrt{1-r^2})$, although this exact value won't concern us. Only the relative factor of $e^{-1/2}$ will matter here. For all points on the $\rho(x, y) = e^{-1/2}\rho(0,0)$ ellipse (we'll call this the "$e^{-1/2}$ ellipse"), $\rho(x, y)$ is smaller than its value at the origin by a factor of $e^{-1/2}$. The exponent in Eq. (6.30) or Eq. (6.34) therefore equals $-1/2$. Any other factor would serve the purpose here just as well, but we're picking $e^{-1/2}$ because it parallels the one-standard-deviation probability density for the single-variable Gaussian distribution, $e^{-x^2/2\sigma^2}/\sqrt{2\pi}\sigma$. If $x = \sigma$ then $\rho(x) = e^{-1/2}\rho(0)$.

What is the value of x at the rightmost point on the $e^{-1/2}$ ellipse? Since we know from above that the line $y = mx$ passes through this point, the second term in the exponent in Eq. (6.30) equals zero. The first term must therefore equal $-1/2$, which means that x is simply σ_x. The same reasoning holds for the leftmost point, so the $e^{-1/2}$ ellipse ranges from $x = -\sigma_x$ to $x = \sigma_x$. We will now take advantage of the fact that Eq. (6.34) is symmetric in x and y. This means that any statement we can make about x, we can also make about y. Therefore, by the same reasoning with x and y switched, the highest point on the ellipse has a y value of σ_y, and the lowest point has a y value of $-\sigma_y$. So the "bounding box" around the $e^{-1/2}$ ellipse is described by the lines $x = \pm\sigma_x$ and $y = \pm\sigma_y$. This box is called the "standard-deviation box" and is shown in Fig. 6.13.[2]

---

[1]It isn't so obvious that given an arbitrary tilted ellipse, the locus of points with this property is in fact a line. But the derivation of $\rho(x, y)$ in Section 6.5 basically proves it. Just work backwards starting with the elliptical distribution in Eq. (6.34), and you will find in Eq. (6.30) that $\rho(x, y)$ decreases symmetrically above and below the $y = mx$ line.

[2]Alternatively, you can find the bounding box by using calculus and taking the differential of the exponent in Eq. (6.34). Setting the result equal to zero gives a relation between nearby points on a given ellipse. Setting $dx = 0$ then gives a relation between x and y at the leftmost and rightmost points, where the slope is infinite. You can plug this relation back into the expression for the $e^{-1/2}$ ellipse to find the x values at the leftmost and rightmost points. Similar reasoning with $dy = 0$ gives the y values at the highest and lowest points.

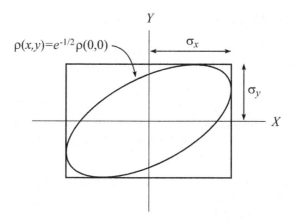

Figure 6.13: The standard-deviation box.

Note that nowhere in the preceding paragraph did we make use of any specific value of the correlation coefficient r (or equivalently, of m). This means that for given values of σ_x and σ_y, the $e^{-1/2}$ ellipse is always bounded by the same box, for any value of r. Different values of r simply determine the shape of the ellipse inside the box. Two examples are shown in Fig. 6.14. They both have the same σ_x and σ_y values (where $\sigma_x = (1.5)\sigma_y$), but the r for the thin ellipse is about 0.93, while the r for the wide ellipse is about 0.19. We will discuss in the next section how to determine r from an ellipse and its standard-deviation box. Note that the two ellipses in Fig. 6.14 have different values of σ_z; they have the same σ_y, so the different values of r lead to different values of $\sigma_z = \sigma_y \sqrt{1 - r^2}$, from Eq. (6.18). The thin ellipse, which has a larger r, has a smaller σ_z.

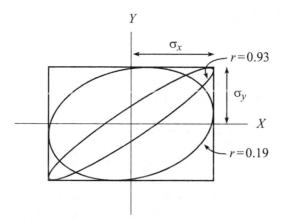

Figure 6.14: The standard-deviation box for an ellipse depends only on σ_x and σ_y, and not on r. Different values of r yield different shapes of ellipses inside the box.

6.7 The regression lines

Given a $\rho(x,y) = e^{-1/2}\rho(0,0)$ ellipse, there are a number of lines that we might reasonably want to draw, as shown in Fig. 6.15.

- We can draw the $y = mx$ line passing through the leftmost and rightmost points on the ellipse, along with the analogous line passing through the highest and lowest points. These are the solid lines shown, and they are called *regression lines*. The reason for this name will become apparent in Section 6.8. These lines are very important.

- We can draw the *standard-deviation line* passing through the corners of the standard-deviation box. This is the long-dashed line, with slope σ_y/σ_x. This line is somewhat important.

- We can draw the line along the major axis of the ellipse. This is the short-dashed line. It might seem like this line should have some importance, being a symmetry axis of the ellipse. However, it actually doesn't have much to do with anything in probability, so we won't be concerned with it.

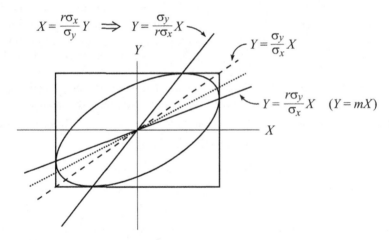

Figure 6.15: The two regression lines (solid), the standard-deviation line (long-dashed), and the unimportant symmetry axis of the ellipse (short-dashed).

How do the slopes of the two regression lines relate to the slope σ_y/σ_x of the standard-deviation line? The slope of the lower[3] regression line is simply m, which equals $r\sigma_y/\sigma_x$ from Eq. (6.18). Equivalently, the slope is the $r\sigma_y/\sigma_x$ coefficient of X in Eq. (6.35). This slope is just r times the σ_y/σ_x slope of the standard-deviation line, which is about as simple a result as we could hope for. Similarly, Eq. (6.36) tells us that if we tilt our head sideways (so that the X axis is now vertical and the Y axis is horizontal), then the slope of the upper regression line is $r\sigma_x/\sigma_y$

---

[3]By "lower" we mean the line that is lower in the first quadrant. Likewise for the "upper" regression line. Of course, these adjectives are reversed in the third quadrant. So perhaps we should be labeling the lines as "shallower" and "steeper." But we'll go with lower and upper, and you'll know what we mean.

(ignoring the sign), because this is the coefficient of Y in Eq. (6.36). This slope (with our head tilted) is simply r times the σ_x/σ_y slope (with our head tilted) of the standard-deviation line, ignoring the sign.

The two regression lines pass through the points of tangency between the ellipse and the standard-deviation box. So from the previous paragraph, we see that the tangency points are the *same* fraction r of the way from each of the coordinate axes to the upper-right corner of the box. This is shown in Fig. 6.16. Determining either of these (identical) fractions therefore gives us the correlation coefficient r. This conclusion checks in the extreme cases of $r = 1$ (perfect correlation, thin ellipse) and $r = 0$ (zero correlation, wide ellipse with axes parallel to the coordinate axes).

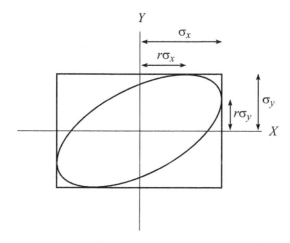

Figure 6.16: The points of tangency between the ellipse and the standard-deviation box are the same fraction r of the way from each of the coordinate axes to the upper-right corner of the box.

Note that the slope of the standard-deviation line is the geometric mean of the slopes of the two regression lines *as they appear on the paper* (with no head tilting). This is true because the slope of the upper regression line (with no head tilting) is the reciprocal of the slope with a tilted head, so it equals $\sigma_y/r\sigma_x$; this is indicated above in Fig. 6.15. The geometric mean of the two slopes as they appear on the paper is then

$$\sqrt{\left(\frac{r\sigma_y}{\sigma_x}\right)\left(\frac{\sigma_y}{r\sigma_x}\right)} = \frac{\sigma_y}{\sigma_x}, \tag{6.37}$$

which is the slope of the standard-deviation line, as desired.

Another way to determine r is the following. What is the y-intercept of the $\rho(x,y) = e^{-1/2}\rho(0,0)$ ellipse in Fig. 6.16? To answer this, we can use the fact that x equals zero on the y axis. So if we want the exponent in Eq. (6.34) to equal $-1/2$, as it does for all points on the ellipse, then since $x = 0$ we see that we need $y = \sigma_y\sqrt{1 - r^2}$. This makes sense, because this is simply σ_z for our original random variable Z, which we labeled as $Z_{X-\text{ind}}^{\sigma_y\sqrt{1-r^2}}$ in Eq. (6.35). Said in another way, when $x = 0$ the exponent in Eq. (6.30) equals $-1/2$ when $y = \sigma_z$.

By the same reasoning, the x-intercept of the $e^{-1/2}$ ellipse is $x = \sigma_x\sqrt{1-r^2}$, which is the σ_z for the random variable $Z_{Y-ind}^{\sigma_x\sqrt{1-r^2}}$ in Eq. (6.36). The intercepts are indicated in Fig. 6.17. Measuring either of these intercepts and dividing by the standard deviation along the corresponding axis gives $\sqrt{1-r^2}$, which gives r. This conclusion checks in the extreme cases of $r = 1$ and $r = 0$.

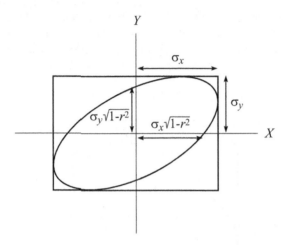

Figure 6.17: The intersection points between the ellipse and the coordinate axes are the same fraction $\sqrt{1-r^2}$ of the way from the origin to the sides of the standard-deviation box.

The critical property of the regression lines is that they are the "lines of averages." We saw in Fig. 6.12 that the lower regression line bisects the vertical span between any two vertically aligned points on a constant-ρ ellipse. This follows from the fact that the $Z_{X-ind}^{\sigma_y\sqrt{1-r^2}}$ distribution in Eq. (6.35) is a (symmetric) Gaussian with zero mean. Similarly, the upper regression line bisects the horizontal span between any two horizontally aligned points on a constant-ρ ellipse. This follows from the fact that the $Z_{Y-ind}^{\sigma_x\sqrt{1-r^2}}$ distribution in Eq. (6.36) is a (symmetric) Gaussian with zero mean. Two vertical and two horizontal pairs of equal distances are shown in Fig. 6.18.

We've been drawing constant-$\rho(x,y)$ ellipses for a while now, so let's return to a scatter plot (generated numerically from Gaussian distributions). Fig. 6.19 illustrates the same idea that Fig. 6.18 does. If we look at an arbitrary *vertical* strip of points, the distribution within the strip is symmetric around the intersection of the strip with the *lower* regression line (at least in the limit of a large number of points). And if we look at an arbitrary *horizontal* strip of points, the distribution within the strip is symmetric around the intersection of the strip with the *upper* regression line (for a large number of points). The intersections are indicated by the large white dots in the figure.

When dealing with, say, vertical strips, remember that it is the *same* $Z_{X-ind}^{\sigma_y\sqrt{1-r^2}}$ distribution that holds for all strips. This follows from Eq. (6.35). The only reason why the spread of points (relative to the regression line) appears to be smaller in the extremes of the plot is that there are fewer points with values of X out there. But

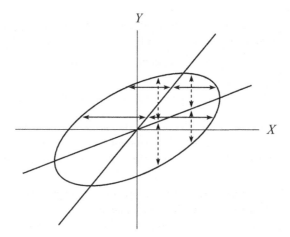

Figure 6.18: The regression lines bisect the horizontal or vertical spans between any two horizontally or vertically aligned points on the ellipse.

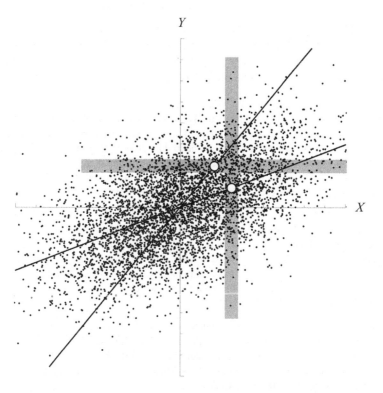

Figure 6.19: The distribution of points within a vertical strip is symmetric around the intersection of the strip and the lower regression line. Likewise for a horizontal strip and the upper regression line.

given a value of X, the value of Y has the same distribution $Z_{X-\text{ind}}^{\sigma_y \sqrt{1-r^2}}$ relative to the $Y = mX$ point on the lower regression line. That is, the probability of obtaining a certain value of $Z_{X-\text{ind}}^{\sigma_y \sqrt{1-r^2}}$ is independent of X, because this Z is assumed to be independent of X.

Let's reiterate where each of the two regression lines is relevant:

- The lower regression line is relevant if you are considering X as the independent variable, with Y being dependent on X; see Eq. (6.35). The lower regression line gives the average value of Y associated with each value of X.

- Conversely, the upper regression line is relevant if you are considering Y as the independent variable, with X being dependent on Y; see Eq. (6.36). The upper regression line gives the average value of X associated with each value of Y.

REMARKS:

1. You might think it odd that the line that cuts through the middle of the vertical strips in Fig. 6.19 is different from the line that cuts through the middle of the horizontal strips. It might seem like a *single* line (perhaps the short-dashed symmetry axis in Fig. 6.15) should do the trick for both types of strips. And indeed, when there is a high correlation ($r \approx 1$), the two regression lines are nearly the same, so one line essentially does the trick. But in the small-correlation limit ($r \approx 0$), it is clear that two lines are needed. In the $r = 0$ case in Fig. 6.5, the lower regression line is the x axis (which cuts through the middle of any vertical strip), and the upper regression line is the y axis (which cuts through the middle of any horizontal strip). These are as different as two lines can be, being perpendicular. There is no way that a single line can cut through the middle of both the vertical and horizontal strips. This fact is true for all r except $r = \pm 1$, although it is most obvious for small r.

2. Consider the lower regression line in Fig. 6.18 or Fig. 6.19. (The upper line would serve just as well.) At first glance, this line might look incorrect, because it doesn't look "balanced" properly, in the sense that the symmetry axis of the ellipse is balanced, and this line is different from the symmetry axis. But that is fine. The important thing is that any vertical strip of points is cut in half by the line. This is *not* the case with the symmetry axis of the elliptical blob of points, which *is* (irrelevantly) balanced within the ellipse.

3. If you are presented with some real data of (x_i, y_i) points in a scatter plot (as opposed to the above numerically-generated plots), you can draw the regression lines by calculating the various quantities in the steps enumerated on page 290. The slopes of the regression lines are given in Fig. 6.15 as $r\sigma_y/\sigma_x$ and $\sigma_y/r\sigma_x$, except with the σ's replaced with the $\tilde{s}$'s from Eq. (3.60).

4. If all of the above figures, we have been assuming that the means μ_x and μ_y are zero. In the more general case where the means are nonzero, the regression lines intersect at the point (μ_x, μ_y), that is, at the middle of the blob of points. Equivalently, you can define new variables by $X' \equiv X - \mu_x$ and $Y' \equiv Y - \mu_y$. The (μ_x, μ_y) point in the X-Y plane becomes the origin in the X'-Y' plane. ♣

6.8 Two regression examples

To get some practice with regression lines, let's do two examples. In both examples, we've chosen the X and Y variables to be IQ (Intelligence Quotient) scores. We've done this partly because IQ scores are easy and standard things to talk about, and partly because we want to draw some analogies between the two examples. However, in dwelling on IQs, we certainly don't mean to imply that they're terribly important. If you want to think of IQ as standing for something else like "Interesting Qualities" or "Illuminati Qualifications," then by all means do!

6.8.1 Example 1: Retaking a test

A specific example

Imagine that a large number of people have just taken an IQ test. Assume that the average score is 100, which is how an IQ test is designed. Consider all of the people who scored 130. The standard deviation of an IQ test is designed to be 15, so 130 is two standard deviations above the mean. If this group of people takes another IQ test (or the same test, if we somehow arrange for them to have amnesia), is their average score expected to be higher than, lower than, or equal to 130?

In answering this question, let's make a model and specify the (reasonable) assumptions of the model. We'll assume that each person's score is partially determined by his/her innate ability and partially determined by random effects (misreading a question, lucky guess, bad day, etc.). Although it's hard to define "innate ability," let's just take it to be a person's average score on a large number of tests. Our model therefore gives a person's actual score Y on the test as their innate score X, plus a random contribution Z which we'll assume is independent of X. So the distribution Z is the same for everyone. A person's score is then given by $Y = X + Z$. This is just our old friend Eq. (6.2), with $m = 1$. For the sake of making some nice scatter plots, we'll assume that X and Z (and hence Y, from Problem 6.4) are Gaussian.

If we take the average of the equation $Y = X + Z$ over a large number of tests taken by a given person whose innate ability has the value $X = x_0$, we obtain $\mu_y = x_0 + \mu_z$, where the μ_y here stands for the average of the given person. But the person's innate ability x_0 is defined simply as their average score μ_y over a large number of tests. We therefore conclude (basically by definition) that the mean of Z is $\mu_z = 0$. And we might as well measure X and Y relative to their population means (which are both 100). So all of $X, Y,$ and Z now have zero means.

Given all of these (quite reasonable) assumptions, what is the answer to the question we posed above? Will the people who scored 130 on their first test score (on average) higher, lower, or the same on their second test? To answer this, let's look at a scatter plot of some numerically-generated X and Y values for the first test, shown in Fig. 6.20. We have plotted 5000 points, relative to the $\mu_x = \mu_y = 100$ averages. Since we've assumed continuous Gaussian distributions for X and Z, our (X,Y) points in the plane don't have integer values, as they would on an actual test. But this won't change our general results.

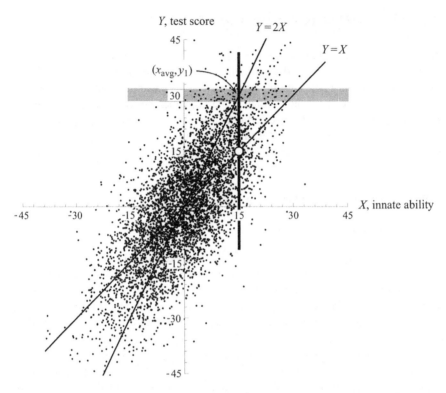

Figure 6.20: The average X value of the people in the shaded strip corresponds to the vertical black line (defined by the intersection of the shaded strip and the upper regression line). The average score of these people on a second test is given by the Y value of the white dot shown on the lower regression line.

To generate the plot in Fig. 6.20, we have arbitrarily assumed $\sigma_z = \sigma_x$. (This is probably too large a value for σ_z. In real life, the standard deviation of the random contribution Z is likely a fair bit smaller than the standard deviation of the innate ability X. But using a large value for σ_z makes things clearer in the plot.) From Eq. (6.5) with $m = 1$ and $\sigma_z = \sigma_x$, we obtain $\sigma_y = \sqrt{2}\sigma_x$. Since IQ tests are designed to have a standard deviation of 15, we have set $\sigma_y = 15$. This then implies that σ_x and σ_z are both equal to $15/\sqrt{2} \approx 10.6$. These values, along with $m = 1$, were used to generate the plot. From Eq. (6.6), the correlation coefficient is

$$r = \frac{m\sigma_x}{\sigma_y} = \frac{1 \cdot \sigma_x}{\sqrt{2}\sigma_x} = \frac{1}{\sqrt{2}} \approx 0.71. \qquad (6.38)$$

The two regression lines are drawn. The slope of the lower line is just $m = 1$. And from Fig. 6.15 the slope of the upper line is $\sigma_y/r\sigma_x = \sqrt{2}/(1/\sqrt{2}) = 2$. As a double check, the geometric mean of these two slopes is $\sqrt{1 \cdot 2}$, which is correctly the slope of the standard-deviation line, $\sigma_y/\sigma_x = \sqrt{2}$, as we noted in Eq. (6.37).

Now consider all of the people who scored 130 on the first test. Since we've subtracted off the mean score of 100 from the Y values, a score of 130 corresponds to $y_1 = 30$, or equivalently $y_1 = 2\sigma_y$. (The subscript "1" refers to the first test.) Since the Y values in our model take on a continuum of values, no one will have a score of *exactly* 30, so let's look at a small interval of scores around 30. This interval is indicated by the shaded strip in Fig. 6.20. There are about 70 points in this strip. So our goal is to predict the average score of the 70 people associated with these points when they take a second test.

What is the average X value (innate ability) of these 70 points? Answering this question is exactly what the upper regression line is good for. As we noted in Fig. 6.19, the upper regression line passes through the mean of any horizontal strip of data points (on average). We are therefore concerned with the X value of the intersection of the horizontal strip and the upper regression line. The slope of the upper regression line is 2, so the intersection point (x_{avg}, y_1) satisfies $y_1/x_{avg} = 2$. Since $y_1 = 30$, this gives $x_{avg} = 15$. Therefore, 15 (or really 115) is the average innate ability X of the 70 people who scored 130 on the first test.

We now claim that when these 70 people take a second test, their average score will simply be their average innate ability, 115. This is true because if we take the average of the $Y = X + Z$ relation over the 70 people in the group, the Z values average out to zero (or rather, the expectation value is zero). So we are left with $y_{avg} = x_{avg}$. (We should probably be using the notation $y_{2,avg}$, to make it clear that we're talking about the second test, but we won't bother writing the 2.)

In the more general case where $Y = mX + Z$, taking the average of this relation yields

$$y_{avg} = mx_{avg}. \tag{6.39}$$

But mx_{avg} is the height of the point on the lower regression line with an X value of x_{avg}. To obtain this result graphically, just draw a vertical line through (x_{avg}, y_1) and look at the intersection of this line with the lower regression line, indicated by the white dot in Fig. 6.20. The Y value of this dot is the desired average score on the second test. (Having determined the *average* second score of the 70 people, we might also want to determine the *distribution* of their scores. This is the task of Problem 6.6.)

The answer to the question posed at the beginning of this section is therefore "lower than 130." Additionally, given the various parameters we arbitrarily chose, we can be quantitative about how much lower the new average score of 115 is. Additively, it is 15 points lower. Multiplicatively, it is $1/2$ as high above the mean, 100, as the original common score, 130, was. Note that since Eq. (6.38) gave $r = 1/\sqrt{2}$ in our setup, we have $r^2 = 1/2$. The agreement of these factors of $1/2$ is no coincidence, as we will show below.

General discussion

Looking at Fig. 6.20, it is clear why the 70 people in the shaded horizontal strip have an average on the second test that is lower than 130. The upper regression line lies to the left of the lower regression line (in the upper righthand quadrant), so the intersection of the horizontal strip with the upper regression line lies to the left of its intersection with the lower regression line. Translation: the average innate ability X of the 70 people (which is given by the intersection of the horizontal strip with the upper line) is smaller than the innate ability that would correspond to a score of 130 if there were no random Z effect (which is given by the intersection of the horizontal strip with the lower line).

In fact, in Fig. 6.20 it happens to be the case that *all* 70 points in the strip lie to the left of, and hence *above*, the lower regression line. That is, they all involve *positive* contributions from Z. Of course, if we were to generate the plot again, there might be some points in the strip that lie to the right of the lower line (with negative contributions from Z). Or if we had 50,000 points instead of 5000, we would undoubtedly have some such points. But for any (large) total number of points, there will be more of them in the shaded strip that have positive Z values than negative Z values.

The preceding observation provides an intuitive way of understanding why the average on the second test is lower than 130. Since $Y = X + Z$, there are two basic possibilities that lead to a score of $Y = 130$ on the first test: A person can have an innate ability X that is *less* than 130 and get *lucky* with a positive value of Z, or they can have an innate ability that is *greater* than 130 and get *unlucky* with a negative value of Z. The first of these possibilities is more likely, on average, because 130 is greater than the mean of 100, which implies that there are more people with an innate ability of $130 - a$ than $130 + a$ (for any positive a), as shown in Fig. 6.21 for $a = 10$. So more of the 130 scorers have an innate ability that is less than 130, than greater than 130, consistent with what we observed in Fig. 6.20. In the end, therefore, the decrease in average score on the second test comes down to the obvious fact that a Gaussian has its peak in the middle and falls off on either side.

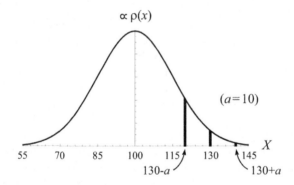

Figure 6.21: There are more people with an innate ability of $130 - a$ than $130 + a$. A score of 130 is therefore more likely to come from a person with a lower innate ability who got lucky, than from a person with a higher innate ability who got unlucky.

Everything we've been saying is relevant to a score that lies above the mean of 100. If we start with a score that is smaller than 100, say $Y = 70$, then all of the above conclusions are reversed. (Note that the number 70 here has nothing to do with the 70 people in the $Y = 130$ shaded strip in Fig. 6.20!) The average on the second test will now *increase* toward the mean. It will be 85 on average, by all the same reasoning. All of the action in Fig. 6.20 will now be in the lower-left quadrant instead of the upper-right quadrant. In this new scenario, there are more people in the $Y = 70$ group who have an innate ability X that is *greater* than 70 but who got *unlucky*, than the other way around. The general conclusion is that for any score on the first test, the average score on the second test will be closer to the mean. This effect is called *regression toward the mean*.

Let's now prove a handy little theorem, which we alluded to above. This theorem allows us to be quantitative about the degree to which averages regress toward the mean.

Theorem 6.1 *Consider the group of people who score y_1 points above the mean on a test. If they take the test (or an equivalent one) again, then their average score on the second test will be $y_{avg} = r^2 y_1$ points above the mean (on average), where r is the correlation coefficient between the actual score and innate ability.*

This theorem is consistent with the above example, because the correlation coefficient was $r = 1/\sqrt{2}$, and the initial score of $y_1 = 30$ points above the mean was reduced (on average) on the second test to $y_{avg} = r^2 \cdot 30 = 15$ points above the mean.

Proof: The proof is quick. We simply need to reapply the reasoning associated with Fig. 6.20, but now with general parameters instead of given numbers. If the shaded strip in Fig. 6.20 has a height y_1, then from Fig. 6.15 the intersection of the strip with the upper regression line has an X value of $x_{avg} = (r\sigma_x/\sigma_y)y_1$. This is the average X value of the people who score y_1 on the first test. On the second test, the Z values will average out to zero, so the lower regression line gives the desired average second-test score of the "y_1 group" of people, via Eq. (6.39). With $m = r\sigma_y/\sigma_x$ from Fig. 6.15 (we'll work with a general m, instead of $m = 1$), Eq. (6.39) gives

$$y_{avg} = mx_{avg} = \left(\frac{r\sigma_y}{\sigma_x}\right) x_{avg} = \left(\frac{r\sigma_y}{\sigma_x}\right)\left(\frac{r\sigma_x}{\sigma_y}\right) y_1 \implies \boxed{y_{avg} = r^2 y_1} \quad \blacksquare \quad (6.40)$$

We stated the theorem in terms of a test-retaking setup with $Y = X + Z$, where m equals 1. But as we just showed, the theorem holds more generally with $Y = mX + Z$, where m is arbitrary. In such cases, the theorem can be stated (in a less catchy manner) as, "Given a large set of data points, consider all of the points whose Y value is y_1. Let the average X value of these points be x_{avg}. Then the average Y value associated with x_{avg} is $r^2 y_1$." Or more succinctly, "The average Y value associated with the average X value of the points with $Y = y_1$ equals $r^2 y_1$."

The theorem checks in two extremes. If $r = 1$, then all scores lie on the $Y = X$ line, or more generally the $Y = mX$ line. The random Z value is always zero, so all scores are exactly equal to the innate ability, or more generally exactly equal to mX. A given person always scores the same every time they take the test. Everyone who scored a 130 on the first test will therefore score a 130 on the second test (and all future tests). So $y_{avg} = (1)^2 y_1$, consistent with Eq. (6.40). In the other extreme where $r = 0$, Eq. (6.6) tells us that either $m = 0$ or $\sigma_x = 0$. So $\sigma_y = \sigma_z$. Basically, everyone's score is completely determined by the random contribution Z, which means that the scores of any given group of people on the second test will be random and will therefore average out to zero. So $y_{avg} = (0)^2 y_1$, again consistent with Eq. (6.40).

The above theorem provides a nice way of determining the correlation coefficient r between the actual score and innate ability, without doing any heavy calculations. Just take a group of people who score y_1 points above the mean on the test, and then have them take the test (or an equivalent one) again. If their new average is y_{avg}, then r is given by

$$y_{avg} = r^2 y_1 \implies r = \sqrt{\frac{y_{avg}}{y_1}}. \tag{6.41}$$

It's rather interesting that r can be determined by simply giving a second test, without knowing anything about σ_x, σ_y, σ_z, or m!

The new property of r in Eq. (6.41) is one of many properties/interpretations of r that we've encountered in this chapter. Let's collect them all together here.

1. Eq. (6.6) tells us that r is (by definition) the fraction of σ_y that can be attributed to X.

2. Eq. (6.18) tells us that the slope m of the lower regression line is $m = r\sigma_y/\sigma_x$. This means that r is the ratio of the slope of the regression line to the slope of the standard-deviation line. This interpretation of r is evident in Figs. 6.15 and 6.16.

   The preceding fact can be restated as: If we consider an X value that is n times σ_x above the mean, then the expected associated Y value is rn times σ_y above the mean.

3. Eq. (6.27) tells us that $1 - r^2$ is the factor by which the "badness" of a prediction of Y is reduced if we take into account the particular value of X. This is the same $1-r^2$ term that appears in the $\sigma_y \sqrt{1 - r^2}$ $(= \sigma_z)$ length in Fig. 6.17.

4. Eq. (6.40) tells us that if we consider the people who score y_1 points above the mean on a test, their average score on a second equivalent test will be $y_{avg} = r^2 y_1$ points above the mean (on average).

6.8.2 Example 2: Comparing IQ's

Consider the following setup. A particular school has an equal number of girls and boys. On a given day, the students form self-selecting girl/boy pairs. Assume that there is a nonzero correlation between the IQ scores within each pair. That is, students with a high (or low) IQ tend to pair up with other students with a high (or low) IQ, on average.[4] This is plausible, because students who are friends with each other (and thus apt to pick each other as partners) might have similar priorities and study habits (or lack thereof).

The question we will pose here is the following. Consider all of the girls who have a particular IQ score, say 130. Will their boy partners have (on average) an IQ score that is higher than, lower than, or equal to 130?

---

[4]By "IQ" or "IQ score" here, we mean a student's innate ability, or equivalently their average score on a large number of IQ tests. In this example, we aren't concerned with the random fluctuations on each test, as we were in the above test-retaking example.

To answer this, let's pick some parameters and make a scatter plot of the IQ scores. We'll assume that both girls and boys have an average IQ of 100 and that the standard deviation for both is 15. And as usual, we'll assume that the underlying distributions are Gaussian.[5] In order to numerically generate a scatter plot, we'll need to pick a value of the correlation coefficient r. Let's pick 0.6. The qualitative answer to the above question won't depend on the exact value. The resulting scatter plot of 5000 points (it's a big school!) is shown in Fig. 6.22. Each point is associated with a girl/boy pair. The x coordinate is the boy's IQ, and the y coordinate is the girl's IQ (relative to the average of 100).

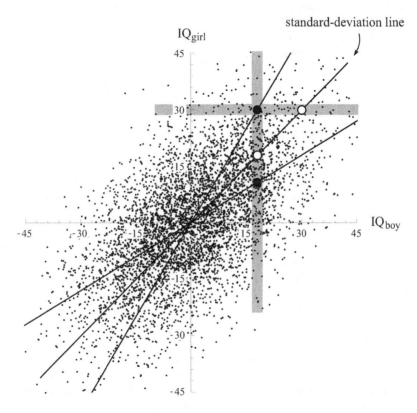

Figure 6.22: IQ scores of girl/boy pairs.

The analysis in the present setup is essentially the same as in the above test-retaking example. The horizontal shaded strip in Fig. 6.22 indicates all pairs in which the girl has an IQ within a small range around 130. (The vertical shaded strip isn't relevant yet; we'll use it below.) To determine the average IQ score (that is, the average x coordinate) of the boys in this group, we simply need to look at the intersection (indicated by the upper large solid dot) of the horizontal shaded strip with the upper regression line. As we know well by now, this is exactly what the

---

[5]We should emphasize that in real life, despite the central limit theorem (see Section 5.5), things are often not as clean as we might lead you to believe by always picking nice Gaussian distributions. But the qualitative results we obtain will generally still hold for messier distributions.

upper regression line is good for. This line passes through (on average) the middle of the distribution of points in any horizontal strip.

From Fig. 6.15, we know that if we tilt our head sideways, the slope of the upper regression line is r times the slope of the standard-deviation line, which is 1 here, because we are assuming $\sigma_x = \sigma_y$ (= 15). So the (tilted-head) slope of the upper regression line is 0.6. Its intersection point (the upper large solid dot) with the horizontal strip at $y = 30$ therefore has an x value of $(0.6)(30) = 18$. Geometrically, in the horizontal strip, the solid dot is r as far to the right as the hollow dot. So 18 (or rather 118) is the desired average IQ of boys who are in pairs where the girl has an IQ of 130. The answer to the question posed above is therefore "lower than 130."

What do we conclude from this? That girls are smarter than boys? Or that girls actively pick boys with a lower IQ? No, neither of these conclusions logically follow from our result. One way of seeing why is to note that we can apply all of the above reasoning to pairs where the girl has an IQ that is *lower* than the overall mean of 100. Let's say 70 instead of 130. The relevant action will then take place in the lower-left quadrant, and we will find that the average IQ of boys in this group is *higher* than 70. It is $100 - 18 = 82$.

The fact of the matter is that there isn't much we can conclude. The lower/higher results that we have found are simply consequences of randomness. There isn't anything deep going on here. The reason why girls with an IQ of 130 are paired with boys with lesser IQs (on average) is the same as the reason why, in the above test-retaking example, people who scored 130 had (on average) an innate ability less than 130. A high number such as 130 is more likely to be the result of a low number (innate ability or boy's IQ) paired with a positive random effect, than a high number paired with a negative random effect. And as we noted in Fig. 6.21, this is due to the simple fact that a Gaussian has its peak in the middle and falls off on either side.

The above "lower than 130" answer when $r = 0.6$ is consistent with the answer in the extreme case where $r = 0$. In this case, if we look at all of the pairs with girls who have an IQ of 130 (or any other number, for that matter), the boys in these pairs will have an average IQ of 100. This is true because if there is no correlation, then knowledge of the girl's IQ is of no help in predicting the boy's IQ; it is completely random. In the other extreme where $r = 1$ (perfect correlation), all of the boys in the "130-girls" pairs will have an IQ of exactly 130, so their average will also be 130. The answer to the question is then "equal to 130." But any degree of non-perfect correlation will change the answer to "lower than 130."

Let's take our setup one step further. We found that girls with an IQ of 130 are paired with boys who have an average IQ of 118. What if we now look at all boys with an IQ of 118? Can we use some sort of symmetry reasoning to say that these boys will be paired with girls whose IQ is 130, on average? No, because when we look at all boys with an IQ of 118 (plus or minus a little), this corresponds to looking at the *vertical* shaded strip in Fig. 6.22. This strip represents a *different set of pairs* from the ones in the horizontal shaded strip, which means that any attempt at a symmetry argument is invalid. We're talking about a different group of pairs, so it's apples vs. oranges.

The average IQ of the girls in the pairs lying in the $x = 18$ (or really 118) vertical shaded strip is given by the intersection (indicated by the lower large solid dot) of the vertical shaded strip with the *lower* regression line. Again, this is exactly what the lower regression line is good for. This line passes through (on average) the middle of the distribution of points in any vertical strip.

From Fig. 6.15, we know that the slope of the lower regression line is r times the slope of the standard-deviation line, which is 1 here. So the slope of the lower regression line is 0.6. Its intersection point (the lower large solid dot) with the vertical strip at $x = 18$ therefore has a y value of $(0.6)(18) = 10.8$. (Note that 10.8 equals $r^2 \cdot 30 = (0.6)^2 \cdot 30$. This is the same factor of r^2 that we found above in the test-retaking example.) Geometrically, in the vertical strip, the lower solid dot has r times the height of the hollow dot. So 10.8 (or rather 110.8) is the desired average IQ of girls who are in pairs where the boy has an IQ of 118. But as above, we can't logically conclude that boys are smarter than girls or that boys actively pick girls with a lower IQ. The smaller average is simply a consequence of the partially random nature of girl/boy pairings.

As mentioned above, the calculations in this example are essentially the same as in the above test-retaking example. This is evidenced by the fact that Fig. 6.22 has exactly the same structure as Fig. 6.20, although we didn't draw the standard-deviation line (with a slope of $\sqrt{2}$) in Fig. 6.20.

6.9   Least-squares fitting

In Section 6.4 we saw that the lower regression line yields the best prediction for Y, given X. We'll now present a different (but very much related) interpretation of the lower regression line.

Assume that we are given a collection of n points (x_i, y_i) in the plane, for example, the 20 points we encountered in Fig. 6.10. How do we determine the "best-fit" line that passes through the points? That is, how do we pick the line that best describes the collection of points? Well, the first thing we need to do is define what we mean by "best." Depending on what definition we use, we might end up with any of a variety of lines, for example, any of the four lines in Fig. 6.15.

We'll go with the following definition: The *best-fit line* is the line that *minimizes the sum of the squares* of the vertical distances from the given points to the line. For example, in Fig. 6.23 the best-fit line of the given 10 points is the line that minimizes the squares of the 10 vertical distances shown. Other definitions of the best-fit line are possible, but this one has many nice properties. The seemingly simpler definition involving just the sum of the distances (not squared) has drawbacks; see the remark in the solution to Problem 6.11.

How do we mathematically determine this "least-squares" line, given a set of n points (x_i, y_i) in the plane? If the line takes the form of $y = Ax + B$, then the vertical distances are $|y_i - (Ax_i + B)|$. So our goal is to determine the parameters A and B that minimize the sum,

$$S \equiv \sum_1^n [y_i - (Ax_i + B)]^2. \tag{6.42}$$

This minimization task involves some straightforward but tedious partial differentiation. If you don't know calculus yet, you can just skip to Eq. (6.46) below; the

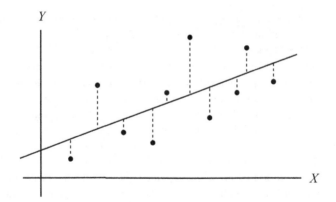

Figure 6.23: The best-fit line is the line that minimizes the sum of the squares of the vertical distances from the given points to the line.

math here isn't terribly important. But the end result in Eq. (6.47) *is* very important. And it is also rather pleasing; we will find that the "least-squares" line is none other than the regression line! More precisely, it is the lower regression line, with slope $m = r\sigma_y/\sigma_x$. See Fig. 6.15.

Let's now do the math. If we expand the square in Eq. (6.42), we can write S as

$$S = \sum y_i^2 - 2 \sum y_i(Ax_i + B) + \sum (Ax_i + B)^2$$
$$= \sum y_i^2 - 2A \sum x_i y_i - 2B \sum y_i + A^2 \sum x_i^2 + 2AB \sum x_i + nB^2. \quad (6.43)$$

All of the sums go from 1 to n. Since the (x_i, y_i) points are given, these sums are all known quantities. S is therefore a function of only A and B. To minimize this function, we must set the partial derivatives of S with respect to A and B equal to zero. This yields the following two equations:

$$\frac{\partial S}{\partial A} = 0 \implies 0 = -\sum x_i y_i + A \sum x_i^2 + B \sum x_i,$$
$$\frac{\partial S}{\partial B} = 0 \implies 0 = -\sum y_i + A \sum x_i + nB. \quad (6.44)$$

Solving for B in the second equation and plugging the result into the first gives

$$0 = -\sum x_i y_i + A \sum x_i^2 + \left(\frac{\sum y_i - A \sum x_i}{n} \right) \sum x_i. \quad (6.45)$$

Solving for A (which is the slope of the $y = Ax + B$ line) gives

$$A = \frac{n \sum x_i y_i - \sum x_i \sum y_i}{n \sum x_i^2 - (\sum x_i)^2}. \quad (6.46)$$

We can make this expression look a little nicer by multiplying both the numerator and denominator by $1/n^2$. We'll use brackets $\langle \; \rangle$ here to denote an average, so $(\sum x_i)/n \equiv \langle x \rangle$, $(\sum x_i y_i)/n \equiv \langle xy \rangle$, etc. (We're changing notation from our usual $\overline{x}$,

$\overline{xy}$, etc. notation for an average, because that would make the following expression for A look rather messy and confusing.) The factors of n all work out nicely, and we obtain the clean result,

$$A = \frac{\langle xy \rangle - \langle x \rangle \langle y \rangle}{\langle x^2 \rangle - \langle x \rangle^2} = \frac{\text{Cov}(x,y)}{\tilde{s}_x^2} = \frac{r \tilde{s}_y}{\tilde{s}_x} \qquad (6.47)$$

The second expression here follows from the results of Problem 6.1, and the third expression follows from Eq. (6.12), which yields $\text{Cov}(x,y) = r\tilde{s}_x \tilde{s}_y$. Our result for A is about as simple as we could have hoped for, given the messiness of the original expression for S in Eq. (6.43).

The slope A in Eq. (6.47) takes the same form as the slope m in Eq. (6.13), with the distribution covariance $\text{Cov}(X,Y)$ replaced with the data-point covariance $\text{Cov}(x,y)$, and with the distribution standard deviations σ_x and σ_y replaced with the data-point standard deviations $\tilde{s}_x$ and $\tilde{s}_y$. The difference between these two results is that the slope m in Eq. (6.13) was based on given distributions for X and Y (equivalently, it was based on an infinite number of data points), whereas the slope A in Eq. (6.47) is based on a finite number of data points. But when the number of points is large, the distributions of x and y values mimic the underlying X and Y distributions, so $\text{Cov}(x,y)$ approaches $\text{Cov}(X,Y)$, and the $\tilde{s}$'s approach the σ's. The slope A of the least-squares line therefore approaches the slope m of the regression line for the complete distribution, as we promised above.

$$\boxed{A \to m} \qquad \text{(large number of data points)} \qquad (6.48)$$

This is a splendid result, although not so surprising. Both the least-squares line in the present section and the (lower) regression line in Section 6.4 minimize the sum of the squared vertical distances from the line. The latter is true because (as we saw in Section 6.4) it is true for any vertical strip.

To solve for B (the y-intercept of the $y = Ax + B$ line), we can plug the A from Eq. (6.46) into either of the equations in Eq. (6.44). The second one makes things a little easier. After simplifying and rearranging the factors of n as we did when producing the A in Eq. (6.47), we obtain

$$B = \frac{\langle y \rangle \langle x^2 \rangle - \langle x \rangle \langle xy \rangle}{\langle x^2 \rangle - \langle x \rangle^2} = \langle y \rangle - A \langle x \rangle \qquad (6.49)$$

The second expression here is derived in Problem 6.8. Note that B is zero if the averages $\langle x \rangle$ and $\langle y \rangle$ are zero. Because of this, we usually aren't too concerned with B, since we can always arrange for it to be zero by measuring x_i and y_i relative to their means (as we have generally been doing with the X and Y distributions throughout this chapter). In this case, the best-fit line passes through the origin, and the only parameter needed to describe the line is the slope A. This line is the same as the lower regression line with slope m (assuming a large number of data points).

In the above derivations of A and B, we treated y as being dependent on x. But what if we're just given a blob of points in the plane, with x and y treated on equal footing? There is then no reason why vertical distances should be given preference over horizontal distances. It is therefore just as reasonable to define the best-fit line as the line that minimizes the sum of the squares of the *horizontal* distances from the given points to the line. Due to all the symmetry we've seen earlier in this chapter, it's a good bet that this line will turn out to be the *upper* regression line. And indeed, if we describe the best-fit line by the equation $x = Cy + D$, then the sum of the squares of the horizontal distances is

$$S \equiv \sum_1^n [x_i - (Cy_i + D)]^2. \tag{6.50}$$

To find the value of C that minimizes this sum, there is no need to go through all of the above mathematical steps again, because all we've done is modify Eq. (6.42) by interchanging x and y, replacing A with C, and replacing B with D. C is therefore obtained by simply letting $x \leftrightarrow y$ in Eq. (6.47), which gives (using the fact that $\mathrm{Cov}(y,x) = \mathrm{Cov}(x,y)$)

$$C = \frac{\mathrm{Cov}(x,y)}{\tilde{s}_y^2} = \frac{r\tilde{s}_x}{\tilde{s}_y}. \tag{6.51}$$

D is found similarly by switching x and y in Eq. (6.49). But if we assume that the means $\langle x \rangle$ and $\langle y \rangle$ are zero, then D equals zero, just as B does. The $x = Cy + D$ line therefore takes the form of $x = (r\tilde{s}_x/\tilde{s}_y)y$, which has the same form as the upper regression line in Figure 6.15, namely $X = (r\sigma_x/\sigma_y)Y$.

So which of the two least-squares lines is the actual best-fit line? Is it the one involving vertical distances, or the one involving horizontal distances? Well, if you're considering y to be dependent on x, then the lower regression line is the best-fit line. It minimizes the variance of the y_i values measured relative to the $Ax_i + B$ values on the line. Conversely, if you're considering x to be dependent on y, then the upper regression line is the best-fit line. It minimizes the variance of the x_i values measured relative to the $Cy_i + D$ values on the line. Note that if you're given an elliptical blob of points in the x-y plane, you might subconsciously think that the best-fit line that serves both of the preceding purposes is the symmetry axis of the ellipse (the short-dashed line in Figure 6.15). But this line in fact serves neither of the purposes.

REMARK: Continuing the discussion following Eq. (6.49), let's talk a little more about measuring x_i and y_i relative to their means. Since the second expression for B in Eq. (6.49) tells us that the $y = Ax + B$ line takes the form of $y = Ax + (\langle y \rangle - A\langle x \rangle)$, it immediately follows that the point $(\langle x \rangle, \langle y \rangle)$ satisfies this equation. That is, the point $(\langle x \rangle, \langle y \rangle)$ lies on the line. (This is no surprise, and you might have just assumed it was true anyway.) Therefore, if we shift the origin to the point $(\langle x \rangle, \langle y \rangle)$, then the least-squares line is the line passing through the origin with a slope given by the A in Eq. (6.47).

Note that measuring x_i and y_i relative to their means doesn't affect A, because A is independent of the averages $\langle x \rangle$ and $\langle y \rangle$. This is intuitively clear; shifting the blob of points and the best-fit line around in the plane doesn't affect the distances to the line, so it doesn't affect our derivation of A. Mathematically, this independence follows from the fact that both $\mathrm{Cov}(x,y)$ and $\tilde{s}_x^2$ in Eq. (6.47) are independent of $\langle x \rangle$ and $\langle y \rangle$. This is true because the expressions in Eq. (6.11) and Eq. (3.60) involve only the differences between x_i values and their mean $\overline{x}$ (likewise for y). And shifting all of the x_i values by a fixed amount changes $\overline{x}$ by this same amount, so the differences $x_i - \overline{x}$ are unaffected. ♣

6.10 Summary

- Let Y be given by $Y = mX + Z$, where X and Z are independent variables. Then the *correlation coefficient r* between X and Y is defined as the fraction of the standard deviation of Y that comes from X. It is given by

$$r \equiv \frac{m\sigma_x}{\sigma_y} = \frac{m\sigma_x}{\sqrt{m^2\sigma_x^2 + \sigma_z^2}}. \tag{6.52}$$

It can also be written as

$$r \equiv \frac{\mathrm{Cov}(X,Y)}{\sigma_x \sigma_y}, \tag{6.53}$$

where the *covariance* is defined as

$$\mathrm{Cov}(X,Y) \equiv E[(X - \mu_x)(Y - \mu_y)]. \tag{6.54}$$

- If you are instead just given a collection of data points in the x-y plane, without knowing the underlying distributions, then Eq. (6.53) turns into

$$r = \frac{\mathrm{Cov}(x,y)}{\tilde{s}_x \tilde{s}_y} = \frac{\sum(x_i - \overline{x})(y_i - \overline{y})}{\sqrt{\sum(x_i - \overline{x})^2}\sqrt{\sum(y_i - \overline{y})^2}}. \tag{6.55}$$

- The higher the correlation, the greater the degree to which knowledge of X helps predict Y. A measure of how much better the prediction is (compared with the naive guess of the mean of Y) is

$$\frac{\sigma_z^2}{\sigma_y^2} = 1 - r^2. \tag{6.56}$$

- Given r, along with σ_x and σ_y, the probability density in the x-y plane is

$$\rho(x,y) = \frac{1}{2\pi\sigma_x\sigma_y\sqrt{1 - r^2}} \exp\left(-\frac{1}{2(1-r^2)}\left(\frac{x^2}{\sigma_x^2} + \frac{y^2}{\sigma_y^2} - \frac{2rxy}{\sigma_x\sigma_y}\right)\right). \tag{6.57}$$

This density is symmetric in x and y (and σ_x and σ_y).

- There are two regression lines. If Y is considered to be dependent on X, then the lower regression line is relevant. This line gives the average value of Y for any given X. Its slope is $m = r\sigma_y/\sigma_x$. If instead X is considered to be dependent on Y, then the upper regression line is relevant. This line gives the average value of X for any given Y. Its slope is $\sigma_y/r\sigma_x$.

- If a group of people score y_1 points above the mean on a test, and if they take the test (or an equivalent one) again, then their average score on the second test will be $y_{\mathrm{avg}} = r^2 y_1$ points above the mean (for a large number of data points), where r is the correlation coefficient between the actual score and innate ability. Since $r \leq 1$, the average new score is closer to the mean than the old score was (except in the $r = 1$ case of perfect correlation, where it is the same). This effect is known as *regression toward the mean*.

- Given a set of data points in the x-y plane, the best-fit line is customarily defined as the *least-squares* line. This line has slope

$$A = \frac{\text{Cov}(x, y)}{\tilde{s}_x^2} = \frac{r\tilde{s}_y}{\tilde{s}_x}, \qquad (6.58)$$

which takes the same form as the slope m given in Eq. (6.13) for the lower regression line.

6.11 Exercises

See **www.people.fas.harvard.edu/~djmorin/book.html** for a supply of problems without included solutions.

6.12 Problems

Section 6.3: The correlation coefficient r

6.1. **Alternative forms of Cov(x,y) and $\tilde{s}$** *

   (a) Show that the $\text{Cov}(x, y)$ defined in Eq. (6.11) can be written as $\langle xy \rangle - \langle x \rangle \langle y \rangle$. ($\langle x \rangle$ means the same thing as $\overline{x}$.)

   (b) Show that the $\tilde{s}^2$ defined in Eq. (3.60) can be written as $\langle x^2 \rangle - \langle x \rangle^2$.

6.2. **Rescaling X** **

   Using Eq. (6.9), we showed in the third remark on page 287 that the correlation coefficient r doesn't change with a uniform scaling of X or Y. Demonstrate this again here by using the expression for r in Eq. (6.6).

6.3. **Uncorrelated vs. independent** **

   If two random variables X and Y are independent, are they necessarily also uncorrelated? If they are uncorrelated, are they necessarily also independent?

Section 6.5: Calculating $\rho(x, y)$

6.4. **Sum of two Gaussians** *** *(calculus)*

   Given two independent Gaussian distributions X and Y with standard deviations σ_x and σ_y, show that the sum $Z \equiv X + Y$ is a Gaussian distribution with standard deviation $\sqrt{\sigma_x^2 + \sigma_y^2}$. You may assume without loss of generality that the means are zero.

6.5. **Maximum $\rho(x, y)$** * *(calculus)*

   For a given y_0, what value of x maximizes the probability density $\rho(x, y_0)$ in Eq. (6.34)?

Section 6.8: Two regression examples

6.6. **Distribution on a second test** ∗∗

Consider the 70 people who scored (roughly) 130 on the IQ test in the example in Section 6.8.1. If these people take a second test, describe the distribution of the results. You can do this by finding the σ_x, σ_y, σ_z, m, r, μ_x, and μ_y values associated with the scatter plot of the (X, Y) values.

6.7. **One standard deviation above the mean** ∗∗

Assume that for a particular test, the correlation coefficient between the score Y and innate ability X is r. Consider a person with an X value that is one standard deviation σ_x above the mean. What is the probability that this person scores at least one standard deviation σ_y above the mean? Assume that all distributions are Gaussian. (To give a numerical answer to this problem, you would need to be given r. And you would need to use a table or a computer. It suffices here to state the value of the standard deviation multiple that you would plug into the table or computer.)

Section 6.9: Least-squares fitting

6.8. **Alternate form of B** ∗

Show that the second expression for B in Eq. (6.49) equals the first.

6.9. **Finding all the quantities** ∗∗

Given five (X, Y) points with values $(2, 1)$, $(3, 1)$, $(3, 3)$, $(5, 4)$, $(7, 6)$, calculate (with a calculator) all of the quantities referred to in the five steps listed on page 290. Also calculate the B in Eq. (6.49), and make a rough plot of the five given points along with the regression (least-squares) line.

6.10. **Equal distances** ∗∗ *(calculus)*

In Section 6.9 we defined the best-fit line as the line that minimizes the sum of the squares of the vertical distances from the given points to the line. Let's kick things down a dimension and look at the 1-D case where we have n values x_i lying on the x axis. We'll define the "best-fit" point as the value of x (call it x_b) that minimizes the sum of the squares of the distances from the n given x_i points to the x_b point.

(a) Show that x_b is the mean of the x_i values.

(b) Show that the sum of all the distances from x_b to the points with $x_i > x_b$ equals the sum of all the distances from x_b to the points with $x_i < x_b$.

6.11. **Equal distances again** ∗∗ *(calculus)*

Returning to 2-D, show that the sum of all the vertical distances from the least-squares line to the points above it equals the sum of all the vertical distances from the line to the points below it. *Hint:* Consider an appropriate partial derivative of the sum S in Eq. (6.42).

6.13 Solutions

6.1. **Alternative forms of Cov(x,y) and $\tilde{s}$**

   (a) Starting with the definition in Eq. (6.11), we have

$$\text{Cov}(x,y) = \frac{1}{n} \sum (x_i - \langle x \rangle)(y_i - \langle y \rangle)$$

$$= \frac{1}{n}\left(\sum x_i y_i - \sum x_i \langle y \rangle - \sum y_i \langle x \rangle + n\langle x \rangle \langle y \rangle \right)$$

$$= \frac{\sum x_i y_i}{n} - \frac{\sum x_i}{n}\langle y \rangle - \frac{\sum y_i}{n}\langle x \rangle + \langle x \rangle \langle y \rangle$$

$$= \langle xy \rangle - \langle x \rangle \langle y \rangle - \langle y \rangle \langle x \rangle + \langle x \rangle \langle y \rangle$$

$$= \langle xy \rangle - \langle x \rangle \langle y \rangle. \tag{6.59}$$

as desired. In the limit of a very large number of data points, the above averages reduce to the expectation values for the underlying distributions. That is, $\langle xy \rangle \to E(XY)$, $\langle x \rangle \to E(X) \equiv \mu_x$, and $\langle y \rangle \to E(Y) \equiv \mu_y$. The above result therefore reduces to Eq. (6.14).

   (b) Starting with the definition in Eq. (3.60), we have

$$\tilde{s}^2 = \frac{1}{n} \sum (x_i - \langle x \rangle)^2$$

$$= \frac{1}{n}\left(\sum x_i^2 - 2 \sum x_i \langle x \rangle + n\langle x \rangle^2 \right)$$

$$= \frac{\sum x_i^2}{n} - 2\frac{\sum x_i}{n}\langle x \rangle + \langle x \rangle^2$$

$$= \langle x^2 \rangle - 2\langle x \rangle^2 + \langle x \rangle^2$$

$$= \langle x^2 \rangle - \langle x \rangle^2, \tag{6.60}$$

as desired. As in part (a), in the limit of a very large number of data points, the above averages reduce to the expectation values for the underlying distributions. The above result therefore reduces to $\sigma_x^2 = E(X^2) - \mu_x^2$, which is equivalent to Eq. (3.50). Eq. (6.60) is a special case of Eq. (6.59), when $x = y$. More precisely, when each y_i equals the corresponding x_i, the covariance reduces to the variance.

6.2. **Rescaling X**

   If we let $X' \equiv aX$ and $Y' \equiv bY$, what form does the $Y = mX + Z$ relation in Eq. (6.2) take when written in terms of X' and Y'? We need to generate some X' and Y' (that is, some aX and bY) terms, so let's multiply $Y = mX + Z$ through by b, and let's also multiply the mX term by 1 in the form of a/a. This gives

$$bY = b\frac{m}{a}aX + bZ \implies Y' = \frac{bm}{a}X' + bZ \implies Y' = m'X' + bZ, \tag{6.61}$$

where $m' \equiv bm/a$. Note that Eq. (3.41) tells us that $\sigma_{x'} = a\sigma_x$ and $\sigma_{y'} = b\sigma_y$. Using the expression for r in Eq. (6.6), the correlation coefficient r' between X' and Y' is then

$$r' = \frac{m'\sigma_{x'}}{\sigma_{y'}} = \frac{(bm/a)(a\sigma_x)}{b\sigma_y} = \frac{m\sigma_x}{\sigma_y} = r, \tag{6.62}$$

as desired. Fig. 6.24 shows a scenario with $a = 2$ and $b = 1$. In the first plot, we have chosen $\sigma_x = 1$, $\sigma_y = 1$, with $r = 0.8$. So the second plot has $\sigma_x = 2$, $\sigma_y = 1$, with r again equaling 0.8.

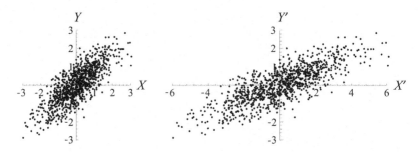

Figure 6.24: The second plot is obtained by stretching the first plot by a factor of 2 in the X direction. The correlation coefficients for the two plots are the same.

6.3. Uncorrelated vs. independent

Assume that the two variables are independent. Then we know from Eq. (3.16) that the expectation value of the product equals the product of the expectation values. The covariance in Eq. (6.14) (the expression in Eq. (6.7) would work just as well) therefore becomes

$$\text{Cov}(X,Y) = E(XY) - \mu_x\mu_y = E(X)E(Y) - \mu_x\mu_y = 0, \qquad (6.63)$$

because $\mu_x \equiv E(X)$ and $\mu_y \equiv E(Y)$. The correlation coefficient r in Eq. (6.9) is then zero. The answer to the first question posed in this problem is therefore "yes." That is, if two random variables X and Y are independent, then they are necessarily also uncorrelated. In short, the logic comes down to the fact that $P(x,y) = P(x)P(y)$ (which is the condition for independence; see Eq. (3.10)) implies via Theorem 3.2 that $E(XY) = E(X)E(Y)$ (which is the condition for $\text{Cov}(X,Y) = 0$; see Eq. (6.14)).

Now assume that the two variables are uncorrelated. It turns out that they are *not* necessarily independent. That is, $E(XY) = E(X)E(Y)$ does not imply $P(x,y) = P(x)P(y)$. The quickest way to see why this is the case is to generate a counterexample. Let X be a discrete random variable taking on the three values of -1, 0, and 1 with equal probabilities of $1/3$. And let $Y = |X|$. Then the three points in the X-Y plane shown in Fig. 6.25 all occur with equal probabilities of $1/3$.

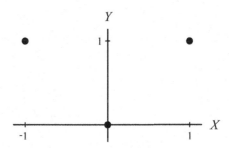

Figure 6.25: If the three points shown have equal probabilities of $1/3$, then X and Y are uncorrelated and dependent.

You can quickly show that $E(XY)$ and $E(X) \equiv \mu_x$ are both equal to zero, which means that the $\text{Cov}(X,Y)$ in Eq. (6.63) is zero. (Consistent with this, we have $E(XY) = E(X)E(Y)$, with the common value being zero.) Therefore $r = 0$. However, the $P(x,y) = P(x)P(y)$ condition for independence is *not* satisfied, because, for example, $P(0,0) = 1/3$ whereas $P(0)P(0) = (1/3)(1/3) = 1/9$. Intuitively, the variables are clearly dependent, because if $X = 0$ then Y is guaranteed to be 0; so Y certainly depends on X. The variables X and Y are therefore (linearly) uncorrelated but dependent.

For a counterexample involving continuous variables, let X be uniformly distributed from -1 to 1, and let $Y = X^2$. Then you can quickly show that $E(XY)$ and $E(X)$ are both equal to zero, which implies that $r = 0$. But X and Y certainly depend on each other.

To sum up: If two random variables are independent, then they are uncorrelated. The contrapositive of this statement is also true: If two random variables are correlated, then they are dependent. However, the converses of the preceding two statements are not valid. That is, if two random variables are uncorrelated, then they are *not* necessarily independent. And if two random variables are dependent, then they are *not* necessarily correlated. These results are summarized in Table 6.2, which indicates which combinations are possible. The only impossible combination is correlated/independent. Remember that throughout this chapter, we are always talking about *linear* correlation.

	Independent	Dependent
Uncorrelated	YES	YES
Correlated	NO	YES

Table 6.2: Relations between (un)correlation and (in)dependence.

6.4. Sum of two Gaussians

There is some overlap between this calculation and the one we did in Section 6.5 when we derived $\rho(x,y)$. We could actually make use of that result to save us some time here, but let's work things out from scratch to get some practice. The solution we'll give here is a standard one involving integration. We'll be a bit pedantic. Many treatments skip the initial material here and effectively just start with Eq. (6.68); see the fourth remark below. If you don't like the following (somewhat involved) solution, we'll present a slick geometric solution in the fifth remark below.

Since X and Y are independent variables, the joint probability of picking an X value that lies in a little span dx around x *and* a Y value that lies in a little span dy around y equals the product of the probabilities, that is, $(\rho_x(x)\,dx)(\rho_y(y)\,dy)$. In other words, the probability $\rho(x,y)\,dx\,dy$ (by the definition of $\rho(x,y)$) of picking X and Y values that lie in a little area $dx\,dy$ around the point (x,y) equals

$$\rho(x,y)\,dx\,dy = \rho_x(x)\rho_y(y)\,dx\,dy. \tag{6.64}$$

Now, a line described by the equation $x + y = C$, where C is a constant, has a slope of -1. Therefore, the shaded strip in Fig. 6.26(a) shows the values of X and Y that yield values of $Z = X + Y$ that lie within a range Δz around a given value of z. (We'll assume that z corresponds to the value at the middle of the strip, although it doesn't matter exactly how z is defined if Δz is small.) The total probability of obtaining a point (x, y) that lies in the shaded strip is found by integrating the above expression for $\rho(x, y)$ over the strip:

$$P(\text{lie in strip}) = \int_{\text{strip}} \rho(x, y)\, dx\, dy = \int_{\text{strip}} \rho_x(x)\rho_y(y)\, dx\, dy. \tag{6.65}$$

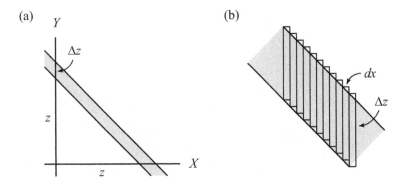

Figure 6.26: (a) The shaded strip indicates the values of X and Y that yield values of $Z = X + Y$ that lie within a range Δz around z. (b) A zoomed-in view of the shaded area, divided into thin rectangles with width dx and height Δz.

Since the probability of obtaining an (x, y) point that lies in the strip is the same as the probability of obtaining a Z value that lies within a range Δz around z, and since the latter probability is $\rho_z(z)\,\Delta z$ by definition (assuming Δz is small), we have

$$P(\text{lie in strip}) = \rho_z(z)\,\Delta z. \tag{6.66}$$

Our goal is therefore to calculate the integral in Eq. (6.65) and then equate it with $\rho_z(z)\,\Delta z$. This will give us the distribution $\rho_z(z)$, which we will find has the desired Gaussian form.

In Fig. 6.26(b) we have divided the shaded strip into thin rectangles, each with width dx and height Δz. We will assume here that dx is much smaller than Δz, so in the $dx \to 0$ limit the thin rectangles exactly cover the strip. Since $Z = X + Y$, the y value in a given rectangle is $y = z - x$. The y value actually varies by Δz within each rectangle, but since Δz is small, we can say that y is essentially equal to $z - x$ over the whole (tiny) rectangle. The integral of $\rho(x, y)$ over each tiny rectangle is therefore equal to the (essentially) uniform value of $\rho_x(x)\rho_y(z - x)$ times the area $dx\,\Delta z$:

$$\int_{\text{rectangle}} \rho(x, y)\, dx\, dy = \rho_x(x)\rho_y(z - x)\, dx\, \Delta z. \tag{6.67}$$

In other words, the integration over y is simply a multiplication by Δz, at least for the way we sliced up the strip into thin vertical rectangles.

We now need to perform the integration over x. That is, we need to integrate the result in Eq. (6.67) over all the little rectangles. This will give us the integral over the entire shaded strip, that is, it will give us $\rho_z(z)\,\Delta z$. Using the explicit Gaussian form of the

ρ's from Eq. (4.42), with the means equal to zero, we obtain

$$\rho_z(z)\,\Delta z = \int_{-\infty}^{\infty} \rho_x(x)\rho_y(z-x)\,dx\,\Delta z$$

$$\implies \rho_z(z) = \frac{1}{\sqrt{2\pi}\sigma_x}\frac{1}{\sqrt{2\pi}\sigma_y}\int_{-\infty}^{\infty} e^{-x^2/2\sigma_x^2}e^{-(z-x)^2/2\sigma_y^2}\,dx. \qquad (6.68)$$

To evaluate this integral, we will complete the square in the exponent. This will require some algebraic manipulation. With $S \equiv \sigma_x^2 + \sigma_y^2$, the exponent equals

$$-\frac{1}{2}\left(\frac{x^2}{\sigma_x^2} + \frac{(z-x)^2}{\sigma_y^2}\right) = -\frac{1}{2\sigma_x^2\sigma_y^2}\left((\sigma_x^2+\sigma_y^2)x^2 - 2\sigma_x^2 zx + \sigma_x^2 z^2\right)$$

$$= -\frac{S}{2\sigma_x^2\sigma_y^2}\left(x^2 - \frac{2\sigma_x^2 z}{S}x + \frac{\sigma_x^2 z^2}{S}\right)$$

$$= -\frac{S}{2\sigma_x^2\sigma_y^2}\left[\left(x - \frac{\sigma_x^2 z}{S}\right)^2 - \left(\frac{\sigma_x^2 z}{S}\right)^2 + \frac{\sigma_x^2 z^2}{S}\right]$$

$$= -\frac{S}{2\sigma_x^2\sigma_y^2}\left(x - \frac{\sigma_x^2 z}{S}\right)^2 - \frac{S}{2\sigma_x^2\sigma_y^2}\left(\frac{-\sigma_x^4 + S\sigma_x^2}{S^2}\right)z^2$$

$$= -\frac{S}{2\sigma_x^2\sigma_y^2}\left(x - \frac{\sigma_x^2 z}{S}\right)^2 - \frac{z^2}{2S}, \qquad (6.69)$$

as you can verify. When we plug this expression for the exponent back into Eq. (6.68), the $z^2/2S$ term is a constant, as far as the x integration is concerned, so we can take it outside the integral. The remaining x integral is a standard Gaussian integral given by Eq. (4.118) in Problem 4.22, with $b \equiv S/(2\sigma_x^2\sigma_y^2)$. (The integral is centered at $\sigma_x^2 z/S$ instead of zero, but that doesn't matter, because the limits are $\pm\infty$.) Eq. (6.68) therefore becomes

$$\rho_z(z) = \frac{1}{\sqrt{2\pi}\sigma_x}\frac{1}{\sqrt{2\pi}\sigma_y}e^{-z^2/2S}\cdot\sqrt{\frac{\pi}{S/(2\sigma_x^2\sigma_y^2)}}$$

$$= \frac{1}{\sqrt{2\pi}\sqrt{\sigma_x^2+\sigma_y^2}}e^{-z^2/2(\sigma_x^2+\sigma_y^2)}. \qquad (6.70)$$

This is a Gaussian distribution with standard deviation $\sqrt{\sigma_x^2 + \sigma_y^2}$, as desired.

REMARKS:

1. If the means μ_x and μ_y aren't zero, we can define new variables $X' \equiv X - \mu_x$ and $Y' \equiv Y - \mu_y$. These have zero means, so by the above reasoning, the sum $Z' = X' + Y'$ is a Gaussian with zero mean. The sum

   $$Z = X + Y = (X' + \mu_x) + (Y' + \mu_y) = Z' + (\mu_x + \mu_y) \qquad (6.71)$$

   is therefore a Gaussian with mean $\mu_x + \mu_y$.

2. Without doing any work, we already knew from Eq. (3.42) that the standard deviation of Z is given by $\sigma_z^2 = \sigma_x^2 + \sigma_y^2$. (Standard deviations add in quadrature, for independent variables.) But it took the above calculation to show that the shape of the Z distribution is actually a Gaussian.

3. The result of this problem also holds for the *difference* of two Gaussians. That is, if X and Y are independent Gaussians with standard deviations σ_x and σ_y, then $Z \equiv X - Y$ is a Gaussian with standard deviation $\sqrt{\sigma_x^2 + \sigma_y^2}$. This follows from writing Z as $X + (-Y)$ and noting that $-Y$ has the same standard deviation as Y, namely σ_y. Note that the standard deviation of $Z \equiv X - Y$ is *not* $\sqrt{\sigma_x^2 - \sigma_y^2}$.

   Consider the special case where Z is the difference between two independent and identically distributed variables X_1 and X_2, each with standard deviation σ_x. Then the preceding paragraph tells us that Z is a Gaussian with standard deviation $\sqrt{2}\sigma_x$. The incorrect $\sqrt{\sigma_x^2 - \sigma_y^2}$ answer mentioned above would yield $\sigma_z = 0$, which certainly can't be correct, because it would mean that Z is guaranteed to take on one specific value.

4. A quicker and less rigorous solution to this problem is to say that if the sum $X + Y$ takes on the particular value z, and if we are given x, then y must equal $z - x$. Integrating over x (to account for all of the different ways to obtain z) yields the second line in Eq. (6.68). So we can basically just start the solution with that equation. However, we chose to include all of the reasoning leading up to Eq. (6.68), because things can get confusing if you don't clearly distinguish between probability densities, such as $\rho_x(x)$ and $\rho(x, y)$, and actual probabilities, such as $\rho_x(x)\,dx$ and $\rho(x, y)\,dx\,dy$. It can also get confusing if you don't distinguish between the different roles of dx and Δz. The former is an infinitesimal integration variable, while the latter is the vertical width of the shaded strip in Fig. 6.26. Although technically the definition of the probability density $\rho_z(z)$ in Eq. (4.2) requires that Δz be infinitesimal, we often think of it as simply being small.

5. There is a slick alternative geometric argument that shows why the sum Z of two independent Gaussian distributions X and Y is again a Gaussian. We'll just sketch the idea here; you can fill in the gaps. Consider first the case where X and Y have the same standard deviation σ. Then

$$\rho(x, y) = \rho_x(x)\rho_y(y) \propto e^{-(x^2+y^2)/2\sigma^2} = e^{-r^2/2\sigma^2}, \qquad (6.72)$$

   where r is the radius in the x-y plane. Since $\rho(x, y)$ depends only on r (and not on the angle θ in the plane), we see that $\rho(x, y)$ has circular symmetry.

   As in our original solution, the values of $\rho_z(z)$ for different values of z are proportional to the integrals of $\rho(x, y)$ over the various thin strips tilted at a $45°$ angle shown in Fig. 6.27(a). We now note that due to the circular symmetry of $\rho(x, y)$, the integrals over the strips are unchanged if we rotate the figure around the origin so that we end up with the vertical strips shown in Fig. 6.27(b). But we know that the integrals over these vertical strips are proportional to the original $\rho_x(x)$ values, because integrating over all the y values in a strip just leaves us with $\rho_x(x)\,dx$, by definition. Therefore, since $\rho_x(x)$ is a Gaussian, $\rho_z(z)$ must be also. To determine σ_z, you can simply invoke Eq. (3.42), or you can use the following reasoning. If the circle shown in Fig. 6.27 has x and y intercepts of $\pm\sigma$, then the rightmost strip in the right figure corresponds to one standard deviation σ_z of the Z distribution, because this strip corresponds to one standard deviation σ_x of the X distribution. But from the left figure, this strip is associated with the z value of $\sqrt{2}\sigma$, because the point $(x, y) = (\sigma/\sqrt{2}, \sigma/\sqrt{2})$ lies in the strip, which means that the corresponding z value is $z = x + y = \sqrt{2}\sigma$. Hence $\sigma_z = \sqrt{2}\sigma$.

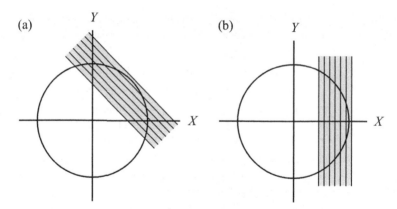

Figure 6.27: (a) Each value of z corresponds to a particular shaded strip. (b) Due to the circular symmetry of $\rho(x,y)$, the integrals of $\rho(x,y)$ over the strips aren't affected by a rotation in the plane. Therefore, since the vertical strips in the right figure yield the Gaussian $\rho_x(x)$ distribution, the diagonal strips associated with $\rho_z(z)$ in the left figure must also yield a Gaussian distribution.

More generally, if X and Y have different standard deviations, then we have elliptical instead of circular symmetry in the plane. But if we stretch/squash one of the axes by the appropriate factor, we obtain circular symmetry, whereupon the above argument holds. It takes a little work to show geometrically that $\sigma_z^2 = \sigma_x^2 + \sigma_y^2$. But again, you can find σ_z by simply invoking Eq. (3.42). ♣

6.5. **Maximum $\rho(x,y)$**

FIRST SOLUTION: Given y, we can maximize $\rho(x,y)$ by taking the partial derivative with respect to x. The exponent in Eq. (6.34) contains all of the dependence on x and y. Taking the partial derivative of the exponent with respect to x, and setting the result equal to zero, gives

$$0 = \frac{2x}{\sigma_x^2} - \frac{2ry}{\sigma_x\sigma_y} \quad \Longrightarrow \quad x = \frac{r\sigma_x}{\sigma_y}y. \tag{6.73}$$

In the case at hand where $y = y_0$, we see that $\rho(x,y)$ is maximized when $x = (r\sigma_x/\sigma_y)y_0$.

SECOND SOLUTION: We claim that the desired value of x is given by the intersection of the horizontal $y = y_0$ line with the upper regression line. Since we know from Fig. 6.15 that the equation for this line is $y = (\sigma_y/r\sigma_x)x$, we obtain

$$y_0 = \frac{\sigma_y}{r\sigma_x}x \quad \Longrightarrow \quad x = \frac{r\sigma_x}{\sigma_y}y_0, \tag{6.74}$$

in agreement with the first solution.

The above claim can be justified as follows. As we saw in Section 6.5, the curves of constant $\rho(x,y)$ are ellipses. Two are shown in Fig. 6.28. The larger the ellipse, the smaller the value of $\rho(x,y)$. The smallest ellipse that contains a point with a y value of y_0 is the inner ellipse shown in the figure. This ellipse is tangent to the horizontal line $y = y_0$. The value of $\rho(x,y)$ at the point B shown is larger than the value at points A and C, because these points lie on a larger ellipse. The point B therefore has the largest value of $\rho(x,y)$ among all points on the horizontal line $y = y_0$; all other points lie on ellipses that are larger than the "*B*" ellipse. Our goal is therefore to find the *x* value of the point *B*. But point *B*, being the highest point on the ellipse on which it lies, is located on the upper regression line, because this line passes through the highest and lowest points of every ellipse. That is how we defined the upper regression line at the beginning of Section 6.7. This proves the above claim.

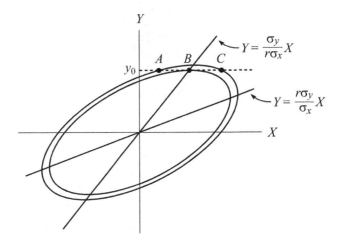

Figure 6.28: The value of $\rho(x, y)$ at *B* is larger than at *A* and *C*, because *B* lies on a smaller ellipse. *B* therefore has the largest $\rho(x, y)$ among all points on the line $y = y_0$.

Since we know from Section 6.7 that the upper regression line gives the average value of *x* associated with a given *y*, we can phrase the result of this problem as: For a given y_0, the value of *x* that maximizes $\rho(x, y_0)$ is the average value of *x* associated with y_0.

6.6. Distribution on a second test

Fig. 6.29 shows what the scores on a second test might look like. We have increased the number of points from 70 to 700, just to smooth out the scatter plot. (You can pretend that there were 50,000 points in Fig. 6.20.) Although the 700 people all scored the same on the first test, they certainly won't all score the same on the second test. However, if these 700 people were to take a third test (or any number of additional tests), their scores would look the same (on average) as they do in Fig. 6.29. We found in Section 6.8.1 that both the average innate ability and the average score of this group of people on the second test is 115. So if we measure *X* and *Y* relative to 100, the blob of points is centered at $(\mu_x, \mu_y) = (15, 15)$.

What standard deviations did we use in numerically generating Fig. 6.29? σ_z still has a value of $15/\sqrt{2} = 10.6$ (from the paragraph preceding Eq. (6.38)). That never changes, since *Z* is independent of *X*. But σ_x^{new} now equals $\sigma_x^{\text{old}} \sqrt{1 - r^2}$, because our 70 (or 700) points all came from a horizontal strip in Fig. 6.20, and the superscript on the *Z* in Eq. (6.36) tells us that $\sigma_x^{\text{old}} \sqrt{1 - r^2}$ is the standard deviation of the spread of points in any horizontal strip. Since we know from Section 6.8.1 that $r = 1/\sqrt{2}$ and $\sigma_x^{\text{old}} = 15/\sqrt{2}$, we have

$$\sigma_x^{\text{new}} = \sigma_x^{\text{old}} \sqrt{1 - r^2} = \frac{15}{\sqrt{2}} \sqrt{1 - \frac{1}{2}} = \frac{15}{2} = 7.5. \tag{6.75}$$

So the values we used in generating Fig. 6.29 were $\sigma_x^{\text{new}} = 7.5$ and $\sigma_z = 10.6$. And

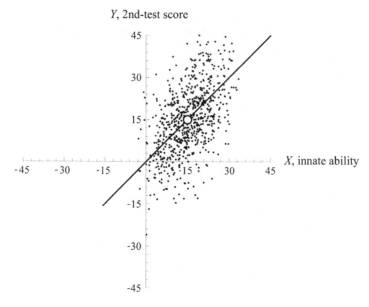

Figure 6.29: The second-test scores of 700 people with the same distribution of innate abilities as the 70 people in the shaded strip in Fig. 6.20.

m still equals 1, because the relation $Y = X + Z$ still holds; the change in the spread of the X values of the people we happen to be looking at doesn't affect this relation. What is the standard deviation of the Y values in Fig. 6.29? From Eq. (6.5) we have

$$\sigma_y^{\text{new}} = \sqrt{m^2(\sigma_x^{\text{new}})^2 + \sigma_z^2} = \sqrt{(1)^2(7.5)^2 + (10.6)^2} = 13. \qquad (6.76)$$

This is smaller that the $\sigma_y = 15$ value in Fig. 6.20, because the smaller spread in the X values affects Eq. (6.5) via σ_x. From Eq. (6.6) the correlation coefficient for Fig. 6.29 is

$$r^{\text{new}} = \frac{m\sigma_x^{\text{new}}}{\sigma_y^{\text{new}}} = \frac{(1)(7.5)}{13} = 0.58. \qquad (6.77)$$

If you work out the numbers exactly, r turns out to be $1/\sqrt{3}$. This is smaller than the $r = 1/\sqrt{2}$ value in Eq. (6.38) for Fig. 6.20, because a smaller fraction of σ_y comes from σ_x (since σ_x is smaller). A larger fraction of σ_y comes from σ_z (which is unchanged). To summarize, in addition to $(\mu_x, \mu_y) = (15, 15)$, the values associated with Fig. 6.29 are

$$\sigma_x^{\text{new}} = 7.5, \qquad \sigma_y^{\text{new}} = 13, \qquad \sigma_z = 10.6, \qquad m = 1, \qquad r^{\text{new}} = 0.58. \qquad (6.78)$$

REMARK: The regression line shown in Fig. 6.29 passes through the origin. Although this isn't the case in general when the blob of points isn't centered at the origin, it is the case here for the following reason. We know that the center of the blob lies on the lower regression line in Fig. 6.20; that's how we found the center, after all. And the regression line in Fig. 6.29 has the same slope (namely $m = 1$) as the lower regression line in Fig. 6.20, because both plots are governed by the same relation, $Y = X + Z$. So the line must pass through the origin in Fig. 6.29. Another way to see why this is true is to recall that the regression line gives the average score for each value of X. And the average score of an $X = 0$ person is still $Y = 0$ (where 0 really means 100 here), because the Z values average out to zero; this doesn't depend on which figure we're looking at. ♣

6.7. One standard deviation above the mean

The expected score of a person with any particular value of X is given by the associated point on the lower regression line. This line takes the form of $Y = mX$, where $m = r\sigma_y/\sigma_x$ from Fig. 6.15. (We'll work with a general m here, even though $m = 1$ in our test-taking setups with $Y = X + Z$.) The expected score (relative to the mean score) of someone with an X value of σ_x (relative to the mean innate ability) is therefore

$$Y = \frac{r\sigma_y}{\sigma_x} \cdot \sigma_x = r\sigma_y. \tag{6.79}$$

This is just the $r\sigma_y$ vertical distance shown in Fig. 6.16. To find the probability that the person achieves a score of *at least* σ_y, note that σ_y exceeds the expected test score of $r\sigma_y$ by

$$\sigma_y - r\sigma_y = \sigma_y(1 - r). \tag{6.80}$$

This is indicated in Fig. 6.30. We have drawn the standard-deviation box for clarity.

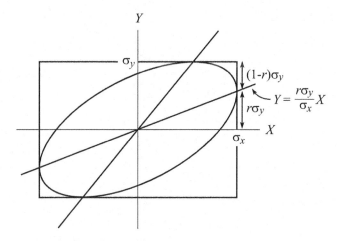

Figure 6.30: The expected Y value associated with an X value of σ_x is $Y = r\sigma_y$, and σ_y exceeds this expected Y value by $(1 - r)\sigma_y$.

Now, since $Y = mX + Z$, the probability distribution of anyone's score is centered on the associated point on the lower regression line and has a standard deviation of σ_z. But $\sigma_z = \sigma_y\sqrt{1 - r^2}$ from Eq. (6.18). So for our given person with $X = \sigma_x$, a score of σ_y exceeds the expected score of $r\sigma_y$ by

$$\frac{\sigma_y(1 - r)}{\sigma_y\sqrt{1 - r^2}} = \sqrt{\frac{1 - r}{1 + r}} \tag{6.81}$$

of the σ_z standard deviations.

To produce a numerical answer to this problem, we must be given a numerical value for r. For example, if $r = 0.5$, then $\sqrt{(1 - r)/(1 + r)} = 0.58$. From a table or computer, it can be shown that the probability of lying outside of 0.58 standard deviations from the mean is 0.56 (assuming a Gaussian distribution). But we must divide by 2 because we're concerned only with the upper tail of the Gaussian. So the desired probability is 0.28. The situation is shown in Fig. 6.31. If $r = 0.5$ then $\sigma_z = \sigma_y \sqrt{1 - (0.5)^2} = (0.87)\sigma_y$. We have arbitrarily chosen $\sigma_y = \sigma_x$ in the figure (and we have set them both equal to 1; the standard-deviation box is shown), but this doesn't affect our results. If $r = 0.5$, then $\sigma_z = (0.87)\sigma_y$, no matter how σ_x and σ_y are related. The $\sigma_z = 0.87$ standard deviation is indicated by the heavy arrows, centered on the expected value given by the lower regression line. A visual inspection of the figure is consistent with the fact that 28% of the dots in the vertical shaded strip are expected to lie above the white dot with height $Y = \sigma_y$. In the present example with $r = 0.5$, both of the $r\sigma_y$ and $(1 - r)\sigma_y$ vertical distances in Fig. 6.30 are equal to $(0.5)\sigma_y$. The upper of these (identical) distances is therefore $(0.5)/(0.87) = 0.58$ times σ_z, as we found above by plugging $r = 0.5$ into Eq. (6.81).

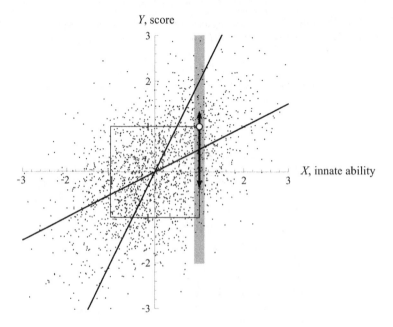

Figure 6.31: If $r = 0.5$, then $\sigma_z = (0.87)\sigma_y$. This standard deviation is indicated by the heavy arrows, centered on the lower regression line. A person with $X = \sigma_x$ has a 28% chance of scoring above $Y = \sigma_y$, indicated by the white dot. The square is the standard-deviation box, with σ_x and σ_y arbitrarily chosen to be 1.

REMARK: We can check some limits of the $\sqrt{(1 - r)/(1 + r)}$ result in Eq. (6.81). If $r = 0$ (no correlation), then Eq. (6.81) reduces to 1 of the σ_z standard deviations. This makes sense, because the $Y = mX + Z$ relation reduces to $Y = Z$ when the correlation coefficient r is zero (which comes about by having $m \to 0$ or $\sigma_z \gg \sigma_x$ in Eq. (6.6)). In this case, the lower regression line has slope zero, which means that it is simply the X axis. So a score of σ_y ($= \sigma_z$) above the overall mean of Y (which is zero) is the same as a score of one σ_z standard deviation above the regression line (the X axis). The desired probability is then 0.16, because this is half of the $1 - 0.68 = 0.32$ probability of lying outside of one standard deviation.

If $r = 1$ (perfect correlation), then Eq. (6.81) reduces to 0 of the σ_z standard deviations. The desired probability is therefore $1/2$, because that is the probability of exceeding the mean of a Gaussian distribution (which is equivalent to lying outside of zero standard deviations from the mean). This result isn't so obvious, because the two relevant quantities in Eq. (6.81) (namely, the distance $\sigma_y(1 - r)$ in Fig. 6.30, and the standard deviation $\sigma_z = \sigma_y\sqrt{1 - r^2}$) both go to zero as r approaches 1. But in the $r \to 1$ limit, the distance $\sigma_y(1 - r)$ goes to zero faster than the standard deviation $\sigma_z = \sigma_y\sqrt{1 - r^2}$. So in Fig. 6.30, σ_y exceeds $r\sigma_y$ by essentially zero of the σ_z standard deviations. ♣

6.8. Alternate form of *B*

If we plug the first expression for A from Eq. (6.47) into the second expression for B in Eq. (6.49), we obtain

$$B = \langle y \rangle - A\langle x \rangle$$

$$= \langle y \rangle - \left(\frac{\langle xy \rangle - \langle x \rangle\langle y \rangle}{\langle x^2 \rangle - \langle x \rangle^2} \right) \langle x \rangle$$

$$= \frac{(\langle y \rangle\langle x^2 \rangle - \langle y \rangle\langle x \rangle^2) - (\langle xy \rangle\langle x \rangle - \langle x \rangle^2\langle y \rangle)}{\langle x^2 \rangle - \langle x \rangle^2}$$

$$= \frac{\langle y \rangle\langle x^2 \rangle - \langle x \rangle\langle xy \rangle}{\langle x^2 \rangle - \langle x \rangle^2}, \tag{6.82}$$

which is correctly the first expression for B in Eq. (6.49).

6.9. Finding all the quantities

The means are

$$\overline{x} = \frac{2 + 3 + 3 + 5 + 7}{5} = 4 \quad \text{and} \quad \overline{y} = \frac{1 + 1 + 3 + 4 + 6}{5} = 3. \tag{6.83}$$

The standard deviations are then

$$\tilde{s}_x = \sqrt{\frac{(2 - 4)^2 + (3 - 4)^2 + (3 - 4)^2 + (5 - 4)^2 + (7 - 4)^2}{5}} = 1.79,$$

$$\tilde{s}_y = \sqrt{\frac{(1 - 3)^2 + (1 - 3)^2 + (3 - 3)^2 + (4 - 3)^2 + (6 - 3)^2}{5}} = 1.90. \tag{6.84}$$

The covariance is

$$\text{Cov}(x, y) = \frac{(2-4)(1-3)+(3-4)(1-3)+(3-4)(3-3)+(5-4)(4-3)+(7-4)(6-3)}{5} = 3.2. \tag{6.85}$$

The correlation coefficient r is then

$$r = \frac{\text{Cov}(x, y)}{\tilde{s}_x \tilde{s}_y} = \frac{3.2}{(1.79)(1.90)} = 0.94. \tag{6.86}$$

The slope m of the lower regression line is

$$m = \frac{r\tilde{s}_y}{\tilde{s}_x} = \frac{(0.94)(1.9)}{1.79} = 1.0. \tag{6.87}$$

Equivalently, Eq. (6.47) gives the slope A (which equals m) as

$$A = \frac{\text{Cov}(x, y)}{\tilde{s}_x^2} = \frac{3.2}{(1.79)^2} = 1.0. \tag{6.88}$$

It turns out that the $A = m$ slope of the regression (least-squares) line is *exactly* equal to 1, as we will see below.

If we want to use the first expression for B in Eq. (6.49), we must calculate $\langle x^2 \rangle$ and $\langle xy \rangle$. You can quickly show that these values are 19.2 and 15.2, respectively. So B equals

$$B = \frac{\langle y \rangle \langle x^2 \rangle - \langle x \rangle \langle xy \rangle}{\langle x^2 \rangle - \langle x \rangle^2} = \frac{(3)(19.2) - (4)(15.2)}{19.2 - 4^2} = -1. \tag{6.89}$$

This result is exact. Alternatively and more quickly, the second expression for B in Eq. (6.49) gives $B = \langle y \rangle - A \langle x \rangle = 3 - (1)(4) = -1$. Fig. 6.32 shows the line $y = Ax + B \implies y = x - 1$ superimposed on the plot of the five given points. We see that the line passes through three of the points. In retrospect, it is clear that we can't do any better than this line when minimizing the sum of the squares of the vertical distances from the points to the line. This is true because for the three points on the line, we can't do any better than zero distance. And for the two points $(3,1)$ and $(3,3)$ off the line, we can't do any better than having the line pass through the point $(3,2)$ midway between them. (As an exercise, you can prove this.) In most setups, however, the location of the least-squares line isn't so obvious. The small number of points in this problem just happened to be located very nicely with respect to each other.

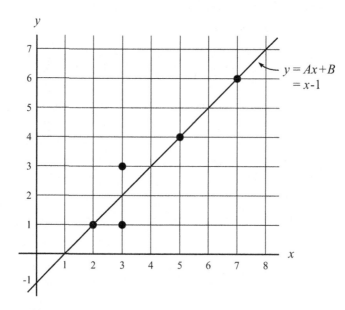

Figure 6.32: The five given points, along with the regression (least-squares) line.

6.10. Equal distances

(a) Given the x_i values, we want to find the value of x_b that minimizes the sum,

$$S \equiv \sum_{1}^{n} (x_i - x_b)^2. \tag{6.90}$$

To do this, we just need to set the derivative dS/dx_b equal to zero. This gives

$$0 = \frac{dS}{dx_b} = -\sum 2(x_i - x_b)$$

$$\implies 0 = -\left(\sum x_i\right) + nx_b$$

$$\implies x_b = \frac{\sum x_i}{n} \equiv \bar{x}. \tag{6.91}$$

(b) The first line in Eq. (6.91) tells us that $\sum(x_i - x_b) = 0$. In words, it tells us that the sum of the *signed differences* from x_b to all of the x_i points equals zero. The points with $x_i > x_b$ yield positive differences, while the points with $x_i < x_b$ yield negative differences. If the sum of the former set of differences is d, then the sum of the latter must be $-d$, so that the sum of all the differences is zero. If we now convert the previous sentence to a statement about *distances* (which are the absolute values of the signed differences, and hence always positive), we see that d is the sum of the distances from x_b to the points with $x_i > x_b$, and d is also the sum of the distances from x_b to the points with $x_i < x_b$. These two sums are therefore equal, as desired.

Combining the results in parts (a) and (b), we see that the mean $\bar{x}$ has two important properties: (1) the sum of the squares of the distances from $\bar{x}$ to the n given points is smaller for $\bar{x}$ than for any other value, and (2) the sum of the distances from $\bar{x}$ to the points above it equals the sum of the distances from $\bar{x}$ to the points below it.

Note that our definition of the "best-fit" point in terms of the minimum sum of the squared distances is essentially the same as the definition we used in Section 6.4 for the "badness" of a prediction. Both definitions involve the variance. But they differ in that the badness definition involves an *expectation value* over points that you will pick in the future, whereas the best-fit point involves an *average* over points that you have already picked; there is no expecting going on in this case. However, if you pick a very large number of points from a given distribution, then the best-fit point $\bar{x}$ will be very close to the mean μ of the distribution (which is the point with the least badness).

REMARK: Why did we define the best-fit point to be the point that minimizes the sum of the *squares* of the distances? Why not define it to be the point that just minimizes the sum of the distances (not squared)? There are two reasons why this latter definition isn't ideal. First, distances involve absolute values like $|x_i - x_b|$, and absolute values are somewhat messy to deal with mathematically. They involve two cases: If z is positive then $|z| = z$, but if z is negative then $|z| = -z$. Squares, on the other hand, are automatically positive (or zero).

Second, the point that minimizes the sum of the distances is simply not the point that most people would consider to be the best-fit point, because this point turns out not to be the *mean*, but rather the *median* (see below). The median is defined to be the point for which half of the x_i lie above it and half lie below it, with no regard for how far the various points are above or below. For example, if we have five x_i values, 2, 3, 5, 90, 100, then the median is 5 and the mean is 40. Most people would probably say that the mean 40 is more indicative of these five numbers than the median 5. The median doesn't take into account the spacing between the numbers.

To show that the above minimum-distance-sum (not squared) definition of the best-fit point leads to the median, we can give a quick proof by contradiction.

Assume that the best-fit point x_b is *not* the median. Then there are n_1 points below x_b and n_2 points above x_b, where $n_1 \neq n_2$. If $n_1 > n_2$ (the $n_1 < n_2$ case proceeds similarly), we can decrease the sum of all the distances by decreasing x_b slightly by, say, d. This will decrease n_1 distances by d and increase n_2 distances by d. And since $n_1 > n_2$, the overall sum of the distances will decrease. This contradicts the fact that x_b was assumed to yield the minimum sum. The only way to escape this contradiction is for n_1 to equal n_2. That is, x_b is the median. If the number of points is odd, then x_b equals the middle point. If the number is even, then x_b can lie anywhere between the middle two points. ♣

6.11. Equal distances again

If we take the partial derivative of the sum in Eq. (6.42) with respect to B, we obtain

$$0 = \frac{\partial S}{\partial B} = -2 \sum_1^n [y_i - (Ax_i + B)]. \qquad (6.92)$$

The $y_i - (Ax_i + B)$ terms here are the signed vertical differences between the given points and the line. The above equation therefore says that the sum of these signed distances is zero. This is exactly analogous to the fact that the sum $\sum (x_i - x_b)$ equaled zero in part (b) of Problem 6.10. So by the same reasoning we used there, we see that the sum of the vertical distances above the line equals the sum of the vertical distances below the line.

Note that the partial derivative of S with respect to A is $-2 \sum x_i [y_i - (Ax_i + B)]$. We can't conclude much from this, due to the x_i factor, which makes the terms in the sum not be the signed vertical differences.

REMARK: As in the remark in the solution to Problem 6.10, minimizing the sum of the distances (instead of their squares) is generally an inferior way to define the best-fit line. By the same reasoning we used in the 1-D case, this definition leads to a line that has half of the given points above it, and half below it, with no regard for how far the various points are above or below. Most people wouldn't consider such a line to be the line that best describes the given set of points. ♣

Chapter 7

Appendices

7.1 Appendix A: Subtleties about probability

In this appendix we will discuss a number of subtle issues with probability. This material isn't necessary for the content in this book, so it can be skipped on a first reading.

Determining probabilities

How do you determine the probability that a given event will occur? There are two ways: You can calculate it theoretically, or you can estimate it experimentally by performing a large number of trials of the process.

We can use a theoretical argument to determine, for example, the probability of obtaining Heads on a coin toss. There is no need to actually perform a coin toss, because it suffices to just think about it and note that the two possibilities of Heads and Tails are equally likely (assuming a fair coin). Each possibility must therefore occur half the time, which means that the probability of each is 1/2. Similar reasoning gives probabilities of 1/6 for each of the six possible rolls of a die (assuming a fair die).

However, there are certainly many situations where we don't have enough information to calculate the probability by theoretical means. In these cases we have no choice but to perform a large number of trials and then assume that the true probability is roughly equal to the fraction of events that occurred. For example, let's say that you take a bus to school or work, and that sometimes the bus is early and sometimes it's late. What is the probability that it is early? There are countless things that influence the bus's timing: traffic, weather, engine issues, delays caused by other passengers, slow service at a restaurant the night before which caused the driver to see a later movie than planned which caused him to go to bed later than usual and hence get up later than usual which caused him to start the route two minutes late, and so on and so forth. It is clearly hopeless to try to incorporate all of these effects into some sort of theoretical reasoning that produces a result that can be trusted. The only option is to observe what happens during a reasonably large number of days, and to assume that the fraction of early arrivals that you observe is

335

roughly the desired probability. If the bus is early on 20 out of 50 days, then we can say that the probability of being early is most likely somewhere around 40%.

Of course, having generated this result of 40%, it just might happen that a construction project on the route starts the next day, which makes the bus late every day for the next three months. So probabilities based on observation should be taken with a grain of salt!

A similar situation arises with, say, basketball free-throw percentages. There is absolutely no hope of theoretically calculating the probability of a certain player hitting a free throw, because it would require knowing everything that's going on from the thoughts in her head to the muscles in her fingers to the air currents on the way to the basket. All we can say is that if the player has hit a certain fraction of the free throws she's already attempted, then that's our best guess for the probability of hitting free throws in the future.

True randomness

We stated above that the probability of a coin toss resulting in Heads is 1/2. The reasoning was that Heads and Tails should have equal probabilities if everything is random, which means that they must each be 1/2. *But is the toss truly random?* What if we know the exact torque and force that you apply to the coin? We can then know exactly how fast it spins and how long it stays in the air (assuming that we know the density and viscosity of the air, etc.). And if we know the makeups of the ground and the coin, we can figure out exactly how the coin bounces, which will allow us to determine which side will land facing up. And even if we *don't* know all these things, they all have definite values, independent of our knowledge of them. So once the coin leaves our hand, the side that will land facing up is completely determined. The "random" nature of the toss is therefore nothing more than a result of our ignorance of the properties of the coin and its surroundings.

The question then arises: *How do we create a process that is truly random?* It's a good bet that if you try to create a random process, you'll discover that it actually isn't random. Instead, it just appears to be random due to your lack of knowledge of various inputs at the start of the process. You might try to make a coin toss random by having a machine flip the coin, where the force and torque that it applies to the coin take on random values. But how do we make *these* things random? All we've done is shift the burden of proof back a step, so we haven't really accomplished anything.

This state of affairs is particularly relevant when computers are used to generate random numbers. By various processes, computers can produce numbers that seem to be random. However, there is no way that they can be truly random, because the output is completely determined by the input. If the input isn't random (we're assuming it isn't, because otherwise we wouldn't need a random number generator!), then the output isn't random either.

In the above coin-toss scenarios, the issue at hand is that our definition of probability in Section 2.1 involved the phrase, "a very large number of *identical* trials." In none of the coin-toss scenarios are the trials identical. They all have (slightly) different inputs. So it's no surprise that things aren't truly random.

This then brings up the question: If we have *truly* identical processes, then shouldn't they give exactly identical results? If we flip a coin in exactly the same manner each time, then we should get exactly the same outcome each time. So our definition of probability seems to preclude true randomness! This makes us wonder if there are actually *any* processes that can be truly identical and at the same time yield different results.

Indeed there are. It turns out that in quantum mechanics, this is exactly what happens. It is possible to have two exactly identical process that yield different results. Things are truly random; you can't trace the different outcomes to different inputs. A great deal of effort has gone into investigating this randomness, and unless our view of the universe is way off-base, there are processes in quantum mechanics that involve true randomness.[1] If you think about this hard enough, it should make your head hurt. Our experiences in everyday life tell us that things happen *because* other things happened. But not so in quantum mechanics. There is no causal structure in certain settings. Some things just happen. Period.

But even without quantum mechanics, there are plenty of physical processes in the world that are *essentially* random, for all practical purposes. The ingredient that makes these processes essentially random is generally either (1) the sheer largeness of the numbers (of molecules, for example) involved, or (2) the phenomenon of "chaos," which turns small uncertainties into huge ones. Using these ingredients, it is possible to create methods for generating nearly random numbers. For example, the noise in the radio frequency range in the atmosphere generates randomness due to the absurdly large number of input bits of data (see www.random.org). And the pingpong balls bouncing around in a box used for picking lottery numbers generate randomness due to the chaotic nature of the ball collisions.

Different information

Let's say that I flip a coin and then look at the result and see a Heads, but I don't show you. Then for you, the probability of the coin being Heads is $1/2$. But for me, the probability is 1. So if someone asks for the probability of the coin showing Heads, which number is it, $1/2$ or 1? Well, there isn't a unique answer to this question, because the question is an incomplete one. The correct question to ask is, "What is the probability of the coin showing Heads, as measured by such-and-such a person?" You have to state who is calculating the probability, because different people have different information, and this affects the probability.

However, you might argue that it's the same process, so it should have a uniquely-defined probability, independent of who is measuring it. But it actually *isn't* the same process for the two of us. The process for me involves looking at the coin, whereas the process for you doesn't. Said in another way, our definition of probability involved the phrase, "a very large number of *identical* trials." As far as you're concerned, if we do 1000 trials of this process, they're all identical to you. But they certainly aren't identical to me, because for some of them I observe Heads, and for some I observe Tails. This is about as nonidentical as they can be. Said in yet another way, we are talking about two fundamentally different probabilities. One

---

[1]Of course, based on induction over the millennia, our view of the universe probably *is* way off-base. But let's not get into that here.

is the probability that the coin shows Heads, given no other information; this probability is 1/2. The other is the *conditional probability* that the coin shows Heads, *given* that it is observed to be Heads; this probability is 1.

"On average"

We now come to the most troublesome issue with probability. At the beginning of Section 2.1, we stated our definition of probability: "Consider a very large number N of identical trials of a certain process. If the probability of a particular event occurring is p, then the event will occur in a fraction p of the trials, on average." There are two related issues here: What do we mean by a "very large" number N of trials, and what do we mean by "on average"? Is $N = 10^9$ (one billion) large? It seems large when talking about coin flips, but it isn't large when talking about an event with $p = 1/10^9$. It turns out that the largeness of N actually isn't a huge issue, due to the words "on average." We can simply consider a very large number N' of sets, each consisting of N trials. (Of course, we're using the words "very large" again here.) However, the words "on average" introduce the following more problematic issue.

First, note that the definition of probability wouldn't make any sense without the words "on average," because there is no guarantee that an event will occur in *exactly* a fraction p of the trials. (Relaxing the condition to involve a small interval around p doesn't help, because there is still no guarantee of ending up in that interval.) Second, given that the words "on average" must appear, we see that we must take an average over a large number N' of sets, each consisting of N trials. (This averaging must be done, independent of the size of N.) In each of the N' sets, the event will occur in a certain fraction of the N trials. If we take the average of these N' fractions, we will obtain p, on average. But since we just said the words "on average" again, we now need to consider a large number N'' of sets, each consisting of N' sets, each consisting of N trials of the process. If we take the average of N'' numbers, each of which is the average of N' fractions (the fractions for the different groups of N trials), then we should obtain p ... on average!

You can see where we're going here. There is no way to end the process. We can never be certain that we will end up with an average of p. Or more precisely, we can never be certain that we will end up with an average that is within, say, 0.0001 (or some other small number of our choosing) of p. Every statement we can make will always end with the words "on average." So we must always tack on one more iteration. Every time we say "on average," we shift the burden of proof to the next step. Our definition of probability is therefore circular. Or perhaps "a never-ending linear chain" would be a more accurate description.

Note that when considering N'' sets of N' sets of N trials, we're simply performing $N''N'N$ trials. So instead of thinking in terms of sets of sets of trials, etc., we can consider one extremely large set of $N''N'N$ trials. It's the same overall set of trials, so we will observe the same fraction of trials in which an events occurs. However, any statement we make about this set will still end with the words "on average." So we're still going to need to consider N''' sets of the number of preceding trials, regardless of how we feel like subdividing that number. At any stage, we will always need to consider a large number of sets of the number of trials we've already done.

Now, you might think that this is all a bit silly, because everyone *knows* that the probability of a fair coin showing Heads is 1/2. You can produce evidence for this statement by flipping a million coins and checking that the percentage of Heads lies between, say, 49% and 51%. Or you can flip a trillion coins and check that the percentage of Heads lies between, say, 49.999% and 50.001%. Or you can flip a larger number of coins and specify a narrower range. In the two cases just mentioned, the calculated probabilities of lying in the given range are the same, with the common value being essentially equal to 1. More precisely, the probability of lying *outside* the range is the ridiculously small number $5 \cdot 10^{-89}$. See Problem 5.3 to get an idea of how small this number really is.

However, even with such a small probability, you *might* get Heads more than 50.001% of the time in a trillion flips. It's certainly unlikely, and to show that it is indeed unlikely, you could consider a large number of sets, each consisting of a trillion coin flips. You will likely find that an extremely small fraction of these sets have Heads occurring more than 50.001% of the time. But since we just said the word "likely," it is understood that we need to consider a large number of sets, each consisting of a large number of sets, each consisting of a trillion trials. And so on. The point is that no matter how many trials you do, you can never be *absolutely sure* that you haven't simply had bad (or good) luck. And, unfortunately, the preceding sentence is one thing you *can* be sure about. There will never be a magical large number for which things abruptly turn from probable to definite. So in that sense, an extremely large number like 10^{1000} is no better than an everyday number like 10. They are fundamentally the same. Any differences are theoretically just a matter of degree.

Having said all this, it would be a monumental mistake to discard the entire theory of probability, just because there are some philosophical issues with its underpinnings (which we have certainly *not* resolved here; our goal in this section was only to make you aware of them). The fact of the matter is that, in practice, probability *works*. Day after day, it proves invaluable in everything from finance to sports to politics to the fact that we don't all spontaneously combust. Therefore, in this book we will take a practical approach, where we intuitively know that the probability of getting Heads on a coin flip is 1/2, the probability of rolling a 5 on a die is 1/6, and so on. Feel free to ponder the philosophy of probability, but don't let that stop you from *using* probability!

7.2 Appendix B: Euler's number, *e*

7.2.1 Definition of *e*

Consider the expression,

$$\left(1 + \frac{1}{n}\right)^n. \tag{7.1}$$

Admittedly, this comes a bit out of the blue, but let's not worry about the motivation for now. After we derive a number of interesting results below, you'll see why we chose to consider this particular expression. Table 7.1 gives the values of $(1+1/n)^n$ for various integer values of n. (Non-integers are fine to consider, too.)

n	1	2	5	10	10^2	10^3	10^4	10^5	10^6
$(1+1/n)^n$	2	2.25	2.49	2.59	2.705	2.717	2.71815	2.71827	2.7182805

Table 7.1: The values of $(1 + 1/n)^n$ approach a definite number, approximately 2.71828, which we call e.

Apparently, the values converge to a number somewhere around 2.71828. This can also be seen in Fig. 7.1, which shows a plot of $(1 + 1/n)^n$ vs. $\log(n)$. The $\log(n)$ here means that the "0" on the x axis corresponds $n = 10^0 = 1$, the "1" corresponds $n = 10^1 = 10$, the "2" corresponds $n = 10^2 = 100$, and so on.

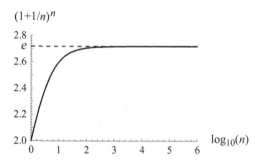

Figure 7.1: The plot of $(1 + 1/n)^n$ approaches e.

It is clear that even before we reach the "6" (that is, $n = 10^6 = 1,000,000$), the curve has essentially leveled off to a constant value. This value happens to be 2.7182818284..... It turns out that the digits in this number go on forever, with no overall pattern. However, the fortuitous double appearance of the "1828" makes it fairly easy to remember to 10 digits, although you'll rarely ever need more accuracy than 2.718. The exact number is known as Euler's number, and it is denoted by the letter e. The precise definition of e in terms of the expression in Eq. (7.1) is

$$e \equiv \lim_{n \to \infty} \left(1 + \frac{1}{n}\right)^n \approx 2.71828 \qquad (7.2)$$

The "lim" notation simply means that we're taking the limit of this expression as n approaches infinity. If you don't like dealing with limits or infinity, just set n equal to a very large number like 10^{10}, and then you pretty much have the value of e.

Remember that Eq. (7.2) is a *definition*. There's no actual content in it. All we did was take the quantity $(1 + 1/n)^n$ and look at what value it approaches as n becomes very large, and then we decided to call the result "e." We will, however, derive some actual results below, which aren't just definitions.

REMARK: If we didn't use a log plot in Fig. 7.1 and instead just graphed $(1 + 1/n)^n$ vs. n, the plot would stretch far out to the right if we wanted to go up to a large number like $n = 10^6$. Of course, we could shrink the plot in the horizontal direction, but then the region of small values of n would be squeezed down to essentially nothing. For example, the region up to $n = 100$ would take up only 0.01% of the plot. We would therefore be left with basically just a horizontal line. Even if we go up to only $n = 10^4$, we end up with the essentially horizontal straight line shown in Fig. 7.2, preceded by an essentially vertical jump from 2.0 to 2.718.

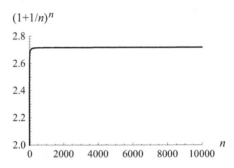

Figure 7.2: The plot of $(1 + 1/n)^n$ vs. n, with n measured on a linear scale.

The features in the left part of the plot in Fig. 7.1 aren't so visible in Fig. 7.2. You can barely see the bend in the curve. Log plots are used to prevent the larger numbers from dominating the plot, as they do in Fig. 7.2. This issue isn't so critical here, since we're concerned only with what $(1 + 1/n)^n$ looks like for large n, but nevertheless it's often more informative to use a log plot in certain settings. ♣

It is quite interesting that $(1 + 1/n)^n$ approaches a definite finite value as n gets larger and larger. On one hand, you might think that because the $1/n$ term gets smaller and smaller (which means that $(1 + 1/n)$ gets closer and closer to 1), the whole expression should get closer and closer to 1, because 1 raised to any power is 1. On the other hand, you might think that because the exponent n gets larger and larger, the whole expression should get larger and larger and approach infinity, because we're raising something to an ever-increasing power. It turns out that $(1 + 1/n)^n$ does neither of these things. Instead, these two effects cancel, and the result ends up somewhere between 1 and ∞, at the particular value of about 2.71828.

As mentioned above, we introduced $(1 + 1/n)^n$ a bit out of the blue. But we've already found one interesting feature of it, namely that it approaches a definite finite number (which we labeled as "e") as n goes to ∞. And there are many other features; so many, in fact, that e ends up being arguably the most important number in mathematics, with the possible exception of π. (But my vote is for e!) From the nearly endless list of interesting facts about e, we include three in the following three subsections.

7.2.2   Raising e to a power

What do we get when we raise e to a power? That is, what is the value of e^x? There are (at least) two ways to answer this. The simple way is to just use your calculator to raise $e = 2.71828$ to the power x. A number will pop out, and that's that.

However, there is another way which turns out to be immensely useful in the study of probability. If we relabel the n in Eq. (7.2) as m (for convenience), and if we then define $n \equiv mx$ in the fourth line below, we obtain

$$e^x = \lim_{m \to \infty} \left(\left(1 + \frac{1}{m} \right)^m \right)^x \qquad \text{(using } m \text{ instead of } n \text{ in Eq. (7.2))}$$

$$= \lim_{m \to \infty} \left(1 + \frac{1}{m} \right)^{mx} \qquad \text{(multiplying exponents)}$$

$$= \lim_{m \to \infty} \left(1 + \frac{x}{mx} \right)^{mx} \qquad \text{(multiplying by 1 in the form of } x/x)$$

$$= \lim_{n \to \infty} \left(1 + \frac{x}{n} \right)^n \qquad \text{(defining } n \equiv mx) \qquad (7.3)$$

In the case where n is large but not infinite, we can replace the "=" sign with a "≈" sign:

$$\boxed{e^x \approx \left(1 + \frac{x}{n} \right)^n} \qquad \text{(for large } n) \qquad (7.4)$$

The bigger the n, the better the approximation. The condition under which the approximation is a good one is

$$x \ll \sqrt{n}. \qquad (7.5)$$

This will usually hold in the situations we'll be dealing with (although there are a few exceptions in Chapter 5). We'll just accept this condition here, but see the second bullet-point case (the $na^2 \ll 1$ one) in Appendix C if you want to know where it comes from.

Eq. (7.4) is a rather nice result. The x that appears in the numerator of the fraction is simply the exponent of e. It almost seems like too simple a generalization of Eq. (7.2) to be correct. (Eq. (7.2) is a special case of Eq. (7.4), with $x = 1$.) Let's check that Eq. (7.4) does indeed hold for, say, $x = 2$. If we pick $n = 10^6$ (which certainly satisfies the $x \ll \sqrt{n}$ condition), we obtain $(1 + x/n)^n = (1 + 2/10^6)^{10^6} = 7.389041$. This is very close to the true value of e^2, which is about 7.389056. Larger values of n will make it even closer.

Example 1 (Compound interest): Assume that you have a bank account for which the interest rate is 5% per year. If this 5% is simply applied as a one-time addition at the end of the year, then after one year you will have 1.05 times the amount of money you started with. However, another way for the interest to be applied is for it to be compounded (applied) daily, with (5%)/365 being the daily rate (which happens to be about 0.014%). That is, your money at the end of each day equals $1 + (0.05)/365$ times what you had at the beginning of that day. In this scenario, by what factor does your money increase after one year?

Solution: Your money gets multiplied by a factor of $1 + (0.05)/365$ each day, so at the end of one year (365 days), it has increased by the factor,

$$\left(1 + \frac{0.05}{365} \right)^{365}. \qquad (7.6)$$

But this has exactly the same form as the expression in Eq. (7.4), with $x = 0.05$ and $n = 365$ (which certainly satisfies the $x \ll \sqrt{n}$ condition). So Eq. (7.4) tells us that after one year, your money increases by the factor $e^{0.05} \approx 1.051$. (Of course, you can also just plug the original expression $(1 + 0.05/365)^{365}$ into your calculator. The result is essentially the same.) Since your money increases by a factor of 1.051, the effective yearly interest rate is 5.1%. That is, someone who has a 5.1% interest rate that is applied as a one-time addition at the end of the year will end up with the same amount of money as you (assuming that the starting amounts were the same).

This effective interest rate of 5.1% is called the *yield*. So an annual rate of 5% has a yield of 5.1%. This yield is larger than 5% because the interest rate each day is being applied not only to your initial amount, but also to all the interest you've received in the preceding days. In short, you're earning interest on your interest.

The increase by 0.1% isn't so much. But if the annual interest rate is instead 10%, and if it is compounded daily, then the above reasoning implies that you will end up with a yearly factor of $e^{0.10} = 1.105$, which means that the yield is 10.5%. And an annual rate of 20% (admittedly rather unrealistic) produces a yearly factor of $e^{0.20} = 1.22$, which means that the yield is 22%.

Example 2 (Doubling your money): In the 5% scenario in the above example, the effect of compound interest (that is, earning interest on the interest) over one year could pretty much be ignored, because it was only 0.1%. However, the effect of compound interest *cannot* be ignored in the following question: If the annual interest rate is 5%, and if it is compounded daily, how many years will it take to double your money?

Solution: First, note the following incorrect line of reasoning: If you start with N dollars, then doubling your money means that you eventually need to increase it by another N dollars. Since it increases by about $(0.05)N$ each year, you need about 20 of these increases (because $20 \cdot (0.05) = 1$) to obtain the desired increase of N. So it takes 20 years. However, this is incorrect, because it ignores the fact that you have more money in each successive year and are hence earning interest on a larger and larger amount of money. The "since it increases by about $(0.05)N$ each year" clause above is therefore incorrect. Even the slightly more correct figure of $(0.051)N$ is still plenty wrong. The correct line of reasoning is the following.

We saw in the previous example that at the end of each year, your money increases by a factor of $e^{0.05}$ compared with what it was at the beginning of the year. So after n years it increases by n of these factors, that is, by $(e^{0.05})^n$ which equals $e^{(0.05)n}$. We want to find the value of n for which this overall factor equals 2. A little trial and error in your calculator shows that $e^{0.7} \approx 2$. (In the language of logs, this is the statement that $\log_e 2 \approx 0.7$, or equivalently $\ln 2 = 0.7$. But this terminology isn't important here.) So we need the $(0.05)n$ exponent to equal 0.7, which in turn implies that $n = (0.7)/(0.05) = 14$. It therefore takes 14 years to double your money.

You can think of this result for n as 70 divided by 5. For a general yearly interest rate of $r\%$, the same reasoning we used above shows that the number of years required to double your money is 70 divided by r. For example, with a 10% rate, your money will double in 7 years. In remembering this general rule, you just need to remember one number: 70. Equivalently, the time it takes to double your money is 70% of the naive answer that ignores the effect of compound interest. From the first paragraph above, this naive answer is 100 divided by r.

Unlike the previous example where the interest earned was small (because we were considering only one year), the interest earned in this example is large; it equals N

dollars by the end. So the effect of earning interest on your interest (that is, the effect of compound interest) cannot be ignored.

Note that even if you don't compound the interest daily (that is, even if you simply apply the 5% at the end of each year), it will still take essentially 14 years to double your money, because $(1.05)^{14} = 1.98 \approx 2$. The extra 0.1% earned each year when the interest is compounded daily doesn't make much of a difference here. ♣

7.2.3   The infinite series for e^x

Eq. (7.3), or equivalently Eq. (7.4), gives an expression for e^x. Another rather interesting expression for e^x that we can derive is

$$e^x = 1 + x + \frac{x^2}{2!} + \frac{x^3}{3!} + \frac{x^4}{4!} + \cdots \qquad (7.7)$$

The first two terms here can be written as $x^0/0!$ and $x^1/1!$, so all of the terms take the form of $x^n/n!$, where n runs from zero to infinity. In calculus language, Eq. (7.7) is known as the *Taylor series* for e^x. But that's just a name, so ignore it if you've never heard of it. We'll give a derivation of Eq. (7.7) below, but let's first look at a few of its consequences.

A special case of Eq. (7.7) occurs when $x = 1$, which yields

$$e = 1 + 1 + \frac{1}{2!} + \frac{1}{3!} + \frac{1}{4!} + \cdots . \qquad (7.8)$$

These terms get very small very quickly, so you don't need to include many of them to get a good approximation to e. Even just going out to the 10! term gives $e \approx 2.71828180$, which is accurate to the seventh digit beyond the decimal point.

A quick corollary to Eq. (7.7) is that if x is small, we can write

$$e^x \approx 1 + x. \qquad (7.9)$$

This is true because if x is small then the $x^2/2!$ term, along with all the higher powers of x in Eq. (7.7), are small compared with x. We can therefore ignore them. You should verify with a calculator that Eq. (7.9) is a good approximation for small x. You can let x be 0.01 or 0.001, etc. The number e is the one special number for which Eq. (7.9) holds. It is *not* the case that $2^x \approx 1 + x$ or $10^x \approx 1 + x$, as you can verify.

Of course, we can also say (by using the exact same reasoning we just used) that if x is small then the x term in Eq. (7.7), along with all the higher powers of x, are small compared with 1. If we ignore all these terms, we obtain the very coarse approximation: $e^x \approx 1$. This is indeed an approximation to e^x for small x, but the question is whether it is good enough for whatever purpose you have in mind. If it isn't, then you need to use the $e^x \approx 1 + x$ expression. And similarly, if that isn't good enough for your purposes, then you need to keep the next term in Eq. (7.7) and write $e^x \approx 1 + x + x^2/2$. And so on. But in many cases the $e^x \approx 1 + x$ approximation gets the job done.

We will now derive Eq. (7.7) by using Eq. (7.3) along with our good old friend, the binomial expansion; see Eq. (1.21). We'll assume that n is an integer here. Letting $a = 1$ and $b = x/n$ in Eq. (1.21), the binomial expansion of Eq. (7.3) gives (expanding the binomial coefficients and rearranging to obtain the third line)

$$e^x = \lim_{n \to \infty} \left(1 + \frac{x}{n}\right)^n \tag{7.10}$$

$$= \lim_{n \to \infty} \left[(1)^n + \binom{n}{1}(1)^{n-1}\left(\frac{x}{n}\right)^1 + \binom{n}{2}(1)^{n-2}\left(\frac{x}{n}\right)^2 + \binom{n}{3}(1)^{n-3}\left(\frac{x}{n}\right)^3 + \cdots\right]$$

$$= \lim_{n \to \infty} \left[1 + x\left(\frac{n}{n}\right) + \frac{x^2}{2!}\left(\frac{n(n-1)}{n^2}\right) + \frac{x^3}{3!}\left(\frac{n(n-1)(n-2)}{n^3}\right) + \cdots\right].$$

This looks roughly like what we're trying to show in Eq. (7.7), if only we could make the terms in parentheses go away. And indeed we can, because in the $n \to \infty$ limit, all of these terms equal 1. This is true because if $n \to \infty$, then both $n - 1$ and $n - 2$ are essentially equal to n (in a multiplicative sense). More precisely, the ratios $(n - 1)/n$ and $(n - 2)/n$ are both equal to 1 if $n = \infty$. So we have

$$\lim_{n \to \infty} \left(\frac{n(n-1)}{n^2}\right) = 1 \quad \text{and} \quad \lim_{n \to \infty} \left(\frac{n(n-1)(n-2)}{n^3}\right) = 1, \tag{7.11}$$

and likewise for the terms associated with higher powers of x. Eq. (7.10) therefore becomes Eq. (7.7) in the $n \to \infty$ limit.[2] If you have any doubts that Eq. (7.7) holds, you should verify with a calculator that it works for, say, $x = 2$. Going out to the 10! term should convince you.

REMARK: Another way to convince yourself that Eq. (7.7) is correct is the following. Consider what e^x looks like if x is a small number, say, $x = 0.0001$. We have

$$e^{0.0001} = 1.0001000050001667\ldots \tag{7.12}$$

This can be written more informatively as

$$e^{0.0001} = 1.0$$
$$+ \ 0.0001$$
$$+ \ 0.000000005$$
$$+ \ 0.0000000000001667\ldots$$

$$= 1 + (0.0001) + \frac{(0.0001)^2}{2!} + \frac{(0.0001)^3}{3!} + \cdots, \tag{7.13}$$

in agreement with Eq. (7.7). If you make x even smaller (say, 0.000001), then the same pattern will form, but with more zeros between the numbers than in Eq. (7.12).

---

[2]For any large but finite n, the terms in parentheses far out in the series in Eq. (7.10) will eventually differ from 1, but by that point the factorials in the denominators will make the terms negligible, so we can ignore them. Even if x is large, so that the powers of x in the numerators become large, the factorials in the denominators will dominate after a certain point in the series, making the terms negligible. But we're assuming $n \to \infty$ anyway, so these issues related to finite n are irrelevant.

Eq. (7.13) shows that if e^x can be expressed as a sum of powers of x (that is, in the form of $a + bx + cx^2 + dx^3 + \cdots$), then a and b must equal 1, c must equal $1/2$, and d must equal $1/6$. If you kept more digits in Eq. (7.12), you could verify the $x^4/4!$ and $x^5/5!$, etc., terms in Eq. (7.7) too. But things aren't quite as obvious for these, because we don't have all the nice zeros that we have among the first 12 digits of Eq. (7.12). ♣

7.2.4   The slope of e^x

Another interesting and important property of e is that if we plot the function $f(x) = e^x$, then the slope of the curve[3] at any point equals the value of the function at that point, namely e^x. For example, in Fig. 7.3 the slope at $x = 0$ is $e^0 = 1$, and the slope at $x = 2$ is $e^2 \approx 7.39$. (Note the different scales on the x and y axes, which make the slopes appear smaller than 1 and 7.39.) The number e is the one special number for which this is true. The same thing is *not* true for, say, 2^x or 10^x. The derivation of this property is by no means necessary for an understanding of the material in this book, but we'll present it in Appendix D, just for the fun of it.

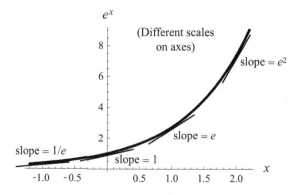

Figure 7.3: For any value of x, the slope of the e^x curve equals e^x. Note the different scales on the axes.

More generally, any function of the form Ae^x (where A is a constant) has the property that the slope at any point equals the value of the function at that point. This is true because both the value and the slope differ by the same factor of A from the corresponding quantities in the e^x case. (You can think about why this is true for the slope.) So if the property holds for e^x (which it does), then it also holds for Ae^x.

7.3   Appendix C: Approximations to $(1 + a)^n$

Expressions of the form $(1 + a)^n$ come up often in mathematics, especially in probability. It turns out that if a is small enough (which is invariably the case in the

---

[3]By "slope" we mean the slope of the line that is tangent to the curve at the given point. You can imagine the curve being made out of an actual piece of wire, and if you press a straight stick up against it, the stick will form the tangent to the curve at the point of contact.

situations we'll be dealing with), then the following approximate formula holds:

$$\boxed{(1 + a)^n \approx e^{na}} \tag{7.14}$$

This relation is equivalent to Eq. (7.4) if we let $a = x/n$.

Eq. (7.14) was critical in our discussion of the exponential and Poisson distributions in Sections 4.6.3 and 4.7.2. However, when we derived the Gaussian approximations to the binomial and Poisson distributions in Sections 5.1 and 5.3, we saw that a more accurate approximation was needed, namely

$$\boxed{(1 + a)^n \approx e^{na} e^{-na^2/2}} \tag{7.15}$$

In the event that a is sufficiently small, the extra factor of $e^{-na^2/2}$ is irrelevant, because it is essentially equal to $e^{-0} = 1$. So Eq. (7.15) reduces to Eq. (7.14). But if a isn't sufficiently small, then the extra factor of $e^{-na^2/2}$ is necessary if we want to have a good approximation. Of course, if a is too large, then even the inclusion of the $e^{-na^2/2}$ factor isn't enough to yield a good approximation. We must tack on another factor, or perhaps many factors, as we'll see in Eq. (7.21) below.

For an example where the $e^{-na^2/2}$ term in Eq. (7.15) is necessary, let's say we have $n = 100$ and $a = 1/10$. Then

$$(1 + a)^n = (1 + 1/10)^{100} \approx 13,781 \quad \text{and} \quad e^{na} = e^{10} \approx 22,026. \tag{7.16}$$

So the $(1 + a)^n \approx e^{na}$ approximation in Eq. (7.14) is a very bad one. However, the $e^{-na^2/2}$ factor in this case equals $e^{-1/2} \approx 0.60653$, which yields

$$e^{na} e^{-na^2/2} \approx (22,026)(0.60653) \approx 13,360. \tag{7.17}$$

The $(1 + a)^n \approx e^{na} e^{-na^2/2}$ approximation in Eq. (7.15) is therefore quite good; 13,360 differs from the actual value of 13,781 by only about 3%.[4] As an exercise, you can show that if we had picked more extreme numbers, say, $n = 10,000$ and $a = 1/100$, then Eq. (7.14) would be a similarly poor approximation, whereas Eq. (7.15) would be an excellent one, off by only 0.3%.

There are various ways to derive Eq. (7.15). The easiest way is to use a little calculus. If you want to avoid using calculus, you can still do the derivation, but it is rather laborious. Furthermore, if you want to generate better approximations by incorporating additional terms, the non-calculus method soon becomes intractable. In contrast, the calculus method gives, in one fell swoop, approximations to whatever accuracy you desire. We'll therefore take that route.

We'll start with the expression for the sum of a geometric series,

$$1 - a + a^2 - a^3 + a^4 - \cdots = \frac{1}{1 + a}. \tag{7.18}$$

---

[4]Whenever we use a "$\approx$" sign, we use it in a *multiplicative* (equivalently, a ratio) sense, and not an *additive* sense. The numbers 13,360 and 13,781 differ by 421, which you might consider to be a large number, but that doesn't matter. The ratio of the numbers is close to 1 (it equals 0.97), so they are "approximately equal" in that sense.

This is valid for $|a| < 1$. (If you plug in, say, $a = 2$, you will get an obviously incorrect statement.) For $|a| < 1$, if you keep enough terms on the left, the sum will essentially be equal to $1/(1 + a)$. If you hypothetically keep an *infinite* number of terms, the sum will be *exactly* equal to $1/(1 + a)$. You can verify Eq. (7.18) by multiplying both sides by $1 + a$. On the lefthand side, the infinite number of cross terms cancel in pairs, so only the "1" survives. Or, as always, you can just plug a small number like $a = 0.01$ or 0.001 into your calculator if you want some reassurance.

Now is where the calculus comes in. If we integrate both sides of Eq. (7.18) with respect to a, we obtain

$$a - \frac{a^2}{2} + \frac{a^3}{3} - \frac{a^4}{4} + \frac{a^5}{5} - \cdots = \ln(1 + a), \tag{7.19}$$

where ln is the natural log, that is, the log base e. We have used the facts that the integral of x^k equals $x^{k+1}/(k + 1)$ and the integral of $1/x$ equals $\ln(x)$. Technically there could be a constant of integration in Eq. (7.19), but it is zero. Eq. (7.19) is the Taylor series for $\ln(1 + a)$, just as Eq. (7.7) is the Taylor series for e^x. Eq. (7.19) can also be derived (as one learns in a calculus class) via the standard way of producing a Taylor series, which involves taking a bunch of derivatives. But the above method involving the geometric series is simpler. As with Eq. (7.18), Eq. (7.19) is valid for $|a| < 1$.

If we now exponentiate both sides of Eq. (7.19), then since $e^{\ln(1+a)} = 1 + a$ by the definition of ln, we obtain (reversing the sides of the equation)

$$1 + a = e^a e^{-a^2/2} e^{a^3/3} e^{-a^4/4} e^{a^5/5} \cdots, \tag{7.20}$$

which again is valid for $|a| < 1$. We have used the fact that the exponential of a sum is the product of the exponentials. Finally, if we raise both sides of Eq. (7.20) to the nth power, we arrive at

$$\boxed{(1 + a)^n = e^{na} e^{-na^2/2} e^{na^3/3} e^{-na^4/4} e^{na^5/5} \cdots} \tag{7.21}$$

This relation is valid for $|a| < 1$. It is exact if we include an infinite number of the exponential factors on the righthand side. However, the question we are concerned with here is how many terms we need to keep in order to obtain a good approximation. (We'll leave "good" undefined for the moment.) Under what conditions do we obtain Eq. (7.14) or Eq. (7.15)? The number of terms we need to keep depends on both a and n. In the following cases, we will always assume that a is small (more precisely, much smaller than 1).

- $na \ll 1$

    If $na \ll 1$, then all of the exponents on the righthand side of Eq. (7.21) are much smaller than 1. The first one (namely na) is small, by assumption. The second one (namely $na^2/2$; we'll ignore the sign) is also small, because it is smaller than na by a factor a (and also by a factor $1/2$), which we are assuming is small. Likewise, all of the other exponents in subsequent terms have additional factors of a and hence are even smaller. Therefore, since all of the exponents in Eq. (7.21) are much smaller than 1, they are, to a good approximation, all equal to zero. The exponential factors are therefore all approximately equal to $e^0 = 1$, so we obtain

$$(1 + a)^n \approx 1 \qquad \text{(valid if } na \ll 1) \qquad (7.22)$$

An example of a pair of numbers that satisfies $na \ll 1$ is $n = 1$ and $a = 1/100$. In this case it is a good approximation to say that $(1 + a)^n \approx 1$. And indeed, the exact value of $(1 + a)^n$ is $(1.01)^1 = 1.01$, so the approximation is smaller by only 1%.

- $na^2 \ll 1$

What if a isn't small enough to satisfy $na \ll 1$, but is still small enough to satisfy $na^2 \ll 1$? In this case we need to keep the e^{na} term in Eq. (7.21), but we can ignore the $e^{-na^2/2}$ term, because it is approximately equal to $e^{-0} = 1$. The exponents in subsequent terms are all also essentially equal to zero, because they are suppressed by higher powers of a. So Eq. (7.21) becomes

$$(1 + a)^n \approx e^{na} \qquad \text{(valid if } na^2 \ll 1) \qquad (7.23)$$

We have therefore derived Eq. (7.14), which we now see is valid when $na^2 \ll 1$. A pair of numbers that doesn't satisfy $na \ll 1$ but does satisfy $na^2 \ll 1$ is $n = 100$ and $a = 1/100$. In this case it is a good approximation to say that $(1 + a)^n \approx e^{na} = e^1 = 2.718$. And indeed, the exact value of $(1 + a)^n$ is $(1.01)^{100} \approx 2.705$, so the approximation is larger by only about 0.5%. The $(1 + a)^n \approx 1$ approximation in Eq. (7.22) is not a good one, being smaller than the approximation in Eq. (7.23) by a factor of e in the present scenario.

A special case of Eq. (7.23) occurs when $n = 1$, which yields $1 + a \approx e^a$. So we have rederived the $e^x \approx 1 + x$ approximation in Eq. (7.9), which we obtained from Eq. (7.7).

As mentioned right after Eq. (7.14), the relation in Eq. (7.4) is equivalent to Eq. (7.14)/Eq. (7.23) when a takes on the value x/n. In this case the $na^2 \ll 1$ condition becomes $n(x/n)^2 \ll 1 \implies x^2 \ll n \implies x \ll \sqrt{n}$, which is the condition stated in Eq. (7.5). But now we know where that condition comes from.

- $na^3 \ll 1$

What if a isn't small enough to satisfy $na^2 \ll 1$, but is still small enough to satisfy $na^3 \ll 1$? In this case we need to keep the $e^{-na^2/2}$ term in Eq. (7.21), but we can ignore the $e^{na^3/3}$ term, because it is approximately equal to $e^0 = 1$. The exponents in subsequent terms are all also essentially equal to zero, because they are suppressed by higher powers of a. So Eq. (7.21) becomes

$$(1 + a)^n \approx e^{na} e^{-na^2/2} \qquad \text{(valid if } na^3 \ll 1) \qquad (7.24)$$

We have therefore derived Eq. (7.15), which we now see is valid when $na^3 \ll 1$. A pair of numbers that doesn't satisfy $na^2 \ll 1$ but does satisfy $na^3 \ll 1$

is $n = 10,000$ and $a = 1/100$. In this case it is a good approximation to say that $(1 + a)^n \approx e^{na}e^{-na^2/2} = e^{100}e^{-1/2} = 1.6304 \cdot 10^{43}$. And indeed, the exact value of $(1+a)^n$ is $(1.01)^{10,000} \approx 1.6358 \cdot 10^{43}$, so the approximation is smaller by only about 0.3%. The $(1 + a)^n \approx e^{na}$ approximation in Eq. (7.23) is not a good one, being larger than the approximation in Eq. (7.24) by a factor of $e^{1/2}$ in the present scenario.

We can continue in this manner. If a isn't small enough to satisfy $na^3 \ll 1$, but is still small enough to satisfy $na^4 \ll 1$, then we need to keep the $e^{na^3/3}$ term in Eq. (7.21), but we can set the $e^{-na^4/4}$ term (and all subsequent terms) equal to 1. And so on and so forth. However, in this book we'll never need to go beyond the two terms in Eq. (7.15)/Eq. (7.24). Theoretically though, if, say, $n = 10^{12}$ and $a = 1/100$, then we need to keep the terms in Eq. (7.21) out to the $e^{-na^6/6}$ term, but we can ignore the $e^{na^7/7}$ terms and beyond, to a good approximation.

In any case, the rough size of the (multiplicative) error is the first term in Eq. (7.21) that is dropped. This is true because however close the first-dropped term is to $e^0 = 1$, all of the subsequent exponential factors are even closer to $e^0 = 1$. In the $n = 10,000$ and $a = 1/100$ case in the third bullet point above, the multiplicative error is roughly equal to the $e^{na^3/3}$ factor that we dropped, which in this case equals $e^{1/300} \approx 1.0033$. This is approximately the factor by which the true answer is larger than the approximate one.[5] This agrees with the results we found above, because $(1.6358)/(1.6304) \approx (1.0033)$. The true answer is larger by about 0.3% (so the approximation is smaller by about 0.3%).

If this factor of 1.0033 is close enough to 1 for whatever purpose we have in mind, then the approximation is a good one. If it isn't close enough to 1, then we need to keep additional terms until it is. In the present example with $n = 10,000$ and $a = 1/100$, if we keep the $e^{na^3/3}$ factor, then the multiplicative error is essentially equal to the next term in Eq. (7.21), which is $e^{-na^4/4} = e^{-1/40,000} = 0.999975$. This is approximately the factor by which the true answer is smaller than the approximate one. The difference is only 0.0025%.

7.4   Appendix D: The slope of e^x

(Note: This Appendix is for your entertainment only. The results here won't be needed anywhere in this book. But the derivation of the slope of the e^x function gives us an excuse to play around with some of the properties of e, and also to present some of the foundational concepts of calculus.)

7.4.1   First derivation

We stated in Section 7.2.4 that the slope of the $f(x) = e^x$ function at any point equals the value of the function at that point, namely e^x. In the language of calculus,

---

[5]The exponent here is positive, which means that the factor is slightly larger than 1. But note that half of the terms in Eq. (7.21) have negative exponents. If one of those terms is the first one that is dropped, then the factor is slightly smaller than 1. This is approximately the factor by which the true answer is smaller than the approximate one.

this is the statement that the *derivative* (the slope) of e^x equals itself, e^x. We will now show why this is true.

There are two main ingredients in the derivation. The first is Eq. (7.9). To remind ourselves that the x in that equation is assumed to be small, let's relabel it as δ, which is a standard letter that mathematicians use for a small quantity. We then have

$$e^\delta \approx 1 + \delta \qquad \text{(for small } \delta) \tag{7.25}$$

The second main ingredient is the strategy of finding the slope of the function $f(x) = e^x$ (or any function, for that matter) at a given point, by first finding an *approximate* slope, and by then making the approximation better and better. This proceeds as follows.

An easy way to make an approximation to the slope of a function at a particular value of x, say $x = 2$, is to find the *average* slope between $x = 2$ and a nearby point, say, $x = 2.1$. The average slope of the function $f(x) = e^x$ between these two points is

$$\text{slope} = \frac{\text{rise}}{\text{run}} = \frac{e^{2.1} - e^2}{0.1} \approx 7.77. \tag{7.26}$$

From Fig. 7.4, however, we see that this approximate slope is larger than the true slope.[6] To produce a better approximation, we can use a closer point, say $x = 2.01$. And then an even better approximation can be generated with $x = 2.001$. These two particular values of x yields slopes of

$$\text{slope} = \frac{\text{rise}}{\text{run}} = \frac{e^{2.01} - e^2}{0.01} \approx 7.43 \quad \text{and} \quad \frac{e^{2.001} - e^2}{0.001} \approx 7.393. \tag{7.27}$$

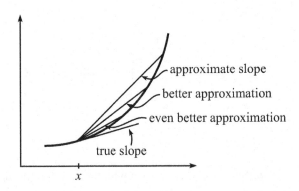

Figure 7.4: Better and better approximations to the true slope of a curve at a given point.

If we kept going with smaller and smaller differences from 2, we would find that the slopes converge to a certain value, which happens to be about 7.389, as you can verify. It is clear from Fig. 7.4 (which, again, is just a picture of a generic-looking curve) that the approximate slopes swing down and get closer and closer to

---

[6]The curve in this figure is an arbitrary curve and not the specific e^x function, but the general features are the same. The curve in the figure is concave upward like the e^x function (although the procedure we're discussing is independent of this property). The reason we're not using the actual e^x function here is that $x = 2.1$ is so close to $x = 2$ that we wouldn't be able to see the important features.

the actual tangent-line slope. This number of 7.389 must therefore be the slope of the e^x curve at $x = 2$.

Now, our goal here is to show that the slope of e^x equals e^x. We just found that the slope at $x = 2$ equals 7.389, so it had better be true that e^2 also equals 7.389. And indeed it does. So at least in the case of $x = 2$, we have demonstrated that the slope of e^x equals e^x.

Having learned how to determine the slope at the specific value of $x = 2$, we can now address the case of general x. To find the slope, we can imagine taking a small number δ and calculating the average slope between x and $x + \delta$ (as we did with 2 and 2.1), and then letting δ become smaller and smaller. Written out explicitly, the formal definition of the slope of a general function $f(x)$ at the value x is

$$\boxed{\text{slope} = \frac{\text{rise}}{\text{run}} = \lim_{\delta \to 0} \left(\frac{f(x + \delta) - f(x)}{\delta} \right)} \qquad (7.28)$$

This might look a little scary, but it's simply saying with an equation what Fig. 7.4 says with a picture: you can get a better and better approximation to the slope by looking at the average slope between two points and having the points get closer and closer together.

For the case at hand where our function $f(x)$ is e^x, we have (with the understanding that we're concerned with the $\delta \to 0$ limit in all of these steps)

$$\begin{aligned}
\text{slope} = \frac{\text{rise}}{\text{run}} &= \frac{e^{x+\delta} - e^x}{\delta} \\
&= e^x \left(\frac{e^\delta - 1}{\delta} \right) \qquad \text{(factoring out } e^x \text{)} \\
&\approx e^x \left(\frac{(1 + \delta) - 1}{\delta} \right) \qquad \text{(using Eq. (7.25))} \\
&= e^x \left(\frac{\delta}{\delta} \right) \\
&= e^x, \qquad\qquad\qquad\qquad\qquad\qquad (7.29)
\end{aligned}$$

as we wanted to show. Since we're concerned with the $\delta \to 0$ limit (that's how the true slope is obtained), the "$\approx$" sign in the third line becomes an "=" sign. So we are correct in saying that the slope of the e^x curve is exactly equal to e^x.

Note that Eq. (7.25) was critical in this derivation. Eq. (7.29) holds only for the special number e, because the $e^\delta \approx 1 + \delta$ result from Eq. (7.25) that we used in the third line holds only for e. The slope of, say, 2^x is *not* equal to 2^x, because Eq. (7.25) doesn't hold if e is replaced by 2 (or any other number).

Given that we're concerned with the $\delta \to 0$ limit, you might be worried about having a δ in the denominator in Eq. (7.29), since division by zero isn't allowed. But there is also a δ in the numerator, so you can cancel them first, and *then* take the $\delta \to 0$ limit, which is trivial because no δ's remain.

7.4.2 Second derivation

In the above derivation, we introduced the strategy of finding the slope by calculating approximate slopes involving x values that differ by a small number δ. Let's use this strategy to find the slope of a general power-law function, $f(x) = x^n$, where n is a nonnegative integer. (Note that x is now the number being raised to a power, as opposed to the power itself, as was the case with e^x.) We'll then use this result to give an alternative derivation of the fact that the slope of e^x equals itself, e^x.

We claim that for any value of x, the slope of the function x^n is given by:

$$\boxed{\text{slope of } x^n \text{ equals } nx^{n-1}} \tag{7.30}$$

(In the language of calculus, this is the statement that the derivative of x^n equals nx^{n-1}.) You can quickly verify Eq. (7.30) for the cases of $n = 0$ and $n = 1$, where the slopes are 0 and 1. To demonstrate Eq. (7.30) for a general nonnegative integer n (although it actually holds for any n), we can (as we did in the first derivation above) find the average slope between x and $x + \delta$, where δ is small. We can then find the true slope by taking the $\delta \to 0$ limit; see Eq. (7.28). To get a feel for what's going on, let's start with a specific value of n, say, $n = 2$. In the same manner as in the first derivation, we have (using Eq. (7.28) along with our trusty friend, the binomial expansion)

$$\text{slope} = \frac{\text{rise}}{\text{run}} = \frac{(x + \delta)^2 - x^2}{\delta}$$
$$= \frac{(x^2 + 2x\delta + \delta^2) - x^2}{\delta}$$
$$= \frac{2x\delta + \delta^2}{\delta}$$
$$= 2x + \delta. \tag{7.31}$$

If we now take the $\delta \to 0$ limit, the δ term goes away, leaving us with only the $2x$ term. So we have shown that the slope of the x^2 function equals $2x$, which is consistent with the nx^{n-1} expression in Eq. (7.30).

Let's try the same thing with $n = 3$. Again using the binomial expansion, we have

$$\text{slope} = \frac{\text{rise}}{\text{run}} = \frac{(x + \delta)^3 - x^3}{\delta}$$
$$= \frac{(x^3 + 3x^2\delta + 3x\delta^2 + \delta^3) - x^3}{\delta}$$
$$= \frac{3x^2\delta + 3x\delta^2 + \delta^3}{\delta}$$
$$= 3x^2 + 3x\delta + \delta^2. \tag{7.32}$$

When we take the $\delta \to 0$ limit, *both* of the $3x\delta$ and δ^2 terms go away, leaving us with only the $3x^2$ term. Basically, anything with a δ in it goes away when we take the $\delta \to 0$ limit. So we have shown that the slope of the x^3 function equals $3x^2$, which is again consistent with the nx^{n-1} expression in Eq. (7.30).

You can see how this works for the case of general n. The goal is to calculate

$$\text{slope} = \frac{\text{rise}}{\text{run}} = \frac{(x + \delta)^n - x^n}{\delta}. \tag{7.33}$$

Using the binomial expansion, the expressions for $(x + \delta)^n$ for the first few values of n are (you'll see below why we've added the parentheses in the second terms on the righthand side):

$$(x + \delta)^0 = x^0,$$
$$(x + \delta)^1 = x^1 + (1)\delta,$$
$$(x + \delta)^2 = x^2 + (2x)\delta + \delta^2,$$
$$(x + \delta)^3 = x^3 + (3x^2)\delta + 3x\delta^2 + \delta^3,$$
$$(x + \delta)^4 = x^4 + (4x^3)\delta + 6x^2\delta^2 + 4x\delta^3 + \delta^4,$$
$$(x + \delta)^5 = x^5 + (5x^4)\delta + 10x^3\delta^2 + 10x^2\delta^3 + 5x\delta^4 + \delta^5. \tag{7.34}$$

When we substitute these expressions into Eq. (7.33), the first term disappears when we subtract off the x^n. Then when we perform the division by δ, we reduce the power of δ by 1 in every term. So at this stage, for each of the expansions in Eq. (7.34), the first term has disappeared, the second term involves no δ's, and the third and higher terms involve at least one power of δ. Therefore, when we take the $\delta \to 0$ limit, the third and higher terms all go to zero, so we're left with only the second term (without the δ). In other words, in each line of Eq. (7.34) we're left with only the term in the parentheses. And this term has the form of nx^{n-1}, as desired. We have therefore proved Eq. (7.30). The multiplicative factor of n here is simply the $\binom{n}{1}$ binomial coefficient, because the general form of all of the $(x + \delta)^n$ expansions in Eq. (7.34) is

$$(x + \delta)^n = x^n + \binom{n}{1}x^{n-1}\delta + \binom{n}{2}x^{n-2}\delta^2 + \cdots. \tag{7.35}$$

We can now provide a second derivation of the fact that the slope of e^x equals itself, e^x. This derivation involves writing e^x in the form given in Eq. (7.7), which we'll copy here,

$$e^x = 1 + x + \frac{x^2}{2!} + \frac{x^3}{3!} + \frac{x^4}{4!} + \cdots. \tag{7.36}$$

We'll find the slope of e^x by applying Eq. (7.30) to each of these $x^n/n!$ terms.

REMARK: In order to make use of Eq. (7.36) in our derivation, we'll need to demonstrate that the slope of the sum of two functions equals the sum of the slopes of the two functions. And also that the "two" here can be replaced by any number. This might seem perfectly believable and not necessary to prove, but let's prove it anyway. We're setting off this proof in a remark, in case you want to ignore it.

Consider a function $F(x)$ that equals the sum of two other functions: $F(x) = f_1(x) + f_2(x)$. We claim that the slope of $F(x)$ at a particular value of x is the sum of the slopes of $f_1(x)$ and $f_2(x)$ at that value of x. This follows from the expression for the slope in

Eq. (7.28). We have

$$\text{slope of } F(x) = \frac{\text{rise}}{\text{run}} = \lim_{\delta \to 0} \left(\frac{F(x+\delta) - F(x)}{\delta} \right)$$

$$= \lim_{\delta \to 0} \left(\frac{(f_1(x+\delta) + f_2(x+\delta)) - (f_1(x) + f_2(x))}{\delta} \right)$$

$$= \lim_{\delta \to 0} \left(\frac{f_1(x+\delta) - f_1(x)}{\delta} \right) + \lim_{\delta \to 0} \left(\frac{f_2(x+\delta) - f_2(x)}{\delta} \right)$$

$$= (\text{slope of } f_1(x)) + (\text{slope of } f_2(x)). \tag{7.37}$$

The main point here is that in the third line we grouped the f_1 terms together, and likewise the f_2 terms. We can do this with any number of functions, of course, so that's why the above "two" can be replaced with any number. We can even have an infinite number of terms, as is the case in Eq. (7.36). ♣

We now know that the slope of e^x equals the sum of the slopes of all the terms in Eq. (7.36), of which there are an infinite number. And Eq. (7.30) tells us how to find the slope of each term. Let's look at the first few.

The slope of the first term in Eq. (7.36) (the 1) is zero. The slope of the second term (the x) is 1. The slope of the third term (the $x^2/2!$) is $(2x)/2! = x$. The slope of the fourth term (the $x^3/3!$) is $(3x^2)/3! = x^2/2!$. For the third and fourth terms, we have used the fact that if A is a numerical constant, then the slope of Ax^n equals Anx^{n-1}. This quickly follows from Eq. (7.28), because the A can be factored outside the parentheses.

We see that when finding the slope, each term in Eq. (7.36) turns into the preceding one; this is due to the factorials in the denominators. So the infinite series that arises after finding the slope is the same as the original infinite series. In other words, the derivative of e^x equals itself, e^x. Written out explicitly, we have

$$\text{Slope of } e^x = \text{Slope of } \left(1 + x + \frac{x^2}{2!} + \frac{x^3}{3!} + \frac{x^4}{4!} + \cdots \right)$$

$$= 0 + 1 + \frac{2x}{2!} + \frac{3x^2}{3!} + \frac{4x^3}{4!} + \frac{5x^4}{5!} + \cdots$$

$$= 0 + 1 + x + \frac{x^2}{2!} + \frac{x^3}{3!} + \frac{x^4}{4!} + \cdots$$

$$= e^x, \tag{7.38}$$

as we wanted to show.

The slope (the derivative) of a function $f(x)$ is commonly written as df/dx or $df(x)/dx$, where the d's indicate infinitesimal (that is, extremely small) changes. The reason for this notation is the following. The numerator in Eq. (7.28) is the change in the function f between two x values (namely x and $x + \delta$). The denominator is the change in the x value. The Greek letter Δ is generally used to denote the change in a quantity, so we can write the quotient in Eq. (7.28) as $\Delta f/\Delta x$, where Δx is simply the δ that we have been using. To find the slope as prescribed by Eq. (7.28), we still need to take the $\delta \to 0$ (or equivalently, the $\Delta x \to 0$) limit. Mathematicians reserve the letter d for this purpose. While a Δ can stand for a change of any size, a

d is used when it is understood that the change is infinitesimally small. So we have

$$\text{slope} = \lim_{\Delta x \to 0} \frac{\Delta f}{\Delta x} \equiv \frac{df}{dx}. \tag{7.39}$$

This is just the rise over run, where df is the infinitesimal rise, and dx is the corresponding infinitesimal run. Both of these quantities are essentially zero, but their ratio (which is the slope) is generally nonzero. In the derivative notation, our above results are

$$\boxed{\frac{d(e^x)}{dx} = e^x \quad \text{and} \quad \frac{d(x^n)}{dx} = nx^{n-1}} \tag{7.40}$$

7.5   Appendix E: Important results

This appendix includes all of the main results in the book. More commentary can be found in the Summary section in each chapter.

Chapter 1

$$\begin{aligned}
\text{Permutations:} \quad & P_N = N! \\
\text{Ordered sets, with repetition:} \quad & N^n \\
\text{Ordered sets, without repetition:} \quad & {}_N P_n = \frac{N!}{(N-n)!} \\
\text{Unordered sets, without repetition:} \quad & {}_N C_n = \frac{N!}{n!(N-n)!} \\
\text{Unordered sets, with repetition:} \quad & {}_N U_n = \binom{n+(N-1)}{N-1}
\end{aligned}$$

Chapter 2

$$\begin{aligned}
\text{Equally likely outcomes:} \quad & p = \frac{\text{number of desired outcomes}}{\text{total number of possible outcomes}} \\
\text{Dependent events:} \quad & P(A \text{ and } B) = P(A) \cdot P(B|A) \\
\text{Independent events:} \quad & P(A \text{ and } B) = P(A) \cdot P(B) \\
\text{Nonexclusive events:} \quad & P(A \text{ or } B) = P(A) + P(B) - P(A \text{ and } B) \\
\text{Exclusive events:} \quad & P(A \text{ or } B) = P(A) + P(B) \\
\text{Independence:} \quad & P(B|A) = P(B) \quad \text{or} \quad P(A|B) = P(A) \quad \text{or} \\
& P(A \text{ and } B) = P(A) \cdot P(B) \\
\text{Bayes' theorem (general form):} \quad & P(A_k|Z) = \frac{P(Z|A_k) \cdot P(A_k)}{\sum_i P(Z|A_i) \cdot P(A_i)} \\
\text{Stirling's formula:} \quad & n! \approx n^n e^{-n} \sqrt{2\pi n}
\end{aligned}$$

Chapter 3

Expectation value:	$E(X) = p_1 x_1 + p_2 x_2 + \cdots + p_m x_m$
For arbitrary variables:	$E(X + Y) = E(X) + E(Y)$
For independent variables:	$E(XY) = E(X) \cdot E(Y)$
Standard deviation:	$\sigma_X \equiv \sqrt{E[(X - \mu)^2]} = \sqrt{E(X^2) - \mu^2}$
For independent variables:	$\sigma_{X+Y}^2 = \sigma_X^2 + \sigma_Y^2$
Biased coin:	$\sigma_{\text{Heads}} = \sqrt{np(1 - p)} \equiv \sqrt{npq}$
Standard deviation of the mean:	$\sigma_{\overline{X}} = \dfrac{\sigma}{\sqrt{n}}$
Variance:	$\text{Var}(X) \equiv E[(X - \mu)^2] = E(X^2) - \mu^2$
For independent variables:	$\text{Var}(X + Y) = \text{Var}(X) + \text{Var}(Y)$
Biased coin:	$\text{Var}(\text{Heads}) = npq$
Variance of a set of numbers:	$\tilde{s}^2 \equiv \text{Var}(S) \equiv \dfrac{1}{n} \sum_1^n (x_i - \overline{x})^2$
Sample variance:	$s^2 \equiv \dfrac{1}{n-1} \sum_1^n (x_i - \overline{x})^2$

Chapter 4

Binomial distribution:	$P(k) = \dbinom{n}{k} p^k (1 - p)^{n-k}$
Exponential distribution:	$\rho(t) = \dfrac{e^{-t/\tau}}{\tau}$ or $\lambda e^{-\lambda t}$
Poisson distribution:	$P(k) = \dfrac{a^k e^{-a}}{k!}$
Gaussian distribution:	$f(x) = \sqrt{\dfrac{b}{\pi}}\, e^{-b(x-\mu)^2}$
	$= \sqrt{\dfrac{1}{2\pi\sigma^2}}\, e^{-(x-\mu)^2/2\sigma^2}$

Chapter 5

Gaussian approx to binomial:	$\dfrac{e^{-x^2/[2np(1-p)]}}{\sqrt{2\pi np(1-p)}}$
Gaussian approx to Poisson:	$\dfrac{e^{-x^2/2a}}{\sqrt{2\pi a}}$

Chapter 6

Linear relation:	$Y = mX + Z$
Covariance:	$\text{Cov}(X, Y) \equiv E\left[(X - \mu_x)(Y - \mu_y)\right]$
For data points:	$\text{Cov}(x, y) \equiv \dfrac{1}{n} \sum (x_i - \overline{x})(y_i - \overline{y})$
Correlation coefficient:	$r \equiv \dfrac{m\sigma_x}{\sigma_y} = \dfrac{m\sigma_x}{\sqrt{m^2\sigma_x^2 + \sigma_z^2}} = \dfrac{\text{Cov}(X, Y)}{\sigma_x \sigma_y}$
For data points:	$r \equiv \dfrac{\text{Cov}(x, y)}{\tilde{s}_x \tilde{s}_y} = \dfrac{\sum (x_i - \overline{x})(y_i - \overline{y})}{\sqrt{\sum (x_i - \overline{x})^2}\, \sqrt{\sum (y_i - \overline{y})^2}}$
Improvement of prediction:	$\dfrac{\sigma_z^2}{\sigma_y^2} = 1 - r^2$
Probability density $\rho(x, y)$:	$\dfrac{1}{2\pi\sigma_x\sigma_y\sqrt{1-r^2}}\exp\left(-\dfrac{1}{2(1-r^2)}\left(\dfrac{x^2}{\sigma_x^2} + \dfrac{y^2}{\sigma_y^2} - \dfrac{2rxy}{\sigma_x\sigma_y}\right)\right)$
Lower regression line slope:	$\dfrac{r\sigma_y}{\sigma_x}$
Upper regression line slope:	$\dfrac{\sigma_y}{r\sigma_x}$
Average retest score:	$y_{\text{avg}} = r^2 y_1$
Slope of least-squares line:	$A = \dfrac{\text{Cov}(x, y)}{\tilde{s}_x^2} = \dfrac{r\tilde{s}_y}{\tilde{s}_x}$

Chapter 7

Euler's number:	$e \equiv \lim\limits_{n\to\infty}\left(1 + \dfrac{1}{n}\right)^n \approx 2.71828$
Taylor series for e^x:	$e^x = 1 + x + \dfrac{x^2}{2!} + \dfrac{x^3}{3!} + \dfrac{x^4}{4!} + \cdots$
For small x:	$e^x \approx 1 + x$
An approximation:	$(1 + a)^n \approx e^{na}$
A better approximation:	$(1 + a)^n \approx e^{na} e^{-na^2/2}$
Two derivatives:	$\dfrac{d(e^x)}{dx} = e^x, \quad \dfrac{d(x^n)}{dx} = nx^{n-1}$

7.6 Appendix F: Glossary of notation

Chapter 1

$$\text{Factorial:} \quad N! = 1 \cdot 2 \cdot 3 \cdot (N-1) \cdot N$$

$$\text{Permutations:} \quad P_N = N!$$

$$\text{Ordered subgroups:} \quad {}_N P_n = \frac{N!}{(N-n)!}$$

$$\text{Unordered subgroups:} \quad {}_N C_n = \frac{N!}{n!(N-n)!}$$

$$\text{Binomial coefficient:} \quad \binom{N}{n} = \frac{N!}{n!(N-n)!}$$

$$\text{Unordered sets with repetitions:} \quad {}_N U_n = \binom{n + (N-1)}{N-1}$$

Chapter 2

$$\text{Probability:} \quad p$$

$$\text{Probability of event } A\text{:} \quad P(A)$$

$$\text{Intersection (joint) probability:} \quad P(A \text{ and } B), \ P(A \cap B)$$

$$\text{Conditional probability:} \quad P(B|A)$$

$$\text{Union probability:} \quad P(A \text{ or } B), \ P(A \cup B)$$

$$\text{Not } A\text{:} \quad {\sim}A$$

Chapter 3

$$\text{Random variable:} \quad X \ \text{(uppercase)}$$

$$\text{Value of random variable:} \quad x \ \text{(lowercase)}$$

$$\text{Expectation value:} \quad E(X), \ \mu_X, \ \mu_x, \ \mu$$

$$\text{Standard deviation:} \quad \sigma_X, \ \sigma_x, \ \sigma$$

$$\text{Standard deviation of the mean:} \quad \sigma_{\overline{X}}, \ \sigma_{\overline{x}}, \ \sigma_{\text{avg}}$$

$$\text{Variance:} \quad \text{Var}(X), \ \sigma_X^2, \ \sigma_x^2, \ \sigma^2$$

$$\text{Set of numbers:} \quad S$$

$$\text{Mean of set } S\text{:} \quad \overline{x}, \ \langle x \rangle \equiv \frac{1}{n} \sum_1^n x_i$$

$$\text{Variance of set } S\text{:} \quad \text{Var}(S), \ \tilde{s}^2 \equiv \frac{1}{n} \sum_1^n (x_i - \overline{x})^2$$

$$\text{Sample variance of set } S\text{:} \quad s^2 \equiv \frac{1}{n-1} \sum_1^n (x_i - \overline{x})^2$$

Chapter 4

Probability:	$P(x)$ (uppercase)
Probability density:	$\rho(x)$, $f(x)$, etc. (lowercase)
Much greater than (multiplicatively):	$\gg$
Much less than (multiplicatively):	$\ll$
Approximately equal (multiplicatively):	$\approx$

Chapter 5

Number of trials in an experiment:	n_t
Number of sets of n_t trials:	n_s

Chapter 6

Slope of (lower) regression line:	m
Correlation coefficient:	r
Covariance of distribution:	$\text{Cov}(X,Y)$
Covariance of data points:	$\text{Cov}(x,y)$
Joint probability density:	$\rho(x,y)$
Slope of least-squares line:	A
y-intercept of least-squares line:	B

Chapter 7

Euler's number:	$e \approx 2.71828$
Derivative (slope) of $f(x)$:	$\dfrac{df(x)}{dx}$

Index